Mathematical Cognition and Learning

Cognitive Foundations for Improving Mathematical Learning

Mathematical Cognition and Learning

Series Editors

Daniel B. Berch
David C. Geary
Kathleen Mann Koepke

VOLUME 5

Cognitive Foundations for Improving Mathematical Learning

Volume Editors

David C. Geary
Daniel B. Berch
Kathleen Mann Koepke

Mathematical Cognition and Learning

Cognitive Foundations for Improving Mathematical Learning

Edited by

David C. Geary
Psychological Sciences
University of Missouri
Columbia, MO, United States

Daniel B. Berch
Curry School of Education
University of Virginia
Charlottesville, VA, United States

Kathleen Mann Koepke
Eunice Kennedy Shriver National Institute of Child Health and Human Development (NICHD)
National Institutes of Health (NIH)
Bethesda, MD, United States

ELSEVIER

ACADEMIC PRESS
An imprint of Elsevier

Academic Press is an imprint of Elsevier
50 Hampshire Street, 5th Floor, Cambridge, MA 02139, United States
525 B Street, Suite 1650, San Diego, CA 92101, United States
The Boulevard, Langford Lane, Kidlington, Oxford OX5 1GB, United Kingdom
125 London Wall, London, EC2Y 5AS, United Kingdom

Library of Congress Cataloging-in-Publication Data
A catalog record for this book is available from the Library of Congress

British Library Cataloguing-in-Publication Data
A catalogue record for this book is available from the British Library

ISBN: 978-0-12-815952-1

ISSN: 2214-2568

For information on all Academic Press publications
visit our website at https://www.elsevier.com/books-and-journals

Publisher: Nikki Levy
Acquisition Editor: Emily Ekle
Editorial Project Manager: Barbara Makinster
Production Project Manager: Anusha Sambamoorthy
Cover Designer: Matthew Limbert

Typeset by SPi Global, India

Contents

Contributors

Pirjo Aunio, Special Needs Education, Faculty of Educational Sciences, University of Helsinki, Helsinki, Finland

Drew Bailey, School of Education, University of California, Irvine, Irvine, CA, United States

Daniel B. Berch, Curry School of Education, University of Virginia, Charlottesville, VA, United States

Talia Berkowitz, Department of Psychology, University of Chicago, Chicago, IL, United States

Douglas H. Clements, Morgridge College of Education, Marsico Institute, University of Denver, Denver, CO, United States

Caitlin Craddock, Department of Special Education, Vanderbilt University, Nashville, TN, United States

Emily N. Daubert, Department of Human Development and Quantitative Methodology, University of Maryland, College Park, MD, United States

Dinorah De León, Centro de Investigación Básica en Psicología, Montevideo, Uruguay

Lydia DeFlorio, College of Education, University of Nevada, Reno, Reno, NV, United States

Dahiana Fitipalde, Centro de Investigación Básica en Psicología, Montevideo, Uruguay

Douglas Fuchs, Department of Special Education, Vanderbilt University, Nashville, TN, United States

Lynn S. Fuchs, Department of Special Education, Vanderbilt University, Nashville, TN, United States

David C. Geary, Psychological Sciences, University of Missouri, Columbia, MO, United States

Dominic J. Gibson, Department of Psychology, University of Chicago, Chicago, IL, United States

Justin Halberda, Johns Hopkins University, Baltimore, MD, United States

Asha K. Jitendra, University of California, Riverside, Graduate School of Education, Riverside, CA, United States

Yemimah King, Department of Human Development and Family Studies, Purdue University, West Lafayette, IN, United States

Alice Klein, Center for Early Learning, STEM Program, WestEd, San Francisco, CA, United States

Josh Langfus, Johns Hopkins University, Baltimore, MD, United States

Susan C. Levine, Department of Psychology; Department of Comparative Human Development; Committee on Education, University of Chicago, Chicago, IL, United States

Melissa E. Libertus, Department of Psychology and Learning Research and Development Center, University of Pittsburgh, Pittsburgh, PA, United States

Alejandro Maiche, Centro de Investigación Básica en Psicología, Montevideo, Uruguay

Álvaro Mailhos, Centro de Investigación Básica en Psicología, Montevideo, Uruguay

Amelia S. Malone, Department of Special Education, Vanderbilt University, Nashville, TN, United States

Kathleen Mann Koepke, National Institutes of Health, *Eunice Kennedy Shriver* National Institute of Child Health & Human Development, Bethesda, MD, United States

Amy R. Napoli, Department of Human Development and Family Studies, Purdue University, West Lafayette, IN, United States

David J. Purpura, Department of Human Development and Family Studies, Purdue University, West Lafayette, IN, United States

Geetha B. Ramani, Department of Human Development and Quantitative Methodology, University of Maryland, College Park, MD, United States

Julie Sarama, Morgridge College of Education, Marsico Institute, University of Denver, Denver, CO, United States

Nicole R. Scalise, Department of Human Development and Quantitative Methodology, University of Maryland, College Park, MD, United States

Pamela M. Seethaler, Department of Special Education, Vanderbilt University, Nashville, TN, United States

Prentice Starkey, Center for Early Learning, STEM Program, WestEd, San Francisco, CA, United States

Xenia Vamvakoussi, Department of Early Childhood Education, University of Ioannina, Ioannina, Greece

John Woodward, School of Education, University of Puget Sound, Tacoma, WA, United States

Foreword: Cognitive Foundations for Improving Mathematical Learning

Ann Dowker
Department of Experimental Psychology, Oxford University, Oxford, United Kingdom

In recent years, there has been increased interest in early mathematical development. This has included both interest in the domain-general and domain-specific underpinnings of early mathematical development (Chu, van Marle, & Geary, 2015; Hannula, Lepola, & Lehtinen, 2010; Hohol, Cipora, Willmes, & Nuerk, 2017), and interest in early interventions to ameliorate and where possible prevent mathematical difficulties (Butterworth, Varma, & Laurillard, 2011; Clements & Sarama, 2011; Cohen Kadosh, Dowker, Heine, Kaufmann, & Kucian, 2013; Dowker, 2017; Fuchs, Fuchs, & Compton, 2012). On the whole, these areas have tended to be studied and discussed separately, though a better understanding of the foundations of early numerical abilities could make it much easier to develop suitable interventions. This book makes a very important contribution to our knowledge and understanding by bringing together and integrating the topics of cognitive foundations and interventions in mathematics and reviewing recent work in these areas.

In the introductory chapter, the editors *David Geary, Daniel Berch,* and *Kathleen Mann Koepke* explore the key issues of the book. They discuss both the domain-general and domain-specific factors that contribute to mathematical development and mathematical difficulties, and how different types of interventions have addressed these. Domain-general factors include fluid intelligence and executive function and working memory. There have been numerous attempts to train children in domain-general abilities, especially working memory, and a few individual results have appeared potentially promising (e.g., Klingberg, 2010). However, as the editors and authors and others point out, there is little consistent evidence that training in domain-general abilities transfers to tasks that are dissimilar to those on which children have been trained (Melby-Verlag, Redick, & Hulme, 2016; Sala & Gobet, 2017). In discussing the chapters,

I will group them not chronologically but thematically as follows: (1) those that propose and discuss cognitive foundations of numeracy; (2) a chapter that discusses informal influences on children's numeracy development, specifically parental input; (3) those that describe and discuss the development and evaluation of intervention programs; and (4) discussions of current problems with interventions and how evaluations of their effectiveness might be improved. Because of the integrative nature of the book, many of the chapters include elements of more than one of these, but I will group the chapters according to what appears to be the main focus.

CHAPTERS THAT DEAL WITH POSSIBLE COGNITIVE FOUNDATIONS FOR NUMERACY

One domain-specific ability that has been proposed to be an important foundation for numeracy is the Approximate Number System (Dehaene, 1997). *Melissa Libertus* discusses the relationship between the Approximate Number System (ANS) and mathematical performance. As she points out, studies suggest a correlation between nonsymbolic magnitude estimation and arithmetic (e.g., Mazzocco, Feigenson, & Halberda, 2011), but it tends to be less strong than the correlations between symbolic magnitude estimation and arithmetic (e.g., De Smedt, Noël, Gilmore, & Ansari, 2013). Moreover, training in symbolic number comparison is more likely than training in nonsymbolic number comparison to lead to improvement in arithmetic (Dillon, Kannan, Dean, Spelke, & Dufio, 2017; Honore & Noel, 2016). The author proposes that the ANS may affect mathematical development not so much directly as through increasing attention to and engagement with mathematical material. This may even have intergenerational effects: recent studies (Braham & Libertus, 2017) suggest that parents' ANS acuity correlates with the amount of mathematical talk that they use when playing with their children, and with their children's mathematical abilities.

With regard to domain-general factors, several authors discuss the importance of both language and visual-spatial representation to arithmetical cognition and learning.

David Purpura, Amy Napoli, and Yemimah King discuss the importance of developing mathematical language. Studies have shown that both language and print knowledge in general are related to informal numeracy and numeral knowledge (Austin, Blevins-Knabe, Ota, Rowe, & Lindauer, 2011; LeFevre et al., 2010; Purpura & Ganley, 2014; Purpura, Hume, Sims, & Lonigan, 2011). Purpura, Napoli, Wehrspann, and Gold (2017) found that instruction in mathematical language through dialogic reading improved children's general numeracy skills, but not their general vocabulary. Analyses showed that it was largely instruction in quantitative language that improved numeracy skills; instruction in spatial language did not have such an effect.

However, visual and spatial representations are also important to mathematical learning. *Asha Jitendra and John Woodward* have studied primary school children of various ages, and have focused on the ways in which visual representations may facilitate word problem solving. They review their own and others' studies, both of how the use of diagrams in text may improve performance, and of how children may be encouraged to develop and use their own visual representations.

Xenia Vamvakoussi discusses the importance in early numerical development of another domain-general ability: analogical reasoning, whereby children may make sense of unfamiliar situations by extrapolating from familiar ones. The author points out that young children frequently use analogical reasoning spontaneously, but often fail to use analogies generated by others such as teachers. For example, children and even adolescents may fail to make productive use of the analogy "numbers are points on the line." The author and colleagues found that secondary school pupils' use of this analogy was improved by explicit instruction as to its implications, and by the use of a bridging analogy of a rubber number line.

Lynn Fuchs, Douglas Fuchs, Amelia Malone, Pamela Seethaler, and *Caitlin Craddock* discuss the importance of understanding and using cognitive, linguistic, and social-emotional processes in treating mathematics learning difficulties. The authors discuss two ways in which this can be applied to interventions. One is to include cognitive, linguistic and/or social-emotional processes within direct skills interventions. For example, Fuchs, Fuchs, Craddock, Seethaler, and Geary (2017) incorporated language comprehension training in an intervention to improve primary school children's word problem solving. Preliminary results have suggested that this adds to the effectiveness of the intervention. The other is to allocate different forms of direct skills interventions to groups of pupils with different profiles of cognitive, linguistic and/or social-emotional processes. For example, Fuchs et al. (2016) found that children with poor working memory and/or poor reasoning abilities benefitted more from a fractions intervention if it included verbal explanations than if it involved word problem training, whereas the reverse was true of children with better working memory and reasoning.

CHAPTER THAT DEALS WITH HOME AND PARENTAL INFLUENCES

Susan Levine, Dominic Gibson, and Talia Berkowitz discuss some parental influences on young children's numeracy development. These include parents' use of numerical language, their use of number gestures (a sometimes neglected topic), and the intergenerational transmission of attitudes toward mathematics. Negative parental attitudes toward mathematics, especially mathematics anxiety, have been found to have adverse effects on children's mathematical attainment. The authors suggest that interventions that involve parents in helping their children with mathematics (see for example the Pre-K Mathematics curriculum by Klein and Starkey, discussed later) may help to break this cycle, and discuss some interventions that they are themselves developing.

CHAPTERS THAT DEAL WITH INTERVENTIONS AND THEIR EFFECTS

Pirjo Aunio discusses general principles of assessment and intervention. Aunio and Räsänen (2016) had developed a model of foundational abilities in arithmetic, with four main groups of crucial abilities: (1) symbolic and non-symbolic number sense; (2) understanding mathematical relations; (3) counting skills; and (4) basic skills in arithmetic. The author emphasizes the importance of identifying children at risk for learning difficulties in mathematics as early as possible. The chapter proposes the use of Curriculum-Based Measurement: the combination of conventional achievement testing and the use of time series (that permit scores for the same individual to be compared at different time points). Reliable assessment may be used to guide suitable interventions. The author discusses in particular the development and use of the "ThinkMath" materials, which target numerical relational skills, counting skills, and simple arithmetical word problems.

Alice Klein, Prentice Starkey, and *Lydia DeFlorio* studied the effects of interventions on American preschool children from disadvantaged backgrounds. The researchers had developed a curricular intervention, Pre-K Mathematics, for children from low-income homes, who were thus at risk of low achievement at school. The intervention includes small-group activities run by teachers using activities with concrete manipulatives to support development of numerical abilities, arithmetic problem solving, spatial abilities and geometrical reasoning, pattern knowledge and informal measurement. It also includes suggestions for parents for games and activities to pursue with children at home. Several studies have shown that the curriculum has positive effects on children's mathematical performance, as measured on the CMA and TEMA tests, at the time of entering kindergarten, with significant improvements from their performance at entry to prekindergarten. However, those who are in the lowest quartile for mathematical performance at the time of entering prekindergarten seem to benefit little from the intervention.

These researchers also examined the effect of giving 541 children from this group either of two additional tutorial interventions. One intervention involved individual tutoring sessions in mathematical tasks related to the overall Pre-K mathematics curriculum. The other combined such tutoring with attention-training tutorials. The children in these individualized intervention groups improved significantly more than those who received just the curricular intervention. The inclusion of attention training did not lead to greater improvement than the mathematics training on its own, again indicating that domain-specific interventions may be more successful than domain-general interventions in improving mathematics performance.

The researchers tested yet another form of intervention: giving the curricular intervention without additional individualized tutoring, but administered over a 2-year period starting at age 3, rather than a 1-year period starting at age 4. This was found to improve outcomes significantly in comparison with children who received only one year of intervention. Thus, it appears that children who appear relatively resistant to a curricular intervention that benefits most children can be helped by intensification of the intervention, either through adding individualized activities or increasing the time devoted to the curricular intervention.

Geetha Ramani, Emily Daubert, and *Nicole Scalise* discuss the role of play and games in building up preschoolers' numerical abilities. For example, they report the work of Geetha Ramani and Robert Siegler (Ramani & Siegler, 2008; Siegler & Ramani, 2008) on the use of board games with nursery school and kindergarten children from disadvantaged backgrounds. They proposed that SES differences in early numeracy may reflect different levels of prior experience with informal numerical activities, such as numerical board games. Most middle-class children will have at least some early experience with traditional board games, such as *Chutes and Ladders*, which involve counting steps on numbered squares, whereas children from disadvantaged backgrounds often do not. The authors found that preschoolers from low-income families performed worse on tests of numerical magnitude estimation than peers from better-off backgrounds. However, playing a simple linear numerical board game for four 15-minute sessions eliminated the differences in numerical estimation proficiency. Playing games that substituted colors for numbers did not improve numerical performance, while a circular numerical board game improved number recognition but not numerical estimation. These findings have been replicated with groups of children in the UK (Whyte & Bull, 2008) and Sweden (Elofsson, Gustafson, Samuelsson, & Träff, 2016). Other number games have also been shown to improve preschoolers' number understanding, such as dominoes (Brankaer, Ghesquière, & De Smedt, 2015) and numerical computer games (Maertens, De Smedt, Sasanguie, Elen, & Reynvoet, 2016; Schacter & Io, 2016; Toll & Van Luit, 2013).

While most of the chapters in this book relate to interventions for preschool and kindergarten children, interventions for early primary school are also described. *Josh Langfus, Alexandro Maiche, Dinorah De Leon, Dahiana Fitipalde, Alvaro Mailhos,* and *Justin Halberda* studied 386 second- and third-grade children in Uruguay. The researchers created a software application comprising simple mathematical games involving magnitude comparisons for approximate numbers of objects, time durations, and surface areas. The games included novel characters such as "monsters," visual rewards for good performance, such as "stars" and online tracking. The children in eight Montevideo classrooms played the games over a period of one month. They were compared with children in eight similar classrooms in the same schools undergoing business-as-usual mathematics instruction. All children in both groups had access to tablets. At the end of the month, all children showed significant improvement in tests of IQ, Vocabulary, Geometry, and especially Arithmetic Ability and Approximate Number System Ability. Overall, the intervention and business-as-usual children did not differ significantly, which may be in part due to the fact that they all experienced positive effects from being in schools, which were taking part in the project, and receiving regular visits from the researchers. There was some suggestion that the intervention had a positive effect on children, who were in the lowest SES group and/or had repeated a grade.

CHAPTERS THAT DEAL WITH PROBLEMS WITH INTERVENTIONS AND THEIR EVALUATION, AND SUGGEST IMPROVEMENTS

Julie Sarama and *Douglas Clements* point out some important weaknesses in the research often carried out on curriculum development. For one thing, it often involves only product evaluation. Researchers may investigate the effectiveness of an intervention or classroom program in comparison with business-as-usual teaching or with other interventions; but it is often not used for the foundations of development of the interventions in the first place. The authors describe their own "Building Blocks" curriculum, which was informed by evidence at all stages of development and evaluation, especially as regards learning trajectories in different components of numerical abilities and how these are influenced by interventions. This chapter focuses in particular on learning trajectories in subitizing.

Drew Bailey discusses reasons why early academic interventions often show significant effects on school readiness, but these effects often fade out during the years after the end of treatment. This may be in part because of the componential nature of arithmetical ability (see Dowker, 2008, 2015): interventions that target components that are important in early schooling may not necessarily affect or transfer to other components that are more important later on. Also, the instruments (tests) used may differ over the years, both in content and in scaling. Another possibility is that the effects of early intervention fade out due to forgetting. Bailey argues that it is important to have more long-term follow-up studies if we are to devise interventions with better long-term impact.

CONCLUSION

To conclude, this book brings together many important and up-to-date theories and findings that will stimulate much further research on the cognitive foundations of numeracy and their applications to interventions to prevent or ameliorate mathematical difficulties. It will also facilitate the development of such interventions and help to ensure that they are truly evidence-based.

REFERENCES

Aunio, P., & Räsänen, P. (2016). Core numerical skills for learning mathematics in children aged five to eight years—a working model for educators. *European Early Childhood Education Research Journal*, *24*(5), 684–704.

Austin, A. M. B., Blevins-Knabe, B., Ota, C., Rowe, T., & Lindauer, S. L. K. (2011). Mediators of preschoolers' early mathematics concepts. *Early Child Development and Care*, *181*, 1181–1198.

Braham, E. J., & Libertus, M. E. (2017). Intergenerational associations in numerical approximation and mathematical abilities. *Developmental Science*, *20*, e12436.

Brankaer, C., Ghesquière, P., & De Smedt, B. (2015). The effect of a numerical domino game on numerical magnitude processing in children with mild intellectual disabilities. *Mind, Brain, and Education*, *9*, 29–39.

Butterworth, B., Varma, S., & Laurillard, D. (2011). Dyscalculia: from brain to education. *Science*, *332*, 1049–1053.

Chu, F. W., van Marle, K., & Geary, D. C. (2015). Early numerical foundations of young children's mathematical development. *Journal of Experimental Child Psychology*, *132*, 1–8.

Clements, D. H., & Sarama, J. (2011). Early childhood mathematics intervention. *Science*, *333*, 968–970.

Cohen Kadosh, R., Dowker, A., Heine, A., Kaufmann, L., & Kucian, K. (2013). Interventions for improving numerical abilities: present and future. *Trends in Neuroscience and Education*, *2*, 85–93.

Dehaene, S. (1997). *The number sense: How the mind creates mathematics*. Oxford: Oxford University Press.

De Smedt, B., Noël, M.-P., Gilmore, C., & Ansari, D. (2013). How do symbolic and non-symbolic numerical magnitude processing relate to individual differences in children's mathematical skills. A review of evidence from brain and behavior. *Trends in Neuroscience and Education*, *2*, 48–55.

Dillon, M. R., Kannan, H., Dean, J. T., Spelke, E., & Dufio, E. (2017). Cognitive science in the field: a preschool intervention durably enhances intuitive but not formal mathematics. *Science*, *357*, 47–55.

Dowker, A. (2008). Individual differences in numerical abilities in preschoolers. *Developmental Science*, *11*, 650–654.

Dowker, A. (2015). Individual differences in arithmetical abilities: the componential nature of arithmetic. In R. Cohen Kadosh & A. Dowker (Eds.), *Oxford handbook of mathematical cognition* (pp. 878–894). Oxford: Oxford University Press.

Dowker, A. (2017). Interventions for primary school children with difficulties in mathematics. In J. Sarama, D. Clements, C. Germeroth, & C. l. Day-Hess (Eds.), *Advances in Child Development and Behavior, Vol. 53: The Development of Early Mathematics Education* (pp. 255–287). New York, NY: Elsevier.

Elofsson, J., Gustafson, S., Samuelsson, J., & Träff, U. (2016). Playing number board games supports 5-year-old children's early mathematical development. *The Journal of Mathematical Behavior*, *43*, 134–147.

Fuchs, L. S., Fuchs, D., & Compton, D. L. (2012). The early prevention of mathematics difficulty: its power and limitations. *Journal of Learning Disabilities*, *45*, 257–269.

Fuchs, L. S., Fuchs, D., Craddock, C., Seethaler, P. M., & Geary, D. C. (2017). In *Embedding word-problem specific language comprehension within word-problem intervention Paper presented at the 2017 math cognition conference*.

Fuchs, L. S., Malone, A., Schumacher, R. F., Namkung, J. M., Hamlett, C. L., Jordan, N. C., et al. (2016). Supported self-explaining during fraction intervention. *Journal of Educational Psychology*, *108*, 493–508.

Hannula, M., Lepola, J., & Lehtinen, E. (2010). Spontaneous focusing on numerosity as a domain-specific predictor of arithmetical skills. *Journal of Experimental Child Psychology*, *107*, 394–406.

Hohol, M., Cipora, K., Willmes, K., & Nuerk, H.-C. (2017). Bringing back the balance: domain general processes are also important in numerical cognition. *Frontiers in Psychology*, *8*, 499.

Honore, N., & Noel, M. P. (2016). Improving preschoolers' arithmetic through number magnitude training: the impact of non-symbolic and symbolic training. *PLoS ONE*, *11*, e0166685.

Klingberg, T. (2010). Training and plasticity of working memory. *Trends in Cognitive Sciences*, *14*(7), 317–324.

LeFevre, J.-A., Fast, L., Skwarchuk, S.-L., Smith-Chant, B. L., Bisanz, J., Kamawar, D., et al. (2010). Pathways to mathematics: longitudinal predictors of performance. *Child Development*, *81*, 1753–1767.

Maertens, B., De Smedt, B., Sasanguie, D., Elen, J., & Reynvoet, B. (2016). Enhancing arithmetic in pre-schoolers with comparison or number line estimation training: does it matter? *Learning and Instruction*, *46*, 1–11.

Mazzocco, M. M. M., Feigenson, L., & Halberda, J. (2011). Pre-schoolers' precision of the approximate number system predicts later school maths performance. *PLoS ONE*, *6*, e23749.

Melby-Verlag, M., Redick, C., & Hulme, C. (2016). Working memory training does not improve performance on measures of intelligence or other measures of "far transfer": evidence from a meta-analytic review. *Perspectives on Psychological Science*, *11*, 512–534.

Purpura, D. J., & Ganley, C. M. (2014). Working memory and language: skill-specific or domain-general relations to mathematics. *Journal of Experimental Child Psychology*, *122*, 104–121.

Purpura, D. J., Hume, L. E., Sims, D. M., & Lonigan, C. J. (2011). Early literacy and early numeracy: the value of including early literacy skills in the prediction of numeracy development. *Journal of Experimental Child Psychology*, *110*, 647–658.

Purpura, D. J., Napoli, A. R., Wehrspann, E. A., & Gold, Z. S. (2017). Causal connections between mathematical language and mathematical knowledge: a dialogic reading intervention. *Journal of Research on Educational Effectiveness*, *10*, 116–137.

Ramani, G. B., & Siegler, R. S. (2008). Promoting broad and stable improvements in low-income children's numerical knowledge through playing number board games. *Child Development*, *79*, 375–394.

Sala, G., & Gobet, F. (2017). Working memory training in typically developing children: a meta-analysis of the available evidence. *Developmental Psychology*, *53*, 671–685.

Schacter, J., & Io, B. (2016). Improving low-income preschoolers mathematics achievement with Math Shelf, a preschool tablet computer curriculum. *Computers in Human Behaviour*, *55*, 223–239.

Siegler, R. S., & Ramani, G. B. (2008). Playing board games promotes low-income children's numerical development. *Developmental Science, Special Issue on Mathematical Cognition*, *11*, 655–661.

Toll, S. W., & Van Luit, J. E. (2013). Accelerating the early numeracy development of kindergartners with limited working memory skills through remedial education. *Research in Developmental Disabilities*, *34*, 745–755.

Whyte, J. C., & Bull, R. (2008). Number games, magnitude representation, and basic number skills in preschoolers. *Developmental Psychology*, *44*, 588–596.

Preface

With *Cognitive Foundations for Improving Mathematical Learning*, we close a five-volume journey through the evolution and early development of number competencies (Volume 1); the brain and genetic foundations for these and more complex abilities (Volume 2); the cognitive bases for the learning of evolutionarily novel mathematics, from arithmetic to trigonometry (Volume 3); the influences of language and culture on mathematical learning and cognition (Volume 4); and now formal and informal instructional approaches for improving children's mathematics learning (Volume 5). With this final volume, we see clear links between topics and discoveries covered in previous volumes and the development of intervention approaches. These include interventions focused on the relation between our evolved number sense and children's early math learning; the influence of home, family, and mathematical language on early math learning; and the integration of cognitive science approaches to mathematical learning into educational interventions. These represent an exciting step forward and a much needed conciliation between basic research in mathematical cognition and mathematics education interventions that we hope will continue well beyond the publication of this volume.

As with the previous volumes, we anticipate this one will be of interest to researchers, graduate students, and undergraduates specializing in cognitive development, cognitive neuroscience, educational psychology, early childhood education, special education, and many other disciplines. More so than the four other volumes, this final volume should be particularly interesting for researchers and practitioners in both mathematics and preschool education. The chapters cover a wide range of approaches and topics. The approaches range from small-scale game-based interventions focused on improving preschoolers' number knowledge to large-scale randomized controlled studies focused on older and academically at-risk children. And, the topics range from fundamental number knowledge, to mathematical language, to use of analogy and visuospatial representations to facilitate mathematical learning. The final chapter addresses the vexing but critically important issue of intervention fade-out effects and the factors that might be underlying these effects.

The chapters herein provide cutting-edge descriptions of the latest approaches to our understanding of the informal (e.g., at home, with parents) and formal (i.e., in educational settings) interventions that can be used to improve young children's and older students' mathematical learning, as well as discussion of potential shortcomings of these approaches and the steps that need to be taken to improve them. This volume could be used as a textbook for several kinds of courses taught in psychology (e.g., educational psychology) and education (e.g., mathematics education, instruction and learning, early childhood education).

In the introduction, we provide an overview of methodological issues and the domain-general and domain-specific cognitive foundations that can influence the effectiveness of various intervention approaches, as well as a brief foray into newly emerging interventions (e.g., brain stimulation) and parental influences on mathematical learning. The early chapters in the volume focus on young children and describe different types of interventions from board games to tablet-based games to large-scale randomized controlled studies, all focused on better preparing children for formal mathematical learning at school entry or as an adjunct to their initial formal learning. The chapters also cover informal influences on children's early learning, including the role of mathematical language (e.g., understanding more versus less) and the ways in which activities with parents can boost children's emerging understanding of number. The later chapters focus on older students' mathematics learning and the integration of insights from cognitive science into intervention design, including the use of visuospatial representations to facilitate students' ability to solve mathematical word problems, analogical reasoning to facilitate students' conceptual understanding of number, and how students' cognitive strengths and weaknesses influence their responsiveness to different intervention approaches. As noted, the volume closes with a sobering but thoughtful discussion of intervention fadeout and provides insights into how we might better address this issue.

We thank the Child Development and Behavior Branch of the *Eunice Kennedy Shriver* National Institute of Child Health and Human Development, NIH for the primary funding of the conference on which this volume is based and Dr. Joan McLaughlin, Commissioner of the National Center for Special Education Research (NCSER), U.S. Department of Education for supplementary funding of the meeting. We are also grateful to Dr. Camilla Benbow, Dean of the Peabody School of Education and Human Development at Vanderbilt University for hosting the conference and to Dr. Lynn Fuchs for her significant contributions to the planning for and successful execution of virtually every phase of this meeting. Finally, we are indebted to the administrative staff of Peabody College for ensuring that the requisite conference facilities were in place and that the meeting ran smoothly and efficiently.

David C. Geary
Daniel B. Berch
Kathleen Mann Koepke

Chapter 1

Introduction: Cognitive Foundations for Improving Mathematical Learning

David C. Geary*, Daniel B. Berch† and Kathleen Mann Koepke‡
**Psychological Sciences, University of Missouri, Columbia, MO, United States*
†*Curry School of Education, University of Virginia, Charlottesville, VA, United States*
‡*National Institutes of Health,* Eunice Kennedy Shriver *National Institute of Child Health & Human Development, Bethesda, MD, United States*

INTRODUCTION

Children's acquisition of some level of competence with mathematics has never been as important as it is today. Mathematical knowledge and skills expand educational and employment opportunities and allow one to better navigate the increasingly number-imbued realities of life in the modern world. As the educational and economic benefits to mathematical knowledge increase, so do the costs of falling behind. Reducing these costs by improving the mathematical development of at-risk children will yield substantive real-world benefits to these individuals and to the communities in which they will reside as adults. As detailed by the authors in this volume, there are many ways to potentially reduce these costs, from increasing parents' engagement in numeracy-related activities with their preschool children to formal interventions to better prepare children for success when they enter school to interventions that target core mathematical competencies (e.g., fractions). These interventions provide some very promising initial steps toward the development of systematic approaches toward ensuring that all children will eventually have the mathematical knowledge needed for functioning in educational, employment, and day-to-day contexts, but much remains to be learned.

The chapters in this volume largely provide reviews of specific interventions or classes of intervention, including issues surrounding fadeout (loss of intervention gains over time). In our introductory chapter, we step back and begin with an overview of intervention work more generally, including

Mathematical Cognition and Learning, Vol. 5. https://doi.org/10.1016/B978-0-12-815952-1.00001-3

discussion of methodological issues and the cognitive mechanisms that support mathematics learning as related to intervention research. The authors of the chapters within this volume have done an excellent job of overviewing current intervention approaches but were not able to cover all of them. Thus we discuss some of the approaches that were not touched upon in these chapters, to provide the reader with a more complete synopsis of the efforts that are being made to improve children's mathematical learning. We close with a brief discussion on how parents can influence their children's mathematical development and suggestions for improving research in this area.

BRIEF HISTORY OF MATHEMATICS INTERVENTION STUDIES

For several millennia, there has been discussion and oftentimes vociferous debate regarding how to best educate youth. The debate reflects a general tension between a teacher-directed classical education with set standards to child-directed Romantic approaches, following Rousseau (1979/1762). The tension between these approaches has fueled much of the controversy in modern education, including the "reading wars" and "math wars" (Klein, 2007; Loveless, 2001). Resolving these tensions is more consequential than your typical "academic feud," as the resolution or resolutions will influence how and what children are taught in school and are likely to be most keenly felt by those students in the most need of well-vetted educational interventions.

Project Follow Through was designed, at least in part, to provide an experimental (actually quasi-experimental) evaluation of the basic tenets of these broad educational approaches and was the largest educational intervention that has ever been conducted (Stebbins, 1977). The project was conducted from 1968 to 1977 across several cohorts of children at risk of poor educational outcomes. In total, about 200,000 children across 180 communities throughout the United States participated in a kindergarten through third grade intervention. Each participating school partnered with a university, research institute, or some other group to develop and implement the four-year intervention grounded in differing educational principles and academic foci. The principles reflected the basic teaching philosophy and broadly included direct instruction, cognitive approaches (e.g., focus on problem solving), and social/affective approaches focusing on, for instance, self-efficacy. The educational goals ranged from a focus on mastery of basic skills to more complex problem solving and conceptual understanding. Relative to Title I comparison groups and with respect to performance on standardized achievement tests, children in the direct instruction programs with a focus on basic skills and mastery had better outcomes on basic skills, problem solving, and had a better academic self-concept than did students in the other programs. There was some fadeout (below) over time, but many of these gains were maintained throughout the elementary school years (Meyer, 1984; Meyer, Gersten, & Gutkin, 1983).

Of course, the results were roundly criticized (House, Glass, McLean, & Walker, 1978) and eventually suppressed in favor of more Romantic approaches to education. The "reading wars" and "math wars" were to follow in the succeeding decades, and in turn were followed by the National Reading Panel (2000), National Mathematics Advisory Panel (2008), and the founding of the Institute of Education Sciences in 2002. The combination, though routinely critiqued and debated (e.g., Benbow & Faulkner, 2008; Boaler, 2008), resulted in more rigorous and scientific standards for evaluating educational programs generally and interventions for at-risk children. Many of the intervention programs described in this volume are the direct result of these changes and all of them represent empirically and scientifically grounded approaches for evaluating and ultimately developing interventions that will be effective.

METHODOLOGICAL ISSUES IN INTERVENTION RESEARCH

As was just noted, there is a long history of debate over how students' mathematical competencies should be assessed—qualitatively or quantitatively—and even deeper divisions about how children learn (e.g., discovery learning versus direct instruction), what should be emphasized during instruction (e.g., concepts versus procedures), and the types of interventions (e.g., randomized controlled trials) that provide the most credible evidence for informing educational practice (e.g., Brickman & Reich, 2015; Cai et al., 2018; Cobb, Yackel, & Wood, 1992; Donaldson, Christie, & Mark, 2015; Geary, 1994). Although we acknowledge that qualitative work can be informative, our focus is on quantitative approaches and research grounded in cognitive psychology, and issues that can compromise the usefulness of these methods for informing educational practice. We describe a few of these most basic issues.

Design Factors

The implementation of effective educational policy at the national or state level and the implementation of instructional programs that consistently promote students' achievement are dependent on the availability of high-quality educational research, including research on interventions (National Research Council, 2002; Pianta, Barnett, Burchinal, & Thornburg, 2009). Researchers must adhere to basic scientific principles in the design of such studies and be alert to potential threats to their validity. The same threats to validity that are of concern to researchers can also be used by consumers (e.g., teachers, policy makers) to evaluate the quality of any particular study, and determine whether and to what extent teaching practices might be adjusted based on these results (Willingham, 2012). We note a few issues regarding the design of educational intervention studies and potential threats to their validity (Gersten et al., 2005).

The What Works Clearing House (2017) provides guidelines for evaluating the quality of intervention research and how this quality influences the strength of the inferences that can be drawn from these studies; the National Mathematics Advisory Panel (2008) adopted very similar standards in their reviews of research related to children's mathematics learning. The best evidence for the effectiveness of interventions comes from randomized controlled trials (RCTs) and quasi-experimental studies. In the former, students (or students clustered within groups, such as classrooms or schools) are randomly assigned to intervention or one of several control groups. Random assignment should ensure that factors that can influence the intervention (e.g., motivation) will be evenly distributed across the intervention and control groups and thus will not bias the results of the study. Quasi-experimental studies, such as Project Follow Through, differ from RCTs in that there is no randomization. Common threats to the validity of both RTCs and quasi-experimental studies include nonequivalence of groups at pretreatment (e.g., groups differ in prerequisite skills) and attrition during the intervention. There are various statistical techniques to adjust for these issues, but the extent to which these improve the quality of the results depends on other factors, such as whether attrition was similar or not in the intervention and control groups.

Quasi-experimental studies come with an additional risk of unmeasured differences between the intervention and control groups. For instance, the intervention group may be composed of students who volunteered for the study, and the control group composed of students who declined to participate. Even if the groups are similar on some dimension (e.g., achievement test scores), this does not ensure they are equivalent on other dimensions (e.g., motivation). As a result, group differences at the end of the intervention could be due to the intervention or some unmeasured differences between the groups; the same issue could occur in RCTs, if there are significant numbers of students, teachers, or schools that decline to participate (Bell, Olsen, Orr, & Stuart, 2016). The What Works Clearing House (2017) also suggests that sample sizes need to be large enough to detect a .25 standard deviation intervention effect with a power of .80. In other words, there needs to be enough participants to ensure detection of a modest intervention effect at least 80% of the time.

Even when these basic issues are addressed, Gersten et al. (2005) note additional factors that could undermine the validity or at least the usefulness (e.g., translating into classroom practices) of RCTs and quasi-experimental studies. These include clear and detailed descriptions of both the content and process of the intervention (e.g., one-on-one, small group, 10 sessions, etc.) and the content and process of the instruction received by the comparison group. The latter often includes students in business-as-usual classrooms; and clear description of what they receive in these classrooms and how it differs from the intervention is critical to the interpretation of the results of such studies (e.g., Fuchs, Geary, et al., 2013, Fuchs, Schumacher, et al., 2013). The fidelity with which the intervention was implemented, as we discuss in the next section, is essential to determining its effectiveness. The content, reliability, and validity of the measures used to assess the intervention are also critical and help to determine whether and the extent to which there are educational effects beyond the specific content covered in the intervention, as we discuss in "Transfer" section.

Fidelity of Implementation

A well-designed intervention, one focusing on enhancing students' conceptual understanding or procedural fluency in a specific area of mathematics is of course critical to the development of effective instructional approaches (National Research Council, 2004). The proper evaluation of the efficacy of lab studies and effectiveness of field studies requires that the program is implemented in ways the developer intended. The expectations for the fidelity of interventions in medicine are well established and have gained considerable traction the past two decades in education, although opportunities for improvement remain (Swanson, Wanzek, Haring, Ciullo, & McCulley, 2013). O'Donnell (2008) notes five basic criteria for ensuring fidelity of an intervention:

> *(a) adherence—whether the components of the intervention are being delivered as designed; (b) duration—the number, length, or frequency of sessions implemented; (c) quality of delivery—the manner in which the implementer delivers the program using the techniques, processes, or methods prescribed; (d) participant responsiveness—the extent to which participants are engaged by and involved in the activities and content of the program; and (e) program differentiation—whether critical features that distinguish the program from the comparison condition are present or absent during implementation*
>
> (O'Donnell, 2008, p. 34)

Demonstrating positive student gains based on interventions that have been implemented with fidelity in lab settings and in restricted educational settings (e.g., intervention implementation monitored by researchers) is just the first step. The results still leave unanswered the question of whether the intervention can be scaled up to large numbers of classrooms and implemented with fidelity without researcher supports (Cai et al., 2017). In the scaling up of interventions to real-world classrooms there are many additional threats to fidelity of implementation, including differences in teacher preparation and autonomy, characteristics of students within each classroom, and general school characteristics (e.g., level of absenteeism), among other factors (McDonald, Keesler, Kauffman, & Schneider, 2006).

Diminishing Intervention Impacts Across Time (Fadeout)

There are many interventions designed to improve children's academic skills and most of them in fact do so, that is, there are demonstrable gains in achievement at the end of such interventions relative to peers who did not participate in the intervention (Barnett, 2011). Unfortunately, nearly all of the effects of these interventions and especially those focused on specific academic skills during the preschool or elementary school years fadeout over time (Bailey, Duncan, Odgers, & Yu, 2017; Puma et al., 2012; Watts, Duncan, Clements, & Sarama, 2018). It is not that all of the gains these children achieved fade, but rather the children who did not receive the intervention tend to catch up within a few years (e.g., Clements, Sarama, Wolfe, & Spitler, 2013). Not all studies show a complete fadeout of early intervention effects, and for some interventions there are often other benefits that are not reflected in achievement scores (e.g., increase in high school graduation rate; Deming, 2009).

Even if there are (sometimes) modest longer term gains to achievement and (also modest) gains in noncognitive areas, the fade-out effects are sobering and indicate a need to rethink the approach to intervention (Bailey, this volume; Fuchs et al., this volume). Bailey et al. (2017) provide a thoughtful discussion of why fade-out effects occur and what might be done to enhance the effectiveness of interventions. Their first point is that interventions designed to build academic skills (e.g., phonemic awareness or counting) that children will be exposed to during normal schooling, whether or not they participate in an intervention, will per force result in some degree of fadeout. This is because children who were not in the intervention will eventually—often the following academic year—be instructed on these same skills and thus will catch up to some extent or completely to the children who participated in the intervention. Skills that are malleable and where the timing of their acquisition is important for subsequent learning or other critical outcomes (e.g., high school graduation) may be an exception.

Cortes and Goodman's (2014) study of the effectiveness of algebra double-dosing—students at risk for failing algebra I enroll in two algebra courses in 9th grade (standard algebra, and algebra with supports)—provides an example. The intervention students receive the same content as other students, but achieving some level of competence in this content has potentially outsized consequences. Passing algebra is critical to later mathematics learning and preparation for the labor market (Rivera-Batiz, 1992) and is also a high school graduation requirement in many states. The double-dose of algebra in 9th grade resulted in modest but significant gains in algebra grades as well as later grades in geometry and trigonometry and, at least for some students, an increase in graduation rates. Bailey et al. (2017) also highlight the importance of interventions that seek to change students' beliefs about learning; changing beliefs that mathematics takes talent that you either have or do not to a belief about the importance of hard work and persistence is one example (Blackwell, Trzesniewski, & Dweck, 2007). Their final point is that the long-term effectiveness of interventions may depend on children's subsequent placement in sustaining environments, such as the quality of the schools they attend after they complete the intervention (Currie & Thomas, 2000).

Transfer

Transfer effects refer to the relation between training or an intervention in one domain on performance in similar (near transfer) or seemingly very different (far transfer) domains. The relation between working memory training—such as practice at tracking and retaining information—on similar working memory tasks (near transfer) or on fluid intelligence or mathematics achievement (far transfer) illustrates different types of transfer. Studies of this type span more than a century, with an early conclusion that:

> *Improvement in any single mental function rarely brings about equal improvements in any other function, no matter how similar ... [the] spread of practice [transfer] occurs only where identical elements are concerned in the influencing and influenced function.*
>
> (Thorndike & Woodworth, 1901, p. 250)

The early evidence thus suggested that transfer can occur but only to the extent to which the trained competencies are embedded within the transfer task. In theory, this could occur for domain-general abilities (below), whereby training to improve working memory might transfer to fluid intelligence or to academic competencies to the extent the latter are dependent on working memory. At this time, the potential for working memory training to transfer to fluid intelligence or to academic competencies is vigorously debated (e.g., Au et al., 2015; Harrison et al., 2013), and little if any far transfer likely occurs (Melby-Lervåg, Redick, & Hulme, 2016). Melby-Lervåg et al. confirmed that working memory training often improves performance on near transfer tasks, but there was very little consistent support for transfer to other competencies (including mathematics), when appropriate controls are used (e.g., an active versus passive control group; see also Sala & Gobet, 2017).

In total, these results are consistent with Thorndike and Woodworth's (1901) early conclusion. The educational implications are that transfer effects will largely occur within the same academic domain and that the strength of transfer will vary directly with the degree of overlap between the training and transfer skills, and this does appear to be the case for mathematics (e.g., Fuchs et al., 2014; Fuchs, Malone, Schumacher, Namkung, & Wang, 2017; Gersten et al., 2008). An example is provided by Fuchs et al.'s (2017) interventions to improve children's fraction competencies. A nice feature of these interventions is their grounding in Siegler, Thompson, and Schneider's (2011) integrated theory of number development; specifically, focusing on children's understanding of fraction magnitudes. The intervention involved activities designed to focus students on the relative magnitudes of fractions, such as identifying the larger of two fractions, ordering a series of fractions from smallest to largest, and placing fractions on a number line. The fractions instruction for the business-as-usual control group focused on part-whole relations (e.g., fractional proportion of shaded areas of a rectangle; see Fuchs, Geary, et al., 2013, Fuchs, Schumacher, et al., 2013).

Relative to the control group, the interventions significantly improved students' performance on tasks that were similar to the intervention (near transfer), specifically, accuracy at placing fractions on a number line (Fuchs et al., 2017). The students also showed substantive gains in competence at procedurally solving fractions problems and on released fractions items from the National Assessment of Educational Progress. The two latter measures are not far transfer, but they are at least intermediate, as they showed that improving students' conceptual understanding of fractions transferred to their ability to solve fractions problems. Using contrasting interventions, Fuchs et al. (2017) identified how different components of the intervention interacted with students' domain-general abilities (below). For instance, students with low working memory capacity especially benefitted from explaining their reasoning behind judgments about fractions magnitudes; this explanation component was not necessary for students with adequate working memory capacity. They also found that instruction on how to solve fractions word problems that involved multiplication transferred to addition word problems, but instruction on addition did not transfer to multiplication. The implication is that the most effective use of instruction time (in terms of promoting transfer) is to focus on multiplication.

WHAT ARE THE COGNITIVE FOUNDATIONS FOR IMPROVING MATHEMATICS LEARNING?

A combination of domain-general abilities, such as working memory, and domain-specific knowledge contribute to the growth of academic competencies but their relative importance is not fully understood, including whether their contributions change over time or with students' level of expertise (Ferrer & McArdle, 2004; Gustafsson & Undheim, 1992; Von Aster & Shalev, 2007). Cattell's (1987) investment theory provides one influential example of the mix of traits that promote academic development. He focused on importance of fluid intelligence (abstract problem solving), in combination with interests and personality, in the development of crystallized intelligence, including domain-specific academic knowledge (see also Ackerman & Beier, 2006). Intelligence influences the ease of learning new information, but interests and personality (e.g., conscientiousness; Poropat, 2009) determine how and where these abilities are invested. Other researchers have demonstrated that current levels of domain-specific knowledge influence the ease of making further gains in this knowledge (Thorsen, Gustafsson, & Cliffordson, 2014; Tricot & Sweller, 2013).

As with other academic domains, the relative contributions of domain-general abilities and domain-specific knowledge to subsequent mathematics achievement are not fully understood and may vary across grade, level of student knowledge, and mathematical content (Bailey, Watts, Littlefield, & Geary, 2014; Friso-van den Bos, van der Ven, Kroesbergen, & van Luit, 2013; Fuchs, Geary, Fuchs, Compton, & Hamlett, 2016; Geary, 2011;

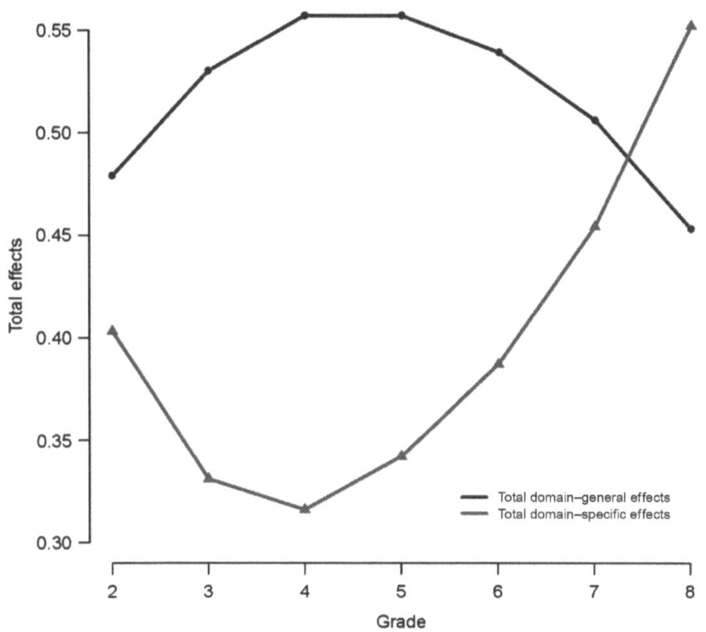

FIG. 1 Changes in the relative contribution of overall domain-general abilities, including intelligence and working memory, and domain-specific knowledge (e.g., recall of basic facts, number and fractions knowledge) on mathematics achievement in the subsequent grade. *(By Geary, D. C., Nicholas, A., Li, Y., & Sun, J.* Journal of Educational Psychology, 109*, 689. Copyright 2017, by American Psychological Association. Reprinted with permission.)*

Siegler et al., 2012; Watts et al., 2015). Two recent longitudinal studies indicated that domain-general abilities are important throughout students' mathematical development, and that the relative importance of domain-specific knowledge increases across successive grades (Geary, Nicholas, Li, & Sun, 2017; Lee & Bull, 2016). The basic pattern is shown in Fig. 1, whereby the combined effect of domain-general abilities, including working memory and intelligence, ranges between .45 and .55 from 2nd to 8th grade, inclusive; a 1 standard deviation increase in domain-general abilities is associated with about a .5 standard deviation increase in math achievement in the following grade, controlling prior math skills. In contrast, prior-grade math competencies on subsequent math achievement are much smaller than the influence of domain-general abilities in the early grades, but become as important as domain-general abilities by middle school.

A more complete understanding of the grade-to-grade contributions of domain-general and domain-specific effects on subsequent mathematics achievement and changes in the relative magnitude of these effects will facilitate our understanding of children's mathematical development generally, and provide insight into when, where, and how to target interventions to improve this development. In the following sections, we briefly overview the core domain-general competencies that contribute to mathematics achievement (and academic achievement more generally) and our current understanding of the effectiveness of domain-general and domain-specific interventions.

Domain-General Components

Intelligence, executive function, and working memory are the most consistently studied domain-general predictors of mathematical development and achievement. The relative importance of these individual abilities may vary with age, students' level of expertise in mathematics, and mathematical content.

Fluid Intelligence

Cattell's and Horn's fluid intelligence indexes people's ability to identify the underlying rules or concepts in novel problem-solving domains (Cattell, 1963; Horn, 1968). As Cattell stated, "Fluid general ability … shows more in tests requiring adaptation to new situations, where crystallized skills [domain-specific knowledge] are of no particular advantage" (Cattell, 1963, p. 3). Mathematics is an evolutionarily novel domain and thus fluid intelligence should be a significant contributor to individual differences in the ease of learning newly introduced mathematical concepts and particularly important as mathematics becomes increasingly abstract in later grades (Geary, 1995, 2005). In theory, however, the relative importance of fluid intelligence will decline once the concept is understood, but with the continuous introduction of new and more abstract concepts in the standard mathematics curriculum, fluid intelligence will remain important.

Most studies of the relation between intelligence and mathematics achievement have used a composite measure (e.g., standardized IQ test) that technically does not provide a direct assessment of fluid intelligence but will be highly correlated with it (Walberg, 1984). One of the largest of these studies included 70,000 students and found that intelligence measured at age 11 years was highly correlated ($r = .77$) with standardized mathematics achievement scores 5 years later (Deary, Strand, Smith, & Fernandes, 2007). In an analysis of almost 5000 children and adolescents, Taub and colleagues separated the contributions of other factors, such as crystallized intelligence, from fluid abilities and found the latter predicts mathematics achievement, albeit the strength of the relation was somewhat lower than that found by Deary et al. ($\beta = .37$ to $.75$) and varied across age (Taub, Keith, Floyd, & McGrew, 2008). Similar to the pattern shown in Fig. 1, contributions of crystallized knowledge and fluid intelligence to the prediction of adolescents' mathematics achievement were about the same, but crystallized knowledge was not important for younger children. Although their measure of crystallized knowledge included many areas, not just mathematics, it should be a reasonable proxy for mathematics knowledge.

This is not to say that domain-specific knowledge is not particularly important for children's early mathematics learning; it depends on the mathematics being assessed. Indeed, a similar overall pattern emerges for the age at which preschool children learn their first mathematical concept—the cardinal value or quantities represented by number words—and for their mathematics achievement generally; a combination of prerequisite domain-specific knowledge (e.g., knowing the list of count words) and intelligence is important (Geary & vanMarle, 2016; Geary, vanMarle, Chu, Hoard, & Nugent, in press). The age at which children learned this mathematical concept and again fluid intelligence independently predict these children's school-entry number knowledge and mathematics achievement 3 years later (Geary, vanMarle, Chu, Hoard, et al., in press, Geary, vanMarle, Chu, Rouder, et al., 2018).

The pattern across all of these studies is consistent with Cattell's (1987) investment theory; "... this year's crystallized ability level is a function of last year's fluid ability level—and last year's interest in school work" (Cattell, 1987, p. 139). In other words, strong fluid abilities accelerate the learning of mathematics, assuming sufficient motivation and engagement with the material, and this domain-specific foundation along with fluid abilities jointly contribute to future learning. Environmental factors such as the quantitative activities at home and quality of curricula material will be important for exposing children to the appropriate mathematics content (LeFevre et al., 2009; Ramani & Siegler, 2008).

Executive Function and Working Memory

The constructs and associated measures of executive function and working memory come from the neuropsychological and cognitive traditions, respectively, but capture many of the same underlying competencies, as we detail later. Performance on measures of fluid intelligence and working memory are at least moderately correlated (Ackerman, Beier, & Boyle, 2002), but appear to assess independent competencies (Embretson, 1995; Jurden, 1995). Our perspective is that the key aspects of working memory as related to mathematics involve attentional control and the ability to update information represented in working memory (Bull & Lee, 2014; Iuculano, Moro, & Butterworth, 2011), as contrasted with the logical problem solving and ease of concept learning captured by fluid intelligence. Whatever the specifics, both types of measures independently predict academic success (Clark, Pritchard, & Woodward, 2010; Taub et al., 2008).

The relations between executive function, working memory, and children's mathematics achievement are assessed somewhat differently for preschool- and school-age children. In the preschool studies, the focus is on a single measure of executive function and performance on such measures consistently predict later mathematics achievement (Blair & Razza, 2007; Bull, Espy, & Wiebe, 2008; Clark et al., 2010). At these ages, executive function captures the ability to maintain attentional focus on the task at hand and to inhibit prepotent responding, that is, inhibit a previously used response that is correct in one context but is not correct in the current context.

Toward the end of the preschool years, these functions differentiate into the central executive, phonological loop, and visuospatial sketchpad components of working memory (Baddeley & Hitch, 1974; Wiebe, Espy, & Charak, 2008). The central executive in turn captures the core abilities of executive function (Allan & Lonigan, 2011) and can be decomposed into the ability to suppress prepotent responses, shift attention between tasks (shifting), and explicitly monitor and update information (updating) represented in the phonological loop or visuospatial sketchpad (Miyake et al., 2000). More recently, the term "working memory" has been used to represent the updating component, and this is how we use working memory hereafter.

The studies of school-age children have consistently revealed a relation between one or several components of Baddeley and Hitch's (1974) system and mathematics achievement (DeStefano & LeFevre, 2004; Geary, Hoard, Byrd-Craven, Nugent, & Numtee, 2007; McLean & Hitch, 1999; Swanson & Sachse-Lee, 2001). The most consistent finding is that the higher the capacity of the central executive the better the performance on measures of mathematics achievement and cognition (Bull et al., 2008; Mazzocco & Kover, 2007; Passolunghi, Vercelloni, & Schadee, 2007). This relation holds across the entire continuum of mathematics achievement and is a core deficit of children with learning difficulties in mathematics (Geary, 2004). In a review of this literature, Bull and Lee (2014) concluded that working memory (the updating component) is particularly important for mathematics learning and commonly found to be associated with mathematical learning difficulties (Geary, Hoard, Nugent, & Bailey, 2012).

The strength of the relation between mathematics achievement and the phonological loop and visuospatial sketchpad is reduced and sometimes eliminated when performance on working memory and intelligence measures are controlled (Fuchs et al., 2010a, 2010b; Geary, Hoard, Nugent, & Bailey, 2013; Holmes & Adams, 2006). The relations that are found tend to be restricted to specific aspects of mathematics (Andersson, 2010; Bull et al., 2008; Geary et al., 2007; Krajewski & Schneider, 2009). The phonological loop contributes to the encoding and processing of number words and numerals and to processes that involve them (e.g., counting "six, seven, eight" to solve 6+2; Geary, 1993; Krajewski & Schneider, 2009; Logie & Baddeley, 1987). The phonological loop may also contribute to skill at solving mathematical word problems, presumably because phonological skills contribute to the comprehension of the reading-related components of word problems (Swanson & Sachse-Lee, 2001).

The visuospatial sketchpad also contributes to the solving of word problems (Johnson, 1984), and some other aspects of mathematics, such as visualizing and representing quantities on the number line (Gunderson, Ramirez, Beilock, &

Levine, 2012). The former effect is likely related to ease of constructing diagrams to represent quantitative relations in the word problems. Overall, however, the strength of the relations between visuospatial memory and mathematics performance tends to get larger across grades, potentially due to changes in the mathematical content covered in higher grades (Li & Geary, 2013, 2017; Mix & Cheng, 2011). There are also different components of visuospatial competencies and some of these may contribute to different aspects of mathematical development (Raghubar, Barnes, & Hecht, 2010). Kyttälä and Lehto (2008) found that the relation between visuospatial ability and high school students' mathematics performance, controlling for fluid intelligence, varied with whether the visuospatial task required simultaneous (maze memory), sequential (e.g., block recall), or active (3-dimensional mental rotation) processing. Memory for simultaneously presented visuospatial information predicted overall mathematics achievement, whereas sequential processing made unique contributions to performance on written word problems and active processing to geometry. Follow-up studies are needed to confirm the specificity of these relations, given the small number of these types of studies.

Although performance on measures of executive function and the central executive requires attentional focus and inhibition of distractors, these measures do not fully capture children's attentive behavior in classroom settings. In-class attentive behavior is rated by teachers and includes items such as, "Gives close attention to detail and avoids careless mistakes" (Swanson et al., 2012). These measures predict mathematics achievement above and beyond the contributions of executive function, the central executive, and fluid intelligence (Fuchs et al., 2006; Geary et al., 2013).

Implications

These types of findings and those reviewed in the next section imply that interventions that are simply focused on addressing deficits in mathematical knowledge will not be sufficient for many children, if they do not also include features that address domain-general deficits. Fuchs et al.'s (2016) approach of integrating domain-specific instruction with supports that help to compensate for any such deficits may be useful in the design of any such intervention (see also Fuchs et al., this volume). In a structured intervention focused on improving elementary student's number knowledge, Fuchs, Geary, et al. (2013), Fuchs, Schumacher, et al. (2013) found that children with below average fluid abilities gained more if the intervention included speeded practice of the instruction material at the end of each session. Children with above-average abilities also gained from this practice, but the relative gains were larger for the below-average students. More generally, meta-analyses conducted by the National Mathematics Advisory Panel (2008)—and consistent with the results from Project Follow Through (Stebbins, 1977)—indicated that students with difficulties in mathematics learning benefit from explicit, teacher-directed instruction (Gersten et al., 2008) that may help to compensate for domain-general deficits.

Domain-General Interventions

As we discussed previously, the majority of studies investigating the role of domain-general factors on mathematical learning and development have focused primarily on fluid intelligence, executive function, and working memory. This being said, it is important to point out that researchers have also begun to study a much wider range of domain-general processes that may also influence mathematical cognition and its development. In fact, a recent special issue of the *Journal of Numerical Cognition* was devoted primarily to the topic of domain-general factors that contribute to numerical and arithmetic processing.

In the introduction to this special issue, Knops, Nuerk, and Gobel (2017) reviewed the surprisingly wide range of domain-general factors examined in this collection of articles, including (in addition to intelligence, executive function, attention, and working memory): general language, mathematical language, ordinality (nonnumerical), perceptual speed, self-regulation, social power, and creativity, among others. Although we will not discuss the findings that explored the extent to which, if any, these factors are associated with mathematical skills, it is worth focusing here on the two studies in this collection that investigated the impact of domain-general interventions.

Honoré and Noël's (2017) study examined whether training visuospatial working memory in 5- to 6-year-olds would not only enhance working memory skills (i.e., visuospatial and/or verbal working memory) but also transfer to numerical abilities; specifically, whether training resulted in improvement in counting and comparison (numerical and collection) skills, number line judgments, and addition. The Cogmed visuospatial working memory training program was administered to these children for 20 min per each school day over 5 weeks. This computerized program is adaptive, in that the difficulty level (i.e., number of items to be remembered) is adjusted on a trial-by-trial basis, thereby taxing working memory capacity to its limit. A control group received the same Cogmed training program except with no adaptive feature, that is, they only had to remember the same, low level number of items they began with—one or two, and were trained for only 10 min per school day.

Interestingly, although the children in the experimental condition exhibited an improvement in visuospatial working memory, there was no effect on verbal working memory. More importantly, even though there was a small improvement in Arabic number comparison, this failed to be sustained 10 weeks following the training (which was also the case for the visuospatial working memory effect). Furthermore, the training did not even show an immediate enhancement of counting, collections comparison, or addition.

The other intervention study in the special issue (Ramani, Jaeggi, Daubert, & Buschkuehl, 2017) is summarized in the chapter by Ramani et al. (this volume). Briefly, although there was some evidence that the numerical magnitude knowledge of Kindergartners was enhanced by playing an adaptive, visuospatial working memory training game on a tablet computer, the only children who exhibited this improvement were those with initially lower, numerical magnitude knowledge. Furthermore, domain-general training was not as effective as domain-specific training with numerical magnitudes. According to Knops et al. (2017), the outcomes of these two studies suggest that significant associations between domain-general factors and numerical or arithmetic skills that have been found in prediction studies do not "easily translate" into improved mathematics learning in intervention studies.

With respect to the current volume, we direct the reader to some of the domain-general interventions that were based on one or two of the other factors discussed by Knops et al. (2017) in their special issue. Namely, both the role of mathematical language in mathematics learning and its influence via an instructional intervention are described by Purpura et al. (this volume). Relatedly, the chapter by Fuchs et al. (this volume) examines the effect of embedding language instruction (both general language and mathematical language) within a word-problem-solving intervention. In both of these cases, the interventions are focused on competencies and knowledge that are directly usable when solving many types of mathematics problems, as contrasted with more general abilities, such as visuospatial ability. Additionally, the chapter by Vamvakoussi (this volume) examines the value of incorporating analogies in mathematics instruction as analogical reasoning can be viewed as a domain-general cognitive process (Alvarez et al., 2017; Forbus, Gentner, Markman, & Ferguson, 2010). Again, the use of analogies is embedded directly in mathematics instruction, and not the focus of an independent intervention to improve general analogical reasoning abilities.

Domain-Specific Components

In their introduction to the special issue discussed previously, Knops et al. (2017) also examined in detail the nature of the domain-specific factors that were investigated in their collection of articles, as well as various meanings that have been attributed to this construct. In their analysis, they point out that this term has been applied to several different components of mathematical cognition studies, including alleged domain-specific *tasks*, *behaviors*, *representations*, and *processing mechanisms*. Rather than exploring these finer distinctions here, we prefer to point out that the domain-specific numerical and mathematical representations, processing mechanisms, and skills investigated to date include, but are not limited to: approximate and exact number judgments; spatial-numerical associations; counting skills, addition and subtraction operations; numerical estimation; fractions knowledge (conceptual and procedural) and arithmetic; and algebraic as well as geometrical reasoning.

Domain-Specific Interventions

Researchers have developed a rather diverse set of interventions for improving domain-specific components of math skills ranging from basic numerical representations and processing mechanisms to higher order mathematical concepts and operations. Examples of such interventions are discussed in many of the following chapters, with the majority focusing on improving the numerical and arithmetic skills of preschool- and early school-age children. In this section, we have chosen to describe a computer-based program for training number line estimations by children with mathematical learning disabilities and their typically achieving peers, as it provides an interesting intervention that targets a numerical skill that is generally considered a critical foundation for subsequent learning of both whole number and rational number arithmetic.

Computer-Based Training of Number Line Judgments

As several of the authors in the present volume describe (Aunio, this volume; Langfus et al., this volume; Ramani et al., this volume; Sarama & Clements, this volume), a growing number of computer-based games have been developed for improving basic numerical and arithmetic skills in children. In this section, we describe a software program that focuses primarily on training number line judgments, known as *Rescue Calcularis*. Development of this program was originally based on neuropsychological and neuroimaging findings from studies of children with developmental dyscalculia (DD) and designed specifically for improving their construction of and access to a mental number line (Kucian et al., 2011). In the first published study making use of this software, 8- to 10-year-old children with DD and an age-matched control group of typically developing (TD) children were shown an Arabic digit, an addition or subtraction problem, or different numbers of dots, and were tasked with positioning the solution to each problem (i.e., a number) on a number line using a joystick. These problems were administered within the context of saving one's home planet with the use of a spaceship. Following 5 weeks of training (15 min a day, 5 days a week), children in both groups showed improved performance on both the number line and arithmetic tests. Furthermore, use of a brain imaging technique—(fMRI, functional magnetic resonance imaging)—revealed reduced recruitment of relevant brain regions (i.e., bilateral parietal) supporting number line processing, which the authors attributed to an automatization of the cognitive processes required for solving number line problems.

In a more recent study employing the Rescue Calcularis game, Michels, O'Gorman, and Kucian (2018) administered number line training to 9.5-year-old children with developmental dyscalculia (DD) over a 5-week period, as well as to TD children to whom the DD participants were matched on age, gender, and handedness. Furthermore, all children were pre- and posttested on a numerical ordinality judgment task during which fMRI and functional connectivity data were collected. The children with developmental dyscalculia exhibited abnormally high hyperconnectivity in frontal, parietal, temporal, and visual regions prior to the training; in other words, they were engaging too many brain regions when attempting to order numerals. The numerical and arithmetic performance of the DD group as well as their number line skills improved significantly following training, but only the latter reached the same level of accuracy as that of the TD group. And of particular interest is that not only did the functional hyperconnectivity of the DD children decline to the point of disappearing, but also a classification analysis of the posttest fMRI data (controlling for age and IQ) could not distinguish between the DD and TD groups. These outcomes led the authors to conclude that number line training normalized the DD children's prior, aberrant neuronal activity and efficiency and generated widespread changes across the distributed brain regions involved in the various stages of information processing needed for them to make correct numerical order judgments.

A recent offshoot of the Rescue Calcularis game is Calcularis—a computer-based training program that like its forerunner focuses on the learning of basic numerical skills, arithmetic operations, and number line estimation (Rauscher et al., 2016). Although this game retains the core features of "Rescue Calcularis"—primarily emphasizing the construction of and access to the mental number line—it includes a more comprehensive training of math skills and is also more flexibly adaptive to each child's learning profile and knowledge state. The program is based upon the robust theoretical principles of both the Triple-Code model (Dehaene, 1992) and a four-step developmental model (Von Aster & Shalev, 2007). Briefly, the two areas of the training program cover: (1) transcoding between alternative representations of number and principles of numerical understanding such as cardinality and ordinality, with the component games ordered hierarchically; and (2) the concepts underlying arithmetic operations and their automatization. Seven- to ten-year-old typically developing children were randomly assigned to a Calcularis training group who received 6–8 weeks of daily (5 days per week) training at home, a computer-based, spelling training control group, or an untrained wait-list control group. The Calcularis group showed significantly greater improvements in number line estimation and subtraction (but not in addition) as compared to both control conditions, with medium to large effect sizes.

As just described, these number line training programs have yielded some important findings. That being said, replications and extensions of such studies are certainly needed before larger scale interventions making use of these programs are warranted. Indeed, a recent critical review and analysis of the numerical training literature by Szucs and Myers (2017) focusing on improving both approximate and exact components of number sense reveal a sizable number of problems with this research. Several of the 10 studies reviewed were considered to have been poorly designed (e.g., poor choice of control condition activities); lacked sufficient power; used inadequate statistical procedures; failed to follow up the most crucial alternative hypotheses of the findings that either had been or could be raised; and there was a severe shortage of critical comments pertaining to the methods, findings, and conclusions of other relevant studies cited. Szucs and Myers conclude by recommending specific research designs, methods, procedures, and statistical analyses; thorough reporting of this information in published articles that should avoid post hoc theorizing and making convoluted arguments when explaining findings; and overcoming citation biases by taking a more critical stance when evaluating evidence from relevant prior studies. This rather devastating critique is certainly worthy of serious attention by researchers planning future training studies.

Domain-Relevant Interventions

In addition to domain-general and domain-specific interventions for improving math skills, we suggest another category that is worthy of consideration: *domain-relevant* interventions. This term represents approaches aimed at enhancing numerical and arithmetic processing indirectly via either training of nonquantitative but still pertinent skills or activation of domain-general cognitive processes that are more relevant for improving numerical judgments than nonquantitative ones (Karmiloff-Smith, 2015; Knops et al., 2017). Examples that we cover here include: (a) providing alerting cues that engage attention toward stimulus features that can enhance the speed of executing enumeration judgments; (b) training nonquantitative, finger differentiation skills that have been shown to be associated with arithmetic abilities; and (c) delivering noninvasive, mild electrical stimulation to cortical brain regions that have been reliably implicated in numerical processing.

Training to Enhance Subitizing Speed

Subitizing refers to the fast and accurate enumeration of small sets of elements (e.g., dots), typically within a range of 1–4 items (Mandler & Shebo, 1982; Trick & Pylyshyn, 1994; see also Sarama & Clements, this volume, for a more detailed examination of subitizing in young children). Although subitizing was historically considered to be a preattentive process, more recent research has shown that if attentional engagement in this enumeration task is reduced (e.g., by concurrent tasks designed to capture attentional resources), the subitizing process is impaired. This finding led Gliksman, Weinbach, and Henik (2016) to ask whether inducing an *increase* in attentional engagement during an enumeration task could lead to improved performance. Alertness denotes the attentional system's state of general readiness to react to a given sensory stimulus that regulates attentional processing to that stimulus as well as the initiation of a response to it (Petersen & Posner, 2012; Wiegand, Petersen, Bundesen, & Habekost, 2017). And as Gliksman et al. point out, researchers can manipulate alerting by presenting brief warning cues, usually an auditory tone or beep that induces a state of high arousal. Importantly, these warning cues are considered neutral in that they provide no information concerning either the identity of the ensuing target stimulus or its spatial location.

Furthermore, subitizing has been shown to be based in part on pattern recognition (Ashkenazi, Mark-Zigdon, & Henik, 2013; Mandler & Shebo, 1982), where quantities within the subitizing range (e.g., three dots) can be identified as familiar whole figures comprised of smaller elements, for example, the spatial configuration of three dots may form a triangle. Taking such evidence into account, Glicksman et al. also proposed that alerting could facilitate the subitizing process by enhancing *global* perceptual processing of geometric shapes such as a triangle and square composed, respectively, of three and four elements in a set; in contrast to a random arrangement of say, seven dots where the individual dots in the array are likely to be processed *locally* (Navon, 1977). Furthermore, these authors noted that strong evidence already exists from other kinds of tasks (e.g., face recognition; letter identification) indicating that alerting cues can induce a global processing bias.

The participants in their first experiment were required to say aloud the number of dots presented in random arrays, consisting of those considered to be within the subitizing range (1–4 dots) and those within the small estimation range (5–9 dots). An alerting cue (auditory tone) appeared briefly before the display of the target array on half of the trials. Although alerting cues facilitated the response times for correctly enumerating the quantities within the subitizing range, they had no such effect on the small estimation range. The authors hypothesized that the facilitative effect of alerting cues on the subitizing process may be attributable to the enhancement of global processing, a process that was previously associated with both alerting and subitizing.

To test this hypothesis, they carried out a second experiment with college students where the procedure was basically the same as that of Experiment 1, except that the arrays containing quantities within the small estimation range were presented in canonical arrangements creating hierarchical shapes—global figures made up of dots such as 6 dots in two vertical, parallel rows of 3 each, as on a die. They reasoned that if alerting regulates the subitizing process by enhancing global processing, then it should likewise improve performance even in the small estimation range, if the quantities are arranged in a canonical array. In other words, small local elements should create a larger global element that would enable participants to apprehend all the displayed dots at the same time rather than having to count each of them in succession. Indeed, this is exactly what happened. Furthermore, Gliksman et al. suggested that future research may be able to investigate whether extant training procedures known to increase alertness and global processing could also enhance enumeration in special populations such as those with developmental dyscalculia.

Training Finger Differentiation: Its Impact on Arithmetic Ability

In the chapter concerning early math development in the home environment, Levine et al. (this volume) discuss numerous studies concerning the role of gesture in general and finger pointing and counting in particular as these may contribute to the early development of numerical and arithmetic skills. One of the interesting findings they mention is an association between children's math skills and the ability to discriminate between their fingers. More specifically, research on this topic begins with measuring the ability to distinguish between one's own fingers when they are touched with a physical object with no visual cues available. This is skill is formally known as *finger gnosis*. To date, several studies have reported significant correlations between finger gnosis and numerical as well as arithmetic abilities.

Fayol, Barrouillet, and Marinthe (1998) found that a measure of finger gnosis administered at age 5 predicted numerical ability up to 3 years later—even better than a measure of general intelligence. Similarly, Noel (2005) found that finger gnosis in first grade predicted the ability to map numerals and their associated magnitudes (but not reading ability) in second grade. Likewise, studies carried out by Penner-Wilger and Anderson (2013) and Penner-Wilger et al. (2009) reported associations between finger gnosis and several kinds of numerical tasks including magnitude comparisons and number line estimation. In contrast, several studies have failed to find any such associations (Long et al., 2016; Newman, 2016; Soylu, Raymond, Gutierrez, & Newman, 2017). Soylu, Lester, and Newman (2018) suggest that evidence demonstrating the codevelopment of finger gnosis, counting, and arithmetic skills might explain some of the conflicting findings on the relation between finger gnosis and arithmetic performance.

Soylu, Lester, and Newman (2018) also provide a detailed analysis of three purported mechanisms underlying these associations: the localization model—close proximity of brain regions (in the parietal cortex) associated with both finger and number processing; the functional model—a functional relation between fingers and number processing that can be attributed to a developmental association originating in early finger counting experiences; and an evolutionary model—finger sensorimotor neural circuits originally evolved for visually guiding fine motor movements are reused to accommodate the representation and processing of numerical quantities.

In the first attempt to test the functional hypothesis (and the only one to do so prior to 2016), Gracia-Bafalluy and Noel, 2008 administered 8 weeks of finger differentiation training to a group of first graders with poor finger gnosis skills. The results showed greater strengthening of finger gnosis for this group as compared with another group of first graders who were trained on story comprehension and a third group of children with initially high finger gnosis skills who spent 8 weeks engaging in business as usual. Furthermore, arithmetic abilities and ordinality judgments also improved significantly.

However, Fischer (2010) criticized this study, arguing that the purported effects of finger differentiation training may be at least partially attributable to a statistical artifact: regression toward the mean.

Testing a larger sample and employing much better controls, Wasner, Nuerk, Martignon, Roesch, and Moeller (2016) found that finger gnosis was a unique albeit weak predictor (accounting for only 1%–2% of the variance) of children's preliminary arithmetic skills at the onset of first grade, independent of the effects of age, gender, general cognitive ability, short-term memory, and antecedents of numerical competencies. Nevertheless, they concluded that their findings are of some theoretical import.

Using Electrical Brain Stimulation to Improve Numerical and Arithmetic Processing

Over the past decade, there has been a growing effort to examine whether weak (low intensity), transcranial electrical stimulation (tES) of the brain can be used for enhancing cognition (Sarkar & Cohen Kadosh, 2016a, 2016b). Indeed, enough studies have been published to have permitted several meta-analyses to be carried out (Simonsmeiera, Grabner, Heina, Krenza, & Schneidera, 2018). Nevertheless, most researchers working in this area would agree that this kind of research is still in its early stages, and that its use for improving numerical cognition in particular is even newer. Sarkar and Cohen Kadosh (2016a, 2016b) have provided a thorough review of such studies, examining the principles underlying this technology and the findings concerning its uses for enhancing numerical and arithmetic processing and skills. Here, we briefly describe two studies that demonstrate some of the potential value of these approaches.

First, it should be noted that the two most frequently used forms of tES are transcranial direct current stimulation (tDCS) and transcranial random noise stimulation (tRNS). Basically, tDCS delivers a constant current through two electrodes attached to the scalp that produces anodal stimulation where it enters the brain (the anode), and cathodal stimulation where it leaves the brain and reenters the electrode (the cathode). The more recently applied form, tRNS, generates randomly assigned amplitudes of current that fluctuate between positive and negative amplitudes, producing "noise" in the targeted brain region, exerting excitatory effects at both electrodes. Furthermore, the rationale for selecting what are considered reliable target regions for brain stimulation is based on evidence drawn from neuroimaging research (including meta-analyses) that denote the locations where cortical activation has taken place during certain kinds of tasks or events. As Sarkar and Cohen Kadosh (2016a) note the parietal lobes and their subregions constitute the two areas of brain that have been implicated in numerical tasks most consistently across a substantial number of studies.

Effects of brain stimulation on numerosity discrimination. Cappelletti, Barth, Fregni, Spelke, and Pascual-Leone (2007) administered tRNS bilaterally to the parietal lobes of healthy adults coupled with intensive cognitive training on a standard numerosity discrimination task—560 trials an hour a day for five successive days to determine whether precision of the approximate number system (ANS) could be improved. This condition was compared with three other conditions: cognitive training without brain stimulation, stimulation with no cognitive training, and cognitive training coupled with stimulation bilaterally to the motor cortex (i.e., a control brain region). The participants in the parietal stimulation plus training group exhibited significantly more improvement than did the participants in all of the other three conditions, and lasted for a longer period of time—up to 16 weeks following training. Furthermore, the improved ANS precision of the combined experimental condition exhibited transfer to other quantity judgment tasks (i.e., spatial and temporal discriminations). According to Sarkar and Cohen Kadosh (2016b), parietal stimulation facilitated the cortical plasticity associated with learning, inducing both the initial improvement in numerical discrimination and its maintenance over time.

Effects of brain stimulation on arithmetic skills. Snowball et al. (2013) carried out a brain stimulation study with adults in which arithmetic skills was trained over 5 consecutive days using two types of problems: (1) drills requiring recall of answers to problems previously presented, and (2) problems requiring calculation by manipulating numerical operands in accord with a specific algorithm. Control tasks were also administered. The experimental group's cognitive training was accompanied by brain stimulation (tRNS) applied bilaterally to the dorsolateral prefrontal cortex (a region implicated in the acquisition of arithmetic), while a control group received sham stimulation during cognitive training. (As Sarkar & Cohen Kadosh, 2016b describe it, sham stimulation consists of applying a current of the same strength and intensity as the treatment condition, but for a significantly shorter duration, e.g., 30 s as compared with 5 min—a typically effective duration, in order to rule out placebo effects.) The authors found that not only did brain stimulation coupled with cognitive training improve performance on both the drill and calculation tasks, but also that these improvements were maintained for a period of 6 months after training.

PARENTAL INFLUENCES ON CHILD COGNITION AND MATHEMATICAL LEARNING

Much of what children learn occurs outside of formal educational settings. This learning includes human universals, such as language, that do not in fact require formal schooling, as well as knowledge that has been viewed by many as the purview of formal education, including symbolic mathematics. We now understand that parents can significantly influence children's initial learning of symbolic mathematics and their readiness for mathematics learning in school. In the following, we provide overviews of parental influences on children's learning in general, as related to numeracy, and associated methodological issues.

A Brief History of Research on Parental Influences

Until recent decades, little was known about mathematical learning prior to formal instruction, and even less about parental and home influences on this learning. This is surprising because from a broader evolutionary perspective, we would expect important early influences on children's learning following from the significant expansion of brain size during hominid evolution, an associated increase in the length of the human developmental period, and almost certainly an increase in the amount of information children must learn to prepare them for adulthood (Bogin, 1999; Foley & Lee, 1991; Geary & Bjorklund, 2000; Wynn, 2002). These shifts were later accompanied by the earliest indications of symbolic representations during our evolutionary history (Street, Navarrete, Reader, & Laland, 2017), and eventually those related to number (Geary, Berch, & Mann Koepke, 2015). Such changes provided the foundation for culture and its cross-generational "evolution" (Richerson & Boyd, 2005), and placed a premium on children's ability to learn (Geary, 2007). The earliest examples of cultural transfer of knowledge no doubt began through children's observation and imitation of adults (Lancy, 2014; Tramacere, Pievani, & Ferrari, 2017) and likely included play with others, including older children, with imitation and emulation at its core (Riede, Johannsen, Hogberg, Nowell, & Lombard, 2018).

Thus it is perhaps surprising that the major and best known theories of cognitive development make only modest reference to parental influences. Despite Piaget's strong interest in evolutionary theories, his own theory of cognitive development placed little emphasis on the role of imitation in learning or to the direct influence parents (or any other adult) may have on that development. Instead, the developmental progression through Piaget's four biologically preordained stages of development required that parents only be aware of the child's incomplete knowledge and then provide an environment that would enable growth of that knowledge. In Piaget's theory, the child is the active agent moving through the environment, constructing new knowledge as she passes through the cognitive development stages.

In contrast, Vygotsky placed much greater emphasis on the social nature of the coconstruction of knowledge and "zones of proximal development" (Vygotsky, 1978, pp. 84–91), highlighting the role of others, including parents in cognitive development. However, in Vygotsky's theory, parents are adjuvants to cognitive development; they have little direct effect and are not discussed at length. Only Bandura's (1986) social learning theory affords parents a substantial and direct role in their child's cognitive development (see also Rosenthal & Zimmerman, 1978). Here, observation and imitation are paramount to cognitive development and parents as "models" have a pivotal role in shaping their child's development through both reinforcing behaviors and modeling. These theories focus our attention on the importance of children's social context for their early learning but have not been thoroughly assessed with respect to children's mathematical development (NMAP, 2008).

Fortunately, more recent investigators have focused on the specifics of how parents can influence their children's early mathematical cognition (e.g., Aunio, Tapola, Mononen, & Niemivirta, 2016; Casey et al., 2018; Kleemans, Segers, & Verhoeven, 2016; Skwarchuk, Sowinski, & LeFevre, 2014). These studies are a crucial step in the right direction, but span very different populations and use diverse assessments, so it is difficult to generalize from them. Further, the studies do not include all the parental variables that may be relevant or of interest; these would include parental math anxiety (del Rio, Susperreguy, Strasser, & Salinas, 2017), parental expectations of child aptitude or achievement (Lefevre, Clarke, & Stringer, 2002), parental valuation of math learning (Di Paola, 2016), or parental beliefs about the aptitude for or importance of mathematics for boys and girls (del Rio et al., 2017; Stoet, Bailey, Moore, & Geary, 2016). Nevertheless, to date, the most consistent parental effects are seen in parental numeracy expectancy and parental engagement.

Research on Children's Home Numeracy Environment

As mentioned, we have achieved important insights into the characteristics of children's home numeracy environment that are associated with their mathematical development and later mathematics achievement (e.g., Blevins-Knabe & Austin, 2016; Elliott & Bachman, 2018; Purpura et al., this volume; Ramani et al., this volume). These characteristics include parental socioeconomic status (SES; Anders et al., 2012; Ginsburg & Russell, 1981); parental education (Anders et al., 2012; Purpura & Reid, 2016); literacy activities (reading) (Anders et al., 2012; Deng, Silinskas, Wei, & Georgiou, 2015); parental number vocabulary (Skwarchuk et al., 2014), frequency, and variety of number words and number-related activities that parents use when engaged with their children (Cankaya & LeFevre, 2016; Casey et al., 2018; Kleemans, Peeters, Segers, & Verhoeven, 2012; Kluczniok, Lehrl, Kuger, & Rossbach, 2013; LeFevre et al., 2009; LeFevre, Polyzoi, Skwarchuk, Fast, & Sowinski, 2010; Zippert & Ramani, 2017; see also Pupura et al., this volume); and home enrichment with numeracy-related materials (e.g., board games) for children to explore (Skwarchuk, Vandermaas-Peeler, & LeFevre, 2016; see also Ramani et al., this volume).

As an example, Casey et al. (2018) found that parents' number talk with their 3-year-old children and especially their bringing attention to number sets (e.g., this is "three" diamonds, ◆◆◆) were predictive of their later mathematics achievement in kindergarten and first grade, controlling a host of child and parental characteristics (e.g., intelligence). Other studies have shown that preschoolers' frequent engagement in board games that involve numbers and counting is associated with later number knowledge and mathematics achievement (Benavides-Varela et al., 2016; Ramani & Siegler, 2008; see also Ramani et al., this volume). Napoli and Purpura (2018) found that parental report of engagement in number-related activities with their children (e.g., counting, pointing out and naming numerals) was correlated with children's later number knowledge (see also LeFevre et al., 2009). Still, much remains to be determined, despite significant progress over the last several decades.

Methodological Issues

The internal consistency and contributions of the previously cited studies are strong, but there are many issues surrounding the definition of numeracy activities and their measurement that remain to be resolved before the consistency of *across*-study findings can be fully evaluated (see Elliott & Bachman, 2018). For example, many studies use a single category of "number-related activities," while other studies break this into distinct categories of "formal activities" and "informal activities." Another inconsistency across studies and perhaps even within some studies is the identification of whether or not there are siblings within the home. Sibling play may increase the time spent in informal number-related activities but decrease the quality of that activity time. Likewise, few studies currently report on the use of full-time or part-time out-of-the-home care; this daycare may include enriched or impoverished, but otherwise unreported number-related activities and serve as an important confound to study results.

Many, if not most, parental and home environment studies rely heavily on parental report. Although checklists can add consistency to this data collection, several concerns remain around self-report reliability, as well as accuracy for delayed recall of frequencies and past events. A few investigators have attempted to overcome concerns of self-report data by conducting naturalistic observations in the home or laboratory-based observations of parent-child interactions. Although these measures can be less biased, more reliable, and eliminate issues around recall of activities, such assessments also add to the cost of the study, may add time and scheduling burdens to the participants and the investigator, and may elicit a Hawthorne effect during periods of direct observation. Finally, the digital world is rapidly approaching saturation levels, yet few studies currently report or include the digital connectivity of homes and the use of electronic devices to engage children in number-related activities (see Ramani et al., this volume). How parents are or will be using digital devices as part their repertoire of number-related activities will be an important measure to include in future research.

CONCLUSION

As outlined in the first four volumes of this series, research on children's mathematical knowledge and its development has blossomed over the past three decades. We now have a much firmer grasp on infants' and young children's intuitive sense of number, the evolved systems that support them, and how these systems interact with early experiences to help children acquire their first understanding of symbolic mathematics (Volume 1). The development of functional and structural brain imaging techniques has led to insights into the neural systems that support people's intuitive understanding of number and that support mathematics learning, including how the brain changes in response to targeted interventions (Volume 2). Advances in the cognitive sciences have led to insights into human learning generally, and the application of these insights to mathematics has resulted in a much deeper understanding of the cognitive mechanisms that can impede or facilitate mathematics learning inside and outside of the classroom (Volume 3). Important insights have also been gained regarding the wider influences of culture (e.g., market economic, mathematics curriculum) and language (e.g., structure of number words) on children's mathematical development (Volume 4).

In this final volume, we focus more directly on educational practices. The latter of course includes the standard curriculum that will be well known to readers of this series, but our focus is on the specifics of how informal parental activities can foster children's learning of mathematics, and on the formal interventions that are needed to support the development of children who are at risk for long-term difficulties in learning mathematics. Much has been learned, but much remains to be done. The reduction of intervention fadeout and the development of interventions for students with domain-general (e.g., working memory) deficits are significant issues that remain to be tackled. Although these are daunting issues, we are confident that advances in the coming decades will make substantive inroads into addressing them.

REFERENCES

Ackerman, P. L., & Beier, M. E. (2006). Determinants of domain knowledge and independent study learning in an adult sample. *Journal of Educational Psychology*, *98*, 366–381.

Ackerman, P. L., Beier, M. E., & Boyle, M. O. (2002). Individual differences in working memory within a nomological network of cognitive and perceptual speed abilities. *Journal of Experimental Psychology: General*, *131*, 567–589.

Allan, N. P., & Lonigan, C. J. (2011). Examining the dimensionality of effortful control in preschool children and its relation to academic and socioemotional indicators. *Developmental Psychology*, *47*, 905–915.

Alvarez, J., Abdul-Chani, M., Deutchman, P., DiBiasie, K., Iannucci, J., Lipstein, R., et al. (2017). Estimation as analogy-making: evidence that preschoolers' analogical reasoning ability predicts their numerical estimation. *Cognitive Development*, *41*, 73–84.

Anders, Y., Rossbach, H. G., Weinert, S., Ebert, S., Kuger, S., Lehrl, S., et al. (2012). Home and preschool learning environments and their relations to the development of early numeracy skills. *Early Childhood Research Quarterly*, *27*, 231–244.

Andersson, U. (2010). Skill development in different components of arithmetic and basic cognitive functions: findings from a 3-year longitudinal study of children with different types of learning difficulties. *Journal of Educational Psychology*, *102*, 115–134.

Ashkenazi, S., Mark-Zigdon, N., & Henik, A. (2013). Do subitizing deficits in developmental dyscalculia involve pattern recognition weakness? *Developmental Science*, *16*, 35–46.

Au, J., Sheehan, E., Tsai, N., Duncan, G. J., Buschkuehl, M., & Jaeggi, S. M. (2015). Improving fluid intelligence with training on working memory: a meta-analysis. *Psychonomic Bulletin & Review*, *22*, 366–377.

Aunio, P., Tapola, A., Mononen, R., & Niemivirta, M. (2016). Early mathematics skill development, low performance, and parental support in the Finnish context. In B. Blevins-Knabe & A. B. M. Austin (Eds.), *Early childhood mathematics skill development in the home environment* (pp. 51–70). New York: Springer.

Baddeley, A. D., & Hitch, G. J. (1974). Working memory. In G. H. Bower (Ed.), *The psychology of learning and motivation: Advances in research and theory* (Vol. 8, pp. 47–90). New York: Academic Press.

Bailey, D., Duncan, G. J., Odgers, C. L., & Yu, W. (2017). Persistence and fadeout in the impacts of child and adolescent interventions. *Journal of Research on Educational Effectiveness*, *10*, 7–39.

Bailey, D. H., Watts, T. W., Littlefield, A. K., & Geary, D. C. (2014). State and trait effects on individual differences in children's mathematical development. *Psychological Science*, *25*, 2017–2026.

Bandura, A. (1986). *Social foundations of thought and action: A social cognitive theory*. Englewood Cliffs, NJ: Prentice-Hall.

Barnett, W. S. (2011). Effectiveness of early educational intervention. *Science*, *333*, 975–978.

Bell, S. H., Olsen, R. B., Orr, L. L., & Stuart, E. A. (2016). Estimates of external validity bias when impact evaluations select sites nonrandomly. *Educational Evaluation and Policy Analysis*, *38*, 318–335.

Benavides-Varela, S., Butterworth, B., Burgio, F., Arcara, G., Lucangeli, D., & Semenza, C. (2016). Numerical activities and information learned at home link to the exact numeracy skills in 5–6 years-old children. *Frontiers in Psychology*, *7*, 94.

Benbow, C. P., & Faulkner, L. R. (2008). Rejoinder to the critiques of the National Mathematics Advisory Panel Final Report. *Educational Researcher*, *37*, 645–648.

Blackwell, L. A., Trzesniewski, K. H., & Dweck, C. S. (2007). Implicit theories of intelligence and achievement across the junior high school transition: a longitudinal study and an intervention. *Child Development*, *78*, 246–263.

Blair, C., & Razza, R. P. (2007). Relating effortful control, executive function, and false belief understanding to emerging math and literacy ability in kindergarten. *Child Development*, *78*, 647–663.

Blevins-Knabe, B., & Austin, A. B. M. (Eds.), (2016). *Early childhood mathematics skill development in the home environment*. New York: Springer.

Boaler, J. (2008). When politics took the place of inquiry: a response to the National Mathematics Advisory Panel's review of instructional practices. *Educational Researcher*, *37*, 588–594.

Bogin, B. (1999). *Patterns of human growth* (2nd ed.). Cambridge, UK: Cambridge University Press.

Brickman, L., & Reich, S. M. (2015). Randomized control trials: a gold standard or gold plated? In S. I. Donaldson, C. A. Christie, & M. M. Mark (Eds.), *Credible and actionable evidence: The foundation for rigorous and influential evaluations* (pp. 83–113). Thousand Oaks, CA: Sage Publications.

Bull, R., Espy, K. A., & Wiebe, S. A. (2008). Short-term memory, working memory, and executive functioning in preschoolers: longitudinal predictors or mathematical achievement at age 7 years. *Developmental Neuropsychology*, *33*, 205–228.

Bull, R., & Lee, K. (2014). Executive functioning and mathematics achievement. *Child Development Perspectives*, *8*, 36–41.

Cai, J., Morris, A., Hohensee, C., Hwang, S., Robison, V., & Hiebert, J. (2017). Making classroom implementation an integral part of research. *Journal for Research in Mathematics Education, 48*, 342–347.

Cai, J., Morris, A., Hohensee, C., Hwang, S., Robison, V., & Hiebert, J. (2018). The role of replication studies in educational research. *Journal for Research in Mathematics Education, 49*, 2–8.

Cankaya, O., & LeFevre, J. A. (2016). The home numeracy environment: what do cross-cultural comparisons tell us about how to scaffold young children's mathematical skills? In B. Blevins-Knabe & A. B. M. Austin (Eds.), *Early childhood mathematics skill development in the home environment* (pp. 87–104). New York: Springer.

Cappelletti, M., Barth, H., Fregni, F., Spelke, E. S., & Pascual-Leone, A. (2007). rTMS over the intraparietal sulcus disrupts numerosity processing. *Experimental Brain Research, 179*, 631–642.

Casey, B. M., Lombardi, C. M., Thomson, D., Nguyen, H. N., Paz, M., Theriault, C. A., et al. (2018). Maternal support of children's early numerical concept learning predicts preschool and first-grade math achievement. *Child Development, 89*, 156–173.

Cattell, R. B. (1963). Theory of fluid and crystallized intelligence: a critical experiment. *Journal of Educational Psychology, 54*, 1–22.

Cattell, R. B. (1987). *Intelligence: Its structure, growth, and action.* Amsterdam: North-Holland.

Clark, C. A. C., Pritchard, V. E., & Woodward, L. J. (2010). Preschool executive functioning abilities predict early mathematics achievement. *Developmental Psychology, 46*, 1176–1191.

Clements, D. H., Sarama, J., Wolfe, C. B., & Spitler, M. E. (2013). Longitudinal evaluation of a scale-up model for teaching mathematics with trajectories and technologies: persistence of effects in the third year. *American Educational Research Journal, 50*, 812–850.

Cobb, P., Yackel, E., & Wood, T. (1992). A constructivist alternative to the representational view of mind in mathematics education. *Journal for Research in Mathematics Education, 23*, 2–33.

Cortes, K. E., & Goodman, J. S. (2014). Ability-tracking, instructional time, and better pedagogy: the effect of double-dose algebra on student achievement. *American Economic Review, 104*, 400–405.

Currie, J., & Thomas, D. (2000). School quality and the longer-term effects of head start. *Journal of Human Resources, 35*, 755–774.

Deary, I. J., Strand, S., Smith, P., & Fernandes, C. (2007). Intelligence and educational achievement. *Intelligence, 35*, 13–21.

Dehaene, S. (1992). Varieties of numerical abilities. *Cognition, 44*, 1–42.

del Rio, M. F., Susperreguy, M. I., Strasser, K., & Salinas, V. (2017). Distinct influences of mothers and fathers on kindergartners' numeracy performance: the role of math anxiety, home numeracy practices, and numeracy expectations. *Early Education and Development, 28*, 939–955.

Deming, D. (2009). Early childhood intervention and life-cycle skill development: Evidence from head start. *American Economic Journal: Applied Economics, 1*, 111–134.

Deng, C. P., Silinskas, G., Wei, W., & Georgiou, G. K. (2015). Cross-lagged relationships between home learning environment and academic achievement in Chinese. *Early Childhood Research Quarterly, 33*, 12–20.

DeStefano, D., & LeFevre, J. (2004). The role of working memory in mental arithmetic. *European Journal of Cognitive Psychology, 16*, 353–386.

Di Paola, B. (2016). Why Asian children outperform students from other countries? Linguistic and parental influences comparing Chinese and Italian children in preschool education. *Mathematics Education, 11*, 3351–3359.

Donaldson, S. I., Christie, C. A., & Mark, M. M. (Eds.), (2015). *Credible and actionable evidence: The foundation for rigorous and influential evaluations*. Thousand Oaks, CA: Sage Publications.

Elliott, L., & Bachman, H. J. (2018). How do parents foster young children's math skills? *Child Development Perspectives*, *12*, 16–21.

Embretson, S. E. (1995). The role of working memory capacity and general control processes in intelligence. *Intelligence*, *20*, 169–189.

Fayol, M., Barrouillet, P., & Marinthe, C. (1998). Predicting arithmetical achievement from neuro-psychological performance: a longitudinal study. *Cognition*, *68*, B63–B70.

Ferrer, E., & McArdle, J. J. (2004). An experimental analysis of dynamic hypotheses about cognitive abilities and achievement from childhood to early adulthood. *Developmental Psychology*, *40*, 935–952.

Fischer, J.-P. (2010). Numerical performance increased by finger training: a fallacy due to regression toward the mean? *Cortex*, *46*, 272–273.

Foley, R. A., & Lee, P. C. (1991). Ecology and energetics of encephalization in hominid evolution. *Philosophical Transactions of the Royal Society of London. Series B, Biological Sciences*, *334*, 223–232.

Forbus, K. D., Gentner, D., Markman, A. B., & Ferguson, R. W. (2010). Analogy just looks like high level perception: why a domain-general approach to analogical mapping is right. *Journal of Experimental & Theoretical Artificial Intelligence*, *10*, 231–257.

Friso-van den Bos, I., van der Ven, S. H., Kroesbergen, E. H., & van Luit, J. E. (2013). Working memory and mathematics in primary school children: a meta-analysis. *Educational Research Review*, *10*, 29–44.

Fuchs, L. S., Fuchs, D., Compton, D. L., Powell, S. R., Seethaler, P. M., Capizzi, A. M., et al. (2006). The cognitive correlates of third-grade skill in arithmetic, algorithmic computation, and arithmetic word problems. *Journal of Educational Psychology*, *98*, 29–43.

Fuchs, L. S., Geary, D. C., Compton, D. L., Fuchs, D., Hamlett, C. L., & Bryant, J. D. (2010a). The contributions of numerosity and domain-general abilities for school readiness. *Child Development*, *81*, 1520–1533.

Fuchs, L., Geary, D. C., Compton, D. L., Fuchs, D., Hamlett, C. L., Seethaler, P. M., et al. (2010b). Do different types of school mathematics development depend on different constellations of numerical versus general cognitive abilities? *Developmental Psychology*, *46*, 1731–1746.

Fuchs, L. S., Geary, D. C., Compton, D. L., Fuchs, D., Schatschneider, C., Hamlett, C. L., et al. (2013). Effects of first-grade number knowledge tutoring with contrasting forms of practice. *Journal of Educational Psychology*, *105*, 58–77.

Fuchs, L. S., Geary, D. C., Fuchs, D., Compton, D. L., & Hamlett, C. L. (2016). Pathways to third-grade calculation versus word-reading competence: are they more alike or different? *Child Development*, *87*, 558–567.

Fuchs, L. S., Malone, A., Schumacher, R. F., Namkung, J. M., Hamlett, C. L., Jordan, N. C., et al. (2016). Supported self-explaining during fraction intervention. *Journal of Educational Psychology*, *108*, 493–508.

Fuchs, L. S., Malone, A. S., Schumacher, R. F., Namkung, J., & Wang, A. (2017). Fraction intervention for students with mathematics difficulties: lessons learned from five randomized controlled trials. *Journal of Learning Disabilities*, *50*, 631–639.

Fuchs, L. S., Powell, S. R., Cirino, P. T., Schumacher, R. F., Marrin, S., Hamlett, C. L., et al. (2014). Does calculation or word-problem instruction provide a stronger route to prealgebraic knowledge? *Journal of Educational Psychology*, *106*, 990–1006.

Fuchs, L. S., Schumacher, R. F., Long, J., Namkung, J., Hamlett, C. L., Cirino, P. T., et al. (2013). Improving at-risk learners' understanding of fractions. *Journal of Educational Psychology, 105*, 683–700.

Geary, D. C. (1993). Mathematical disabilities: cognitive, neuropsychological, and genetic components. *Psychological Bulletin, 114*, 345–362.

Geary, D. C. (1994). *Children's mathematical development: Research and practical applications.* Washington, DC: American Psychological Association.

Geary, D. C. (1995). Reflections of evolution and culture in children's cognition: Implications for mathematical development and instruction. *American Psychologist, 50*, 24–37.

Geary, D. C. (2004). Mathematics and learning disabilities. *Journal of Learning Disabilities, 37*, 4–15.

Geary, D. C. (2005). *The origin of mind: Evolution of brain, cognition, and general intelligence.* Washington, DC: American Psychological Association.

Geary, D. C. (2007). Educating the evolved mind: conceptual foundations for an evolutionary educational psychology. In J. S. Carlson & J. R. Levin (Eds.), *Psychological perspectives on contemporary educational issues: Vol. 2. Educating the evolved mind.* Greenwich, CT: Information Age [pp. 1–99, 177–202].

Geary, D. C. (2011). Cognitive predictors of individual differences in achievement growth in mathematics: a five-year longitudinal study. *Developmental Psychology, 47*, 1539–1552.

Geary, D. C., Berch, D. B., & Mann Koepke, K. (2015). The evolution of number systems. In D. C. Geary, D. B. Berch, & K. Mann Koepke (Eds.), *Evolutionary origins and early development of number processing* (pp. 335–353). San Diego, CA: Elsevier/Academic Press.

Geary, D. C., & Bjorklund, D. F. (2000). Evolutionary developmental psychology. *Child Development, 71*, 57–65.

Geary, D. C., Hoard, M. K., Byrd-Craven, J., Nugent, L., & Numtee, C. (2007). Cognitive mechanisms underlying achievement deficits in children with mathematical learning disability. *Child Development, 78*, 1343–1359.

Geary, D. C., Hoard, M. K., Nugent, L., & Bailey, D. H. (2012). Mathematical cognition deficits in children with learning disabilities and persistent low achievement: a five-year prospective study. *Journal of Educational Psychology, 104*, 206–223.

Geary, D. C., Hoard, M. K., Nugent, L., & Bailey, H. D. (2013). Adolescents' functional numeracy is predicted by their school entry number system knowledge. *PLoS One, 8*(1), e54651.

Geary, D. C., Nicholas, A., Li, Y., & Sun, J. (2017). Developmental change in the influence of domain-general abilities and domain-specific knowledge on mathematics achievement: an eight-year longitudinal study. *Journal of Educational Psychology, 109*, 680–693.

Geary, D. C., & vanMarle, K. (2016). Young children's core symbolic and non-symbolic quantitative knowledge in the prediction of later mathematics achievement. *Developmental Psychology, 52*, 2130–2144.

Geary D.C., van Marle K., Chu F., Hoard M.K. and Nugent L., (in press). Predicting age of becoming a cardinal principle knower, *Journal of Educational Psychology.*

Geary, D. C., vanMarle, K., Chu, F., Rouder, J., Hoard, M. K., & Nugent, L. (2018). Early conceptual understanding of cardinality predicts superior school-entry number system knowledge. *Psychological Science, 29*, 191–205.

Gersten, R., Ferrini-Mundy, J., Benbow, C., Clements, D. H., Loveless, T., Williams, V., et al. (2008). Report of the task group on instructional practices. In National Mathematics Advisory Panel (Ed.), *Reports of the task groups and subcommittees* (pp. 6-i–6-224). Washington, DC: United States Department of Education.

Gersten, R., Fuchs, L. S., Compton, D., Coyne, M., Greenwood, C., & Innocenti, M. S. (2005). Quality indicators for group experimental and quasi-experimental research in special education. *Exceptional Children, 71*, 149–164.

Ginsburg, H. P., & Russell, R. L. (1981). Social-class and racial influences on early mathematical thinking. *Monographs of the Society for Research in Child Development, 46*, 1–69.

Gliksman, Y., Weinbach, N., & Henik, A. (2016). Alerting cues enhance the subitizing process. *Acta Psychologica, 170*, 139–145.

Gracia-Bafalluy, M., & Noel, M. P. (2008). Does finger training increase young children's numerical performance? *Cortex, 44*, 368–375.

Gunderson, E. A., Ramirez, G., Beilock, S. L., & Levine, S. C. (2012). The relation between spatial skill and early number knowledge: the role of the linear number line. *Developmental Psychology, 48*, 1229–1241.

Gustafsson, J. E., & Undheim, J. O. (1992). Stability and change in broad and narrow factors of intelligence from ages 12 to 15 years. *Journal of Educational Psychology, 84*, 141–149.

Harrison, T. L., Shipstead, Z., Hicks, K. L., Hambrick, D. Z., Redick, T. S., & Engle, R. W. (2013). Working memory training may increase working memory capacity but not fluid intelligence. *Psychological Science, 24*, 2409–2419.

Holmes, J., & Adams, J. W. (2006). Working memory and children's mathematical skills: implications for mathematical development and mathematics curricula. *Educational Psychology, 26*, 339–366.

Honoré, N., & Noël, M.-P. (2017). Can working memory training improve preschoolers' numerical abilities? *Journal of Numerical Cognition, 3*, 516–539.

Horn, J. L. (1968). Organization of abilities and the development of intelligence. *Psychological Review, 75*, 242–259.

House, E., Glass, G., McLean, L., & Walker, D. (1978). No simple answer: critique of the follow through evaluation. *Harvard Educational Review, 48*, 462–464.

Iuculano, T., Moro, R., & Butterworth, B. (2011). Updating working memory and arithmetical attainment in school. *Learning and Individual Differences, 21*, 655–661.

Johnson, E. S. (1984). Sex differences in problem solving. *Journal of Educational Psychology, 76*, 1359–1371.

Jurden, F. H. (1995). Individual differences in working memory and complex cognition. *Journal of Educational Psychology, 87*, 93–102.

Karmiloff-Smith, A. (2015). An alternative to domain-general or domain-specific frameworks for theorizing about human evolution and ontogenesis. *AIMS Neuroscience, 2*, 91–104.

Kleemans, T., Peeters, M., Segers, E., & Verhoeven, L. (2012). Child and home predictors of early numeracy skills in kindergarten. *Early Childhood Research Quarterly, 27*, 471–477.

Kleemans, T., Segers, E., & Verhoeven, L. (2016). Towards a theoretical framework on individual differences in numerical abilities: role of home numeracy experiences. In B. Blevins-Knabe & A. B. M. Austin (Eds.), *Early childhood mathematics skill development in the home environment* (pp. 71–86). New York: Springer.

Klein, D. (2007). A quarter century of US 'math wars' and political partisanship. *BSHM Bulletin, 22*, 22–33.

Kluczniok, K., Lehrl, S., Kuger, S., & Rossbach, H. G. (2013). Quality of the home learning environment during preschool age—domains and contextual conditions. *European Early Childhood Education Research Journal, 21*, 420–438.

Knops, A., Nuerk, H.-C., & Gobel, S. (2017). Domain-general factors influencing numerical and arithmetic processing. *Journal of Numerical Cognition, 3*, 112–132.

Krajewski, K., & Schneider, W. (2009). Exploring the impact of phonological awareness, visual–spatial working memory, and preschool quantity–number competencies on mathematics achievement in elementary school: findings from a 3-year longitudinal study. *Journal of Experimental Child Psychology*, *103*, 516–531.

Kucian, K., Grond, U., Rotzer, S., Henzi, B., Schönmann, C., & Plangger, F. (2011). Mental number line training in children with developmental dyscalculia. *NeuroImage*, *57*, 782–795.

Kyttälä, M., & Lehto, J. E. (2008). Some factors underlying mathematical performance: the role of visuospatial working memory and non-verbal intelligence. *European Journal of Psychology of Education*, *23*, 77–94.

Lancy, D. F. (2014). *The anthropology of childhood: Cherubs, chattel, changelings* (2nd ed.). Cambridge, NY: Cambridge University Press.

Lee, K., & Bull, R. (2016). Developmental changes in working memory, updating, and math achievement. *Journal of Educational Psychology*, *108*, 869–882.

Lefevre, J. A., Clarke, T., & Stringer, A. P. (2002). Influences of language and parental involvement on the development of counting skills: comparisons of French- and English-speaking Canadian children. *Early Child Development and Care*, *172*, 283–300.

LeFevre, J. A., Polyzoi, E., Skwarchuk, S. L., Fast, L., & Sowinski, C. (2010). Do home numeracy and literacy practices of Greek and Canadian parents predict the numeracy skills of kindergarten children? *International Journal of Early Years Education*, *18*, 55–70.

LeFevre, J. A., Skwarchuk, S. L., Smith-Chant, B. L., Fast, L., Kamawar, D., & Bisanz, J. (2009). Home numeracy experiences and children's math performance in the early school years. *Canadian Journal of Behavioural Science*, *41*, 55–66.

Li, Y., & Geary, D. C. (2013). Developmental gains in visuospatial memory predict gains in mathematics achievement. *PLoS One*, *8*(7), e70160.

Li, Y., & Geary, D. C. (2017). Children's visuospatial memory predicts mathematics achievement through early adolescence. *PLoS One*, *12*(2), e0172046.

Logie, R. H., & Baddeley, A. D. (1987). Cognitive processes in counting. *Journal of Experimental Psychology: Learning, Memory, and Cognition*, *13*, 310–326.

Long, I., Malone, S. A., Tolan, A., Burgoyne, K., Heron-Delaney, M., Witteveen, K., et al. (2016). The cognitive foundations of early arithmetic skills: it is counting and number judgment, but not finger gnosis, that count. *Journal of Experimental Child Psychology*, *152*, 327–334.

Loveless, T. (Ed.), (2001). *The great curriculum debate: How should we teach reading and math?* Washington, DC: Brookings Institute.

Mandler, G., & Shebo, B. J. (1982). Subitizing: an analysis of its component processes. *Journal of Experimental Psychology: General*, *111*, 1–22.

Mazzocco, M. M. M., & Kover, S. T. (2007). A longitudinal assessment of executive function skills and their association with math performance. *Child Neuropsychology*, *13*, 18–45.

McDonald, S. K., Keesler, V. A., Kauffman, N. J., & Schneider, B. (2006). Scaling-up exemplary interventions. *Educational Researcher*, *35*, 15–24.

McLean, J. F., & Hitch, G. J. (1999). Working memory impairments in children with specific arithmetic learning difficulties. *Journal of Experimental Child Psychology*, *74*, 240–260.

Melby-Lervåg, M., Redick, T. S., & Hulme, C. (2016). Working memory training does not improve performance on measures of intelligence or other measures of "far transfer": evidence from a meta-analytic review. *Perspectives on Psychological Science*, *11*, 512–534.

Meyer, L. A. (1984). Long-term academic effects of the direct instruction project follow through. *The Elementary School Journal*, *84*, 380–394.

Meyer, L. A., Gersten, R. M., & Gutkin, J. (1983). Direct instruction: a project follow through success story in an inner-city school. *The Elementary School Journal*, *84*, 241–252.

Michels, L., O'Gorman, R., & Kucian, K. (2018). Functional hyperconnectivity vanishes in children with developmental dyscalculia after numerical intervention. *Developmental Cognitive-Neuroscience, 30*, 291–303.

Mix, K. S., & Cheng, Y. L. (2011). The relation between space and math: developmental and educational implications. In J. B. Benson (Ed.), *Advances in child development and behavior* (Vol. 42, pp. 197–243). San Diego, CA: Academic Press.

Miyake, A., Friedman, N. P., Emerson, M. J., Witzki, A. H., Howerter, A., & Wager, T. D. (2000). The unity and diversity of executive functions and their contributions to complex "frontal lobe" tasks: a latent variable analysis. *Cognitive Psychology, 41*, 49–100.

Napoli, A. R., & Purpura, D. J. (2018). The home literacy and numeracy environment in preschool: cross-domain relations of parent–child practices and child outcomes. *Journal of Experimental Child Psychology, 166*, 581–603.

National Mathematics Advisory Panel. (2008). *Foundations for success: Final report of the National Mathematics Advisory Panel.* Washington, DC: United States Department of Education.

National Reading Panel. (2000). *Teaching children to read: An evidence-based assessment of the scientific research literature on reading and its implications for reading instruction.* Washington, DC: National Institute of Child Health and Human Development, National Institutes of Health.

National Research Council. (2002). *Scientific research in education.* Washington, DC: National Academies Press.

National Research Council. (2004). *On evaluating curricular effectiveness: Judging the quality of K-12 mathematics evaluations.* Washington, DC: National Academies Press.

Navon, D. (1977). Forest before trees: the precedence of global features in visual perception. *Cognitive Psychology, 9*, 353–383.

Newman, S. (2016). Does finger sense predict addition performance? *Cognitive Processing, 17*, 139–146.

Noel, M. P. (2005). Finger gnosia: a predictor of numerical abilities in children? *Child Neuropsychology, 11*, 413–430.

O'Donnell, C. L. (2008). Defining, conceptualizing, and measuring fidelity of implementation and its relationship to outcomes in K–12 curriculum intervention research. *Review of Educational Research, 78*, 33–84.

Passolunghi, M. C., Vercelloni, B., & Schadee, H. (2007). The precursors of mathematics learning: working memory, phonological ability and numerical competence. *Cognitive Development, 22*, 165–184.

Penner-Wilger, M., & Anderson, M. L. (2013). The relation between finger gnosis and mathematical ability: why redeployment of neural circuits best explains the finding. *Frontiers in Psychology, 4*, 877.

Penner-Wilger, M., Fast, L., LeFevre, J., Smith-Chant, B. L., Skwarchuk, S., Kamawar, D., et al. (2009). Subitizing, finger gnosis, and the representation of number. In N. Taatgen & H. van Rijn (Eds.), *Proceedings of the 31st annual cognitive science society* (pp. 520–525). Austin, TX: Cognitive Science Society.

Petersen, S. E., & Posner, M. I. (2012). The attention system of the human brain: 20 years after. *Annual Review of Neuroscience, 35*, 73–89.

Pianta, R. C., Barnett, W. S., Burchinal, M., & Thornburg, K. R. (2009). The effects of preschool education: what we know, how public policy is or is not aligned with the evidence base, and what we need to know. *Psychological Science in the Public Interest, 10*, 49–88.

Poropat, A. E. (2009). A meta-analysis of the five-factor model of personality and academic performance. *Psychological Bulletin, 135*, 322–338.

Puma, M., Bell, S., Cook, R., Heid, C., Broene, P., Jenkins, F., et al. (2012). *Third grade followup to the Head Start Impact Study: Final report* (OPRE Report No. 2012–45). Retrieved from http://eric.ed.gov/?idDED539264.

Purpura, D. J., & Reid, E. E. (2016). Mathematics and language: individual and group differences in mathematical language skills in young children. *Early Childhood Research Quarterly, 36*, 259–268.

Raghubar, K. P., Barnes, M. A., & Hecht, S. A. (2010). Working memory and mathematics: a review of developmental, individual difference, and cognitive approaches. *Learning and Individual Differences, 20*, 110–122.

Ramani, G. B., Jaeggi, S. M., Daubert, E. N., & Buschkuehl, M. (2017). Domain-specific and domain-general training to improve kindergarten children's mathematics. *Journal of Numerical Cognition, 3*, 468–495.

Ramani, G. B., & Siegler, R. S. (2008). Promoting broad and stable improvements in low-income children's numerical knowledge through playing number board games. *Child Development, 79*, 375–394.

Rauscher, L., Kohn, J., Käser, T., Mayer, V., Kucian, K., McCaskey, U., et al. (2016). Evaluation of a computer-based training program for enhancing arithmetic skills and spatial number representation in primary school children. *Frontiers in Psychology, 7*, 913.

Richerson, P. J., & Boyd, R. (2005). *Not by genes alone: How culture transformed human evolution.* Chicago, IL: University of Chicago Press.

Riede, F., Johannsen, N. N., Hogberg, A., Nowell, A., & Lombard, M. (2018). The role of play objects and object play in human cognitive evolution and innovation. *Evolutionary Anthropology, 27*, 46–59.

Rivera-Batiz, F. (1992). Quantitative literacy and the likelihood of employment among young adults in the United States. *Journal of Human Resources, 27*, 313–328.

Rosenthal, T. L., & Zimmerman, B. J. (1978). *Social learning and cognition.* New York, NY: Academic Press.

Rousseau, J.-J. (1979). *Emile: Or, on education* [A. Bloom, Trans.]. New York: Basic Books [Original work published 1762].

Sala, G., & Gobet, F. (2017). Does far transfer exist? Negative evidence from chess, music, and working memory training. *Current Directions in Psychological Science, 26*, 515–520.

Sarkar, A., & Cohen Kadosh, R. (2016a). Transcranial electrical stimulation and numerical cognition. *Canadian Journal of Experimental Psychology, 70*, 41–58.

Sarkar, A., & Cohen Kadosh, R. (2016b). Transcranial electrical stimulation and the enhancement of numerical cognition. In D. B. Berch, D. C. Geary, & K. Mann Koepke (Eds.), *Development of mathematical cognition: Neural substrates and genetic influences* (pp. 245–296). San Diego, CA: Elsevier/Academic Press.

Siegler, R. S., Duncan, G. J., Davis-Kean, P. E., Duckworth, K., Claessens, A., Engel, M., et al. (2012). Early predictors of high school mathematic achievement. *Psychological Science, 23*, 691–697.

Siegler, R. S., Thompson, C. A., & Schneider, M. (2011). An integrated theory of whole number and fractions development. *Cognitive Psychology, 62*, 273–296.

Simonsmeiera, B. A., Grabner, R. H., Heina, J., Krenza, U., & Schneidera, M. (2018). Electrical brain stimulation (tES) improves learning more than performance: a meta-analysis. *Neuroscience and Biobehavioral Reviews, 84*, 171–181.

Skwarchuk, S. L., Sowinski, C., & LeFevre, J.-A. (2014). Formal and informal home learning activities in relation to children's early numeracy and literacy skills: The development of a home numeracy model. *Journal of Experimental Child Psychology, 121*, 63–84.

Skwarchuk, S. L., Vandermaas-Peeler, M., & LeFevre, J. A. (2016). Optimizing the home numeracy environments of 3- to 6-year-old children in the USA and Canada. In B. Blevins-Knabe & A. B. M. Austin (Eds.), *Early childhood mathematics skill development in the home environment* (pp. 127–146). New York: Springer

Snowball, A., Tachtsidis, I., Popescu, T., Thompson, J., Delazer, M., Zamarian, L., et al. (2013). Long-term enhancement of brain function and cognition using cognitive training and brain stimulation. *Current Biology*, *23*, 987–992.

Soylu, F., Lester, F. K., & Newman, S. D. (2018). You can count on your fingers: the role of fingers in early mathematical development. *Journal of Numerical Cognition*, *4*, 107–135.

Soylu, F., Raymond, D., Gutierrez, A., & Newman, S. (2017). The differential relationship between finger sense, and addition and subtraction: An fMRI study. *3*, 694–715.

Stebbins, L. B. (1977). *Education as experimentation: A planned variation model* (Vol. 4). Lanham, MD: University Press of America.

Stoet, G., Bailey, D. H., Moore, A. M., & Geary, D. C. (2016). Countries with higher levels of gender equality show larger national sex differences in mathematics anxiety and relatively lower parental mathematics valuation for girls. *PLoS One*, *11*(4), e0153857.

Street, S. E., Navarrete, A. F., Reader, S. M., & Laland, K. N. (2017). Coevolution of cultural intelligence, extended life history, sociality, and brain size in primates. *Proceedings of the National Academy of Sciences of the United States of America*, *114*(30), 7908–7914.

Swanson, E., Wanzek, J., Haring, C., Ciullo, S., & McCulley, L. (2013). Intervention fidelity in special and general education research journals. *The Journal of Special Education*, *47*, 3–13.

Swanson, H. L., & Sachse-Lee, C. (2001). Mathematical problem solving and working memory in children with learning disabilities: both executive and phonological processes are important. *Journal of Experimental Child Psychology*, *79*, 294–321.

Swanson, J. M., Schuck, S., Porter, M. M., Carlson, C., Hartman, C. A., Sergeant, J. A., et al. (2012). Categorical and dimensional definitions and evaluations of symptoms of ADHD: History of the SNAP and the SWAN rating scales. *The International Journal of Educational and Psychological Assessment*, *10*, 51–70.

Szucs, D., & Myers, T. (2017). A critical analysis of design, facts, bias and inference in the approximate number system training literature: a systematic review. *Trends in Neuroscience and Education*, *6*, 187–203.

Taub, G. E., Keith, T. Z., Floyd, R. G., & McGrew, K. S. (2008). Effects of general and broad cognitive abilities on mathematics achievement. *School Psychology Quarterly*, *23*, 187–198.

Thorndike, E. L., & Woodworth, R. S. (1901). The influence of improvement in one mental function upon the efficiency of other functions. *Psychological Review*, *8*, 247–261.

Thorsen, C., Gustafsson, J. E., & Cliffordson, C. (2014). The influence of fluid and crystallized intelligence on the development of knowledge and skills. *British Journal of Educational Psychology*, *84*, 556–570.

Tramacere, A., Pievani, T., & Ferrari, P. F. (2017). Mirror neurons in the tree of life: mosaic evolution, plasticity and exaptation of sensorimotor matching responses. *Biological Reviews*, *92*, 1819–1841.

Trick, L. M., & Pylyshyn, Z. W. (1994). Why are small and large numbers enumerated differently? A limited-capacity preattentive stage in vision. *Psychological Review*, *101*, 80–102.

Tricot, A., & Sweller, J. (2013). Domain-specific knowledge and why teaching generic skills does not work. *Educational Psychology Review*, *26*, 265–283.

Von Aster, M. G., & Shalev, R. S. (2007). Number development and developmental dyscalculia. *Developmental Medicine and Child Neurology*, *49*, 868–873.

Vygotsky, L. S. (1978). In M. Cole, V. John-Steiner, S. Scribner, & E. Souberman (Eds.), *Mind in society: The development of higher psychological processes.* Cambridge, MA: Harvard University Press.

Walberg, H. J. (1984). Improving the productivity of America's schools. *Educational Leadership, 41*, 19–27.

Wasner, M., Nuerk, H.-C., Martignon, L., Roesch, S., & Moeller, K. (2016). Finger gnosis predicts a unique but small part of variance in initial arithmetic performance. *Journal of Experimental Child Psychology, 146*, 1–16.

Watts, T. W., Duncan, G. J., Chen, M., Claessens, A., Davis-Kean, P. E., Duckworth, K., et al. (2015). The role of mediators in the development of longitudinal mathematics achievement associations. *Child Development, 86*, 1892–1907.

Watts, T. W., Duncan, G. J., Clements, D. H., & Sarama, J. (2018). What is the long-run impact of learning mathematics during preschool? *Child Development, 89*, 539–555.

What Works Clearing House. (2017). *Standards handbook, version 4.0.* Washington, DC: U.S. Department of Education. Retrieved from https://ies.ed.gov/ncee/wwc/Docs/ReferenceResources/wwc_standards_handbook_v4_draft.pdf.

Wiebe, S. A., Espy, K. A., & Charak, D. (2008). Using confirmatory factor analysis to understand executive control in preschool children: I. Latent structure. *Developmental Psychology, 44*, 575–587.

Wiegand, I., Petersen, A., Bundesen, C., & Habekost, T. (2017). Phasic alerting increases visual attention capacity in younger but not in older individuals. *Visual Cognition, 25*, 343–357.

Willingham, D. T. (2012). *When can you trust the experts?: How to tell good science from bad in education.* San Francisco, CA: Jossey-Bass.

Wynn, T. (2002). Archaeology and cognitive evolution. *Behavioral and Brain Sciences, 25*, 389–402.

Zippert, E. L., & Ramani, G. B. (2017). Parents' estimations of preschoolers' number skills relate to at-home number-related activity engagement. *Infant and Child Development, 26*, 24.

Chapter 2

The Effects of SES, Grade-Repeating, and IQ in a Game-Based Approximate Math Intervention

Josh Langfus*, Alejandro Maiche†, Dinorah De León†, Dahiana Fitipalde†, Álvaro Mailhos† and Justin Halberda*

**Johns Hopkins University, Baltimore, MD, United States*

†Centro de Investigación Básica en Psicología, Montevideo, Uruguay

INTRODUCTION

What tools can be used to close the gap between low- and high-achieving students? While there may be many ways to intervene—for example, teacher training, improving facilities, updating curricula—a great deal of attention has focused on the potential benefits of using technology as a teaching tool in the classroom. Nonprofit organizations like One Laptop Per Child (Trucano, 2011) and the World Computer Exchange (2016) have partnered with schools around the world to give students access to computers, tablets, and the internet. To the extent that digital literacy will be a valuable skill for tomorrow's workforce, providing early access to technology might be a useful investment. Less clear is the role that technology can play in bolstering instruction in traditional subject areas such as science, math, and reading. Some exploratory work has examined the effectiveness of technology-based classroom interventions in math education, with results suggesting benefits of training that focuses on specific skills, for example, manipulating decimals (Zhang, Trussell, Gallegos, & Asam, 2015) and multiplication and division (Pilli & Aksu, 2013). In contrast, other recent work has explored the effectiveness of nontechnology interventions in locations with restricted access to computers, though it is not clear that improvement on these intervention activities transfers to other math abilities (Dillon, Kannan, Dean, Spelke, & Duflo, 2017).

In the context of recent interest in technology's role in education, we present some initial findings from an ongoing classroom technology intervention in

Mathematical Cognition and Learning, Vol. 5. https://doi.org/10.1016/B978-0-12-815952-1.00002-5

elementary schools in and surrounding Montevideo, Uruguay—the country's capital and largest city. This effort is unique because Uruguay stands out as one of the first countries in the world to commit to providing public access to information and communications technology (ICT) on a nation-wide scale. Not only has the Uruguayan government invested in establishing free internet connections in schools and public places across the country, but also since 2008 it has partnered with the nonprofit organization One Laptop Per Child to provide every school-aged child with their own computer to be used in the classroom and at home. The joint initiative is called Plan Ceibal ("Conectividad Educativa de Informática Básica para el Aprendizaje en Línea" or "Educational Connectivity/Basic Computing for Online Learning"), and over the last decade, the program has expanded from its start in the nation's primary schools to include every student in Uruguay from preschool through elementary and middle school.

The dramatic increase in access to ICT has created a need for content on these platforms that educators can use in the classroom. Our team has partnered with a dedicated group of teachers at public elementary schools in the Montevideo area to create educational software for the tablets and evaluate its effectiveness. The present study is the latest from this ongoing project; a previous study engaged 503 first-graders in a number estimation task (Odic et al., 2016), and in the work presented here we introduced 386s- and third-graders from a wide range of backgrounds to three tablet-based magnitude training games. We used a pretest, intervention, posttest design, along with a Business-As-Usual (BAU) Control group, to examine the effectiveness of playing these games on several measures, including a standardized test of school math ability (speeded arithmetic). The context of Uruguay also affords us the opportunity to look at whether such interventions helps some students more than others, e.g., as a function of Socioeconomic Status (SES) and Grade-Repeating status. As we will discuss in more detail later, our project includes students from schools across the SES spectrum, and therefore allows us to study the relationships between SES and measures of achievement and cognitive abilities, as well as how SES might moderate the impact of our intervention software.

In addition to SES, the large proportion of students in Uruguayan schools that repeat a grade allowed us to examine whether our intervention was particularly helpful for these students, and more generally to ask what kinds of students benefit most from our technology-based intervention. Our initial hope was that students who are most vulnerable or disadvantaged might stand to gain the most from our intervention games which focus on simple, intuitive, "core" magnitude discrimination. To explore how the current study may begin to answer these questions, it is important to understand the context of schools in Uruguay.

THE URUGUAYAN CONTEXT

One hope of the educational community in Uruguay is that providing equivalent technology to all children may help to bridge the gaps between schools in high- and low-SES communities. As mentioned previously, the schools in Uruguay provide a unique test case for exploring the potential of technological interventions, not only because of the willingness of teachers and students to engage with technology, but also because this technology is accessible to schools in communities across the socioeconomic spectrum. Table 1 summarizes the number of students per school per SES quintile included in the current study. The Uruguayan National Public Education Administration (ANEP) uses a variety of factors to classify each school's SES context: household education level, socioeconomic indicators (e.g., percentage of households with access to potable water), and social integration (e.g., head of household's employment status and percentage of household children attending school). Using an aggregate of these factors, ANEP classifies schools into quintiles (with level 1 as the lowest).

Uruguay's national government tracks a wide range of data about the nation's public education system, including enrollment and completion rates across many demographics, and these data highlight differences across socioeconomic contexts. A 2017 report from the National Institute of Education Assessment (INEEd) showed that, while access to education has improved over the last 10 years, striking disparities still exist across levels of SES in terms of enrollment, retention, and grade-repetition. These disparities can be seen from the beginning of enrollment in school through graduation. For example, between 2013 and 2015, preschoolers (age 3) in the least advantaged schools were less likely to meet minimum standards for required attendance (at least 141 days in the year) compared to students at the most advantaged schools; 28% of students in SES Quintile 5 did not meet this standard, compared to 46% in Quintile 1 (INEEd, 2017). Though overall attendance increases, and the gap narrows as students progress to higher grades, the data show a consistent 10% difference in meeting the standard for minimum attendance between the highest and lowest SES schools from first grade through sixth grade (INEEd, 2017). Socioeconomic differences are also apparent in the number of students who repeat a grade, which can occur for several reasons including lack of attendance or poor performance. Already in the earliest grades there are differences between higher and lower SES Quintiles with respect to children being on-target to progress through school at the appropriate age. For example, at 7 years of age only 4% of students in Quintile 5 schools had repeated or were repeating a grade, while the figure jumps to 12% for children in Quintile 1 schools, according to country-wide data (INEEd, 2017). And, most dramatically, by the time students graduated from secondary school, 83% of students in the least advantaged schools had repeated at least one grade, compared to 24% in the most advantaged programs (INEEd, 2017).

Arguably the most striking differences across SES manifest in terms of the number of students who leave education at either the primary-, middle-, or high-school level. While Uruguay has made significant strides in ensuring that nearly every student, regardless of SES, completes primary education, only

TABLE 1 Condition Assignment by School With SES Quintile Information (1 = Low, 5 = High)

School		A		B	C		D		E		F		G		H		
SES Quintile		*1*		*1*	2		2		*3*		*4*		*5*		*5*		
Classrooms		*a*	*b*	*a*	*a*	*b*	*a*	*b*	*a*	*b*	*a*	*b*	*a*	*b*	*a*	*b*	c
$n \approx$		17	18	20	27	31	20	22	26	27	24	27	29	28	26	19	27
Grade	2				X	X	X	X	X	X			X	X	X	X	
	3	X	X	X							X	X					X
Condition	BAU control	X		X	X		X		X		X		X		X		
	Game intervention		X			X		X		X		X		X		X	X

Note that for all schools except for B, every school had 2 year-matched classrooms—one for each condition (Intervention and Control).

50% of least advantaged students graduate from middle school, compared to 95% of students from Quintile 5 schools, and the trend continues through the end of high school, where the gap is 15% vs. 71% (INEEd, 2017).

In terms of the effect of SES on access to education, and foreshadowing some of our own result, data from our study show the same patterns as the INEEd report. On both the pre- and postintervention evaluation days of the current study, more students in the lower SES schools were absent compared to students at higher SES schools (see Table 2). Students at poorer schools are also more likely to have repeated a grade; within our sample of 386 children, the percentage of students whose birthdays fell after the legal cutoff date for their grade level was 5.4% at Quintile 5 schools compared to 41.8% at Quintile 1 schools (see Table 3)—a rate that is even more dramatic than what INEEd has reported for the national percentages in Quintile 1 (INEEd, 2017). While these disparities are disheartening, the fact that the schools who have partnered with us for this project seem to represent a microcosm

TABLE 2 Attendance Data by SES Quintile for the Testing Days

Quintile	1	2	3	4	5	Overall
N	55	100	53	49	129	386
% Present both	53	40	55	90	77	62
% Absent pre	18	26	15	8	12	16
% Absent post	22	26	25	2	9	16
% Absent both	7	8	6	0	3	5
Total %	100	100	100	100	100	100

Number of students enrolled in each Quintile is given at the top, and percent of students who were present for both, one, or neither test are given below. Students in lower SES classrooms (Quintiles 1, 2, and 3) were overall less likely to be in school testing days.

TABLE 3 Number of Students per SES Quintile Counted as Repeaters

Quintile	Repeaters	Total Students	Percent Repeating
1	23	55	41.8
2	27	100	27.0
3	9	53	17.0
4	4	49	8.2
5	7	129	5.4

of the issues affecting the education system as a whole in Uruguay allows this and future studies to identify interventions that stand a greater chance of scaling up to meet those broader challenges.

BACKGROUND FOR THE CURRENT STUDY

Socioeconomic Status in Education

Children from low SES households typically underperform their middle- and high-SES peers in school achievement (Sirin, 2005; Valle-Lisboa et al., 2016). These differences may be the result of a range of factors associated with SES (Wilkinson & Pickett, 2010). Performance differences have been observed in many specific domains, including language development (Hart & Risley, 1975; Hoff, 2003), IQ (Turkheimer, Haley, Waldron, D'Onofrio, & Gottesman, 2003), cognitive ability (Larson, Russ, Nelson, Olson, & Halfon, 2015), spatial knowledge (Levine, Vasilyeva, Lourenco, Newcombe, & Huttenlocher, 2005; Verdine, Irwin, Golinkoff, & Hirsh-Pasek, 2014), as well as mathematics (Klibanoff, Levine, Huttenlocher, Vasilyeva, & Hedges, 2006). Based on this literature, we expect to find differences in the initial cognitive and curricular abilities of children as a function of SES. Given these differences in baseline abilities, we might expect that SES differences will differentially affect the impact of our training games and test-retest improvement.

Classroom Geometric and Arithmetic Abilities

As with other content areas, math achievement varies as a function of SES in the United States (Sirin, 2005), Uruguay (Valle-Lisboa et al., 2016) and other South American countries (INEEd, 2017) for all ages. These differences can be tracked across a variety of domains, e.g., intuitive number sense, spatial reasoning. In the present work, we focused on assessing two domains of math achievement: arithmetic and geometry. Our measure of geometry was used primarily as a pilot and consisted of an assessment developed by the Spelke lab and, in a collaboration, translated into Spanish by one of our authors (A.M.) and used with permission (E. Spelke, personal communication, June 23, 2017). A typical question on this measure showed students several shapes and asked them, for example, to circle all that had sides of equal length. Another question in this pilot task assessed students' understanding of symmetry by asking whether a given shape was the same on either side of a dotted line. This measure will serve as a control to contrast with our main measure of interest: arithmetic ability.

To measure arithmetic ability, we used a version of the math fluency subtest of the Woodcock-Johnson Tests of Achievement (Woodcock, McGrew, & Mather, 2001/2007) adapted for use in Spanish-speaking contexts: the Batería III Woodcock-Muñoz (Woodcock et al., 2001/2007) (see Fig. 1). The test measures the number of addition, subtraction, and single-digit multiplication problems children can complete in 3 min. It is a reliable measure of arithmetic achievement and has been widely administered in many countries.

1 − 1	0 + 3	2 + 2	4 − 2	2 + 1	3 − 3	0 + 0	3 − 0	2 − 1	2 + 4
5 + 0	3 − 1	1 + 6	4 + 4	5 − 0	1 + 1	6 − 1	3 + 5	4 − 1	5 − 2
3 − 2	5 + 1	6 − 3	2 − 2	7 + 1	4 − 4	1 + 8	4 − 3	7 + 2	4 + 1
2 + 5	8 − 1	5 − 4	3 + 3	10 − 2	3 + 6	7 − 2	2 + 8	3 + 1	9 − 4
6 − 2	4 + 6	9 + 3	8 − 6	7 + 5	10 − 10	2 + 6	5 − 3	6 − 6	3 + 4
5 + 5	8 − 3	5 − 1	8 + 0	7 − 4	1 + 9	10 − 6	8 + 4	6 + 8	9 − 9
1 × 1	4 + 5	7 + 7	2 × 3	10 − 5	1 × 2	3 + 2	8 − 8	9 + 5	5 × 1
3 + 7	1 × 4	9 − 2	9 + 8	3 × 3	6 − 5	2 × 2	9 − 3	3 × 1	6 + 6

FIG. 1 An example page from the Woodcock-Muñoz speeded arithmetic assessment. Students had 3 min to complete as many problems as possible.

Because norming data for this test do not exist for Uruguay, here we report the raw scores (i.e., the number of problems completed correctly), age-normalized scores (normed to our sample), and regressions controlled for age, geometry, IQ, and vocabulary where appropriate. We expect that our children will differ in math ability (geometry and arithmetic) as a function of SES. We will also look at how this relates to IQ and ANS ability and Repeater status, and how these factors may influence the effectiveness of the intervention.

Approximate Number Abilities

Our magnitude training intervention approach was inspired by work in developmental psychology that has focused on the relationship between classroom mathematics and an intuitive number sense mediated by the approximate number system (ANS: Halberda & Feigenson, 2008). The ANS is a core knowledge system (Feigenson, Dehaene, & Spelke, 2004) that is present in newborns (Izard, Sann, Spelke, & Streri, 2009) and improves over the course of development (Halberda & Feigenson, 2008; Halberda, Ly, Wilmer, Naiman, & Germine, 2012; Odic, Libertus, Feigenson, & Halberda, 2013). The system supports people's intuitive understanding of the approximate quantity of collections of items or events and supports the ability to add and subtract these quantities. Such a system might be specific to number, or the ANS may be part of a more general magnitude system that includes, e.g., surface area, time, and length (Lourenco, Bonny, Fernandez, & Rao, 2012; Sokolowski, Fias, Ononye, & Ansari, 2017; Walsh, 2003). Because we want to remain open to the possibility of a generalized magnitude system, we created mini-games for surface area, time, and number. However, in previous work we have found that each of these dimensions uniquely correlates with school math ability, suggesting some independence among these dimensions (Odic et al., 2016). Our own proposal is that the shared aspects of these representations may derive from the computations that allow one to compare magnitudes (e.g., to determine that one duration is longer than another, or that one size is larger than another), rather than from shared representations of quantity (Odic, Pietroski, Hunter, & Lidz, 2013). For the current paper, we are interested in improvements in this or any other parts of these systems, and in the possibility that training in magnitude games can transfer to speeded arithmetic performance.

A critical component of these ongoing debates is whether ANS precision plays a causal role in shaping formal math abilities. Intervention studies provide a tool to assess this hypothesis by intervening to improve ANS precision and observing if this leads to an improvement in formal math ability. A handful of research projects have looked at classroom interventions using computerized training programs to improve the precision of cognitive abilities, such as the ANS, that are associated with school math performance (DeWind & Brannon, 2012; Hyde, Khanum, & Spelke, 2014). There have been mixed successes, and more research is needed to understand what kinds of interventions are most successful and for which students. One possibility is that these types of ANS interventions may be particularly helpful in fostering foundational understandings of basic concepts or in situations where a child is struggling to attain proficiency with early skills.

In the present work, our measure of ANS ability will be the number of problems solved in a timed paper-and-pencil dot comparison task (Fig. 2). This is a pilot version of a task that we are actively developing which aims to measure ANS ability in a group- or individual-administered, paper-and-pencil format (Mailhos et al., 2018). We expect to find differences in ANS ability across SES in pretest, and we also expect to replicate results showing a correlation between ANS ability and arithmetic ability (DeWind & Brannon, 2012; Hyde et al., 2014; Odic et al., 2016; Zosh, Verdine, Halberda, Hirsh-Pasek, & Golinkoff, 2018). Lastly, we will also look at whether SES,

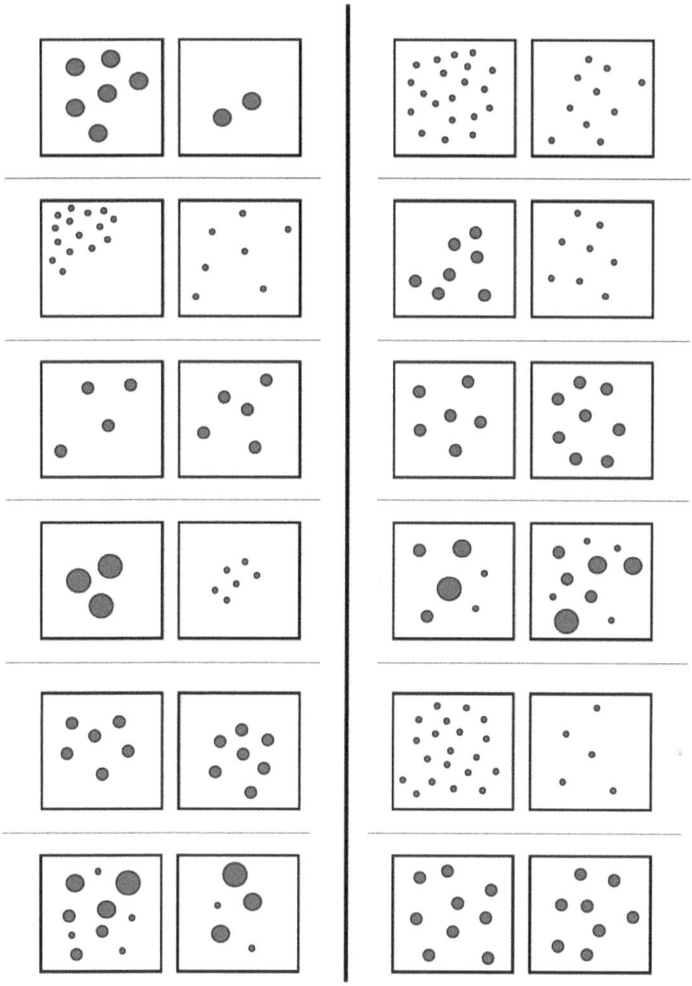

FIG. 2 Sample page from modified dots discrimination task. Students were instructed to consider each pair of squares and place an X in the box with more dots in it. They were given 3 min and told to begin at the top left and continue down each column until time ran out.

Grade-Repeating, and IQ affect the relationships between ANS ability, arithmetic ability, and intervention transfer.

THE PRESENT INTERVENTION STUDY

Our primary goal in partnering with communities in Montevideo was to create a resource that could improve students' math performance and that students could use in the classroom and at home. To do this, we aimed to develop an app that students would be motivated to engage with over an extended period of time and that trained approximate magnitude comparison over various domains.

FIG. 3 (A) The tablet children used in the intervention and sample images from trials of each of the three training games: ANS (B), area (C), and time (D).

To make the intervention engaging for students, we developed a completely novel app that included extensive original artwork and characters (e.g., "monsters"), rewards for good performance (e.g., "stars"), and online tracking of gameplay. The app included three mini-games (see Fig. 3B–D), each focusing on one of the three abilities that have been discussed in the literature as being relevant to general magnitude representation: time, area, number. We sought to improve students' math abilities through training on these magnitude comparison tasks over the course of a 1-month intervention.

Teachers' Responses to the Software

Successfully translating from the lab to the classroom requires building bridges between researchers and educators, and our project cannot succeed without the work of teachers invested in partnering with us for the long term. Not only are teachers responsible for implementing the intervention in their classrooms, but their feedback is our primary means of understanding how students interact with the app. Nearly all the teachers whose classrooms participated in the current study were also with us for a separate pilot phase, and their feedback was particularly valuable in helping us to understand how to motivate the children to continue engaging with the games. For example, teachers suggested that students be able to personalize their game sessions by selecting an image as an avatar. They also suggested introducing a progress-tracking system into the game that would allow students to see how they were doing and compare their performance with their peers. We implemented these ideas in the version of the game for the current study, and we heard from teachers that these changes improved student engagement. At the end of the intervention, the teachers met with our team in Uruguay and offered final comments on the intervention, summarized by our team here:

> *El juego en términos generales, fue evaluado positivamente por parte de las maestras, la mayoría de ellas transmitieron el entusiasmo de los niños por jugar. La incorporación de las estrellas como premio fue de gran motivación para los niños, por ejemplo se veía que comparaban su progreso en los distintos mini-juegos con sus compañeros a la vez que los motivaba a seguir jugando. La temática del juego fue divertida y muy aceptada por los niños, con respecto a esto se observó que cada niño tenía un monstruo preferido, incluso nos compartieron dibujos de los monstruos. Para finalizar, luego de terminada la intervención los niños nos preguntaban si íbamos a volver con las tablets y el juego.*

> *The game, in general, was seen very positively by the teachers, the majority of whom communicated the children's enthusiasm for playing the game. The incorporation of the stars as a reward system was a great motivator for the kids, for example they saw students comparing progress in the different mini-games with their friends which motivated them to continue playing. The theme of the game was fun and very well accepted by the children, and with respect to this it was observed that each child had a favorite monster, and they shared with us drawings of these monsters. Finally, after the end of the intervention, the children asked us if we were going to use the tablets and play the games again*

These comments suggest that students enjoyed engaging with the training app and teachers were easily able to incorporate it into their normal lessons. These are key components of a classroom-based intervention.

Design of the Current Study

As mentioned previously, the study implemented a pretest, intervention, posttest design with a Business-As-Usual (BAU) Control group. We assessed formal Arithmetic Abilities (i.e., the number of single- and double-digit addition, subtraction and multiplication problems students could complete in 3 min), IQ [Raven's Progressive Matrices (Raven, Raven, & Court, 1998)], Approximate Number Abilities (the ability to rapidly determine the larger of two approximate quantities on our paper assessment), Vocabulary, and Geometry Ability. In our analyses, we focus on the diversity of our Uruguayan school sample in terms of SES, Grade-Repeating, IQ, and Arithmetic Ability, and we present improvement scores from pre- to posttest for both our Game Intervention group and our BAU Control group. Figure 4 illustrates the study's overall design. The boxes on the left and right represent the pre- and postintervention evaluations, respectively. All students participated in these evaluations. Classrooms at each of the participating schools were randomly assigned to either the Game

FIG. 4 Study design. Both groups (Intervention and Control) participated in pre- and postintervention evaluations of cognitive abilities. The intervention group was encouraged to play the three tablet games during the 5-week intervention period, during which the control group received Business-as-Usual instruction.

Intervention condition or the BAU Control condition for the five-week intervention period. During that time, students in the Game Intervention group interacted with the intervention software which consisted of three magnitude training games, each focusing on a different ability: area discrimination, time discrimination, and approximate number discrimination. Note that, consistent with Uruguayan school policy, students in the BAU Control condition also had access to tablets during this time as part of their normal curriculum, but they did not have access to the games we developed for the intervention.

Methods

Participants

Sixteen classrooms in eight public elementary schools in and around Montevideo participated in the study, including 386 s- and third-graders overall. Half of the classrooms were assigned to the intervention condition and the training game was installed on tablet computers provided by Plan Ceibal. During the 5-week intervention period, students in this group were encouraged approximately once per week by their teachers to play the game at school for a little while and at home as much as they liked. Otherwise, children in the Game Intervention group received Business-As-Usual (BAU) math instruction. The other eight classrooms comprised a BAU Control group and received normal math instruction. These students had access to tablets but not the training game and, as is usual for schools in Uruguay, use of the tablets was left to the discretion of the classroom teacher. For this study, it was important to us that the teachers in each group felt that they could instruct as normal, along with the small addition of access to the game for the Game Intervention group.

In Fig. 5 we see a map of Montevideo and the surrounding areas where our schools were located. Each school is lettered and the metrics for each school are summarized in Table 1. In this study, SES information was only available

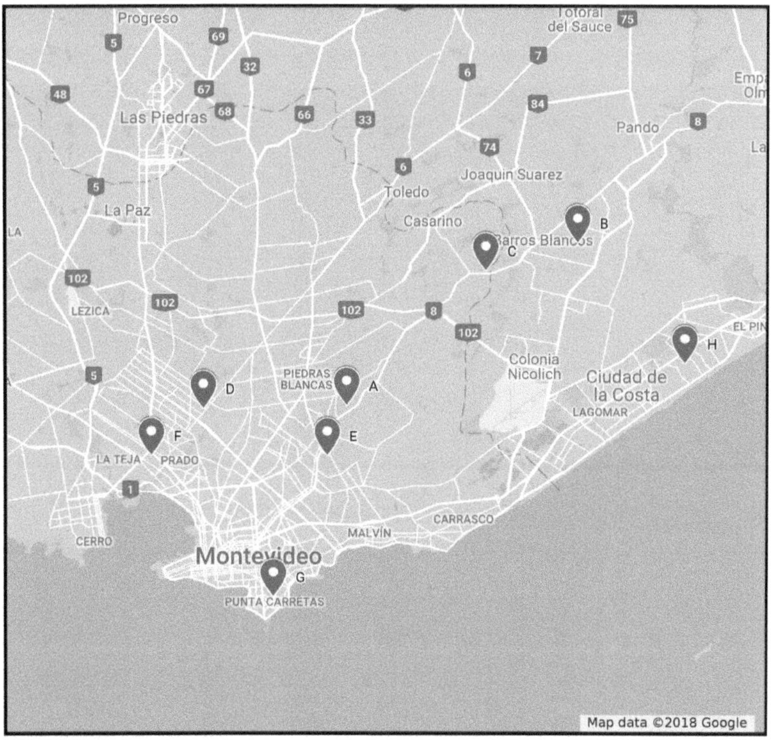

FIG. 5 Locations of our schools. *From Google Maps image used with in accordance with Google's fair-use guidelines: https://www.google.com/permissions/geoguidelines/attr-guide/.*

at the school level (as mentioned previously). In all but two schools, each intervention classroom was matched with a same-year (second- or third-grade) control classroom at the same school. All children spoke Spanish as their first language and all tasks were administered in Spanish. The posttest evaluations were administered by a researcher blind to the experimental conditions of the classrooms (J.L.).

Of the 386 students enrolled in classrooms that participated in the study, 19 students (5%) did not attend class on both pre- and postintervention testing days, and some students (overall 32%) were in attendance on one of these days but not both. As mentioned earlier, attendance at school depends to an extent on SES Quintile, and this is reflected in which students tended to be present for our pre- and posttest measures, as summarized in Table 2. Additionally, descriptive statistics about our sample (e.g., age ranges by condition and Grade) can be found in Table 4. Analyses were conducted using data from all students who completed the relevant measures for each statistical test.

Evaluation of repeater status. Our analyses considered the interaction between Repeater Status on cognitive abilities (such as IQ) and achievement (e.g., arithmetic ability). We considered Repeaters to be any children who had already repeated or were currently repeating a grade, and we determined

TABLE 4 Descriptive Statistics for Students Included in Analyses

Class	Condition	Mean Age (mo)	SD Age	Range (mo)	Females	Total
Second grade	Control	95.24	6.91	85.90–117.10	59	128
Second grade	Intervention	94.22	6.65	85.50–117.40	72	127
Third grade	Control	107.17	7.90	97.73–127.63	31	61
Third grade	Intervention	106.55	6.63	97.73–127.50	35	70
Overall	Control intervention	99.13	9.14	85.90–127.63	90	189
		98.79	8.92	85.50–127.50	107	197

All ages are reported here as-of the pretest date before the 5-week intervention period.

this by comparing each student's birthday with the state-mandated cutoff date for their grade. By law, Uruguayan schools enforce strict cutoff dates for matriculation; it is illegal for parents to choose to hold a child back and start them later in school, a practice that occurs in some other countries like the United States. While our records did not indicate for each child the reason why they were a year older than their peers, the teachers reported that nearly all of these older children had been asked to repeat a grade because of poor performance. In our dataset, we had roughly the same percentage of Repeaters in Grades 2 and 3 (24% and 27%, respectively).

Materials

During the 5-week intervention period, students engaged with the training app using touchscreen tablets specifically designed for children, which they had access to throughout the training period. As mentioned previously, the training app consisted of three mini-games, each designed to probe a distinct modality of magnitude discrimination: time, area, and approximate number. The games could be played without a connection to the internet. Because use of the app was left to the discretion of the teachers in school, and the children themselves while at home, the number of times each child played each game was free to vary; however, in practice, children played each game multiple times over the course of the intervention; see *Engagement with the intervention* in "Results" section.

The cognitive and achievement tests were administered to all students in the study before and after the intervention period by trained researchers following a written protocol. As mentioned previously, we measured ANS Ability, Geometric Ability, Arithmetic Ability, Vocabulary size, and IQ (see Fig. 4). We also tracked SES at a school level using government-reported data (ANEP [Administración Nacional de Educación Primaria], 2012).

Intervention Games

Upon starting the intervention software, students saw the app's home page and could select one of the three mini-games by tapping on that game's icon. In each mini-game, students saw 12 trials in which the ratios of magnitudes (i.e., durations of time, numbers of items, and surface areas) were systematically varied, and they were asked to tap on the larger magnitude. Students received feedback after each trial indicating whether they made a correct response, and the games responded to three consecutive incorrect responses by repeating the game's instructions. After 12 trials the game ended, and a gold star appeared beside the number of correct answers made in that block. Tapping the screen returned the student to the app's home page, where the total number of correct responses across all blocks was displayed over the icon for each game.

Time discrimination. In the time discrimination game, students had to decide which of two sounds had a longer duration. The screen showed two monsters: a green one on the left and a purple one on the right (see Fig. 3D). Students tapped the screen to start a trial in which each monster took its turn to make a singing-like sound for a specific amount of time. To provide additional visual cues, the monsters' mouths moved, and they moved their hands to cover their mouths for 1 s after completing their sound. Students tapped the monster that they thought sang for longer and received feedback based on whether they answered correctly. After correct answers, the monsters would laugh and smile; incorrect responses elicited a "No, no!" sound while the monsters shook their heads. The singing sound varied from trial to trial to make the game more interesting.

Area discrimination. In the area discrimination game, students had to decide which of two monsters blew the largest "bubble" (see Fig. 3C). As in the time discrimination game, students saw two monsters: a blue one on the right and a yellow one on the left. After tapping the screen to begin the trial, they saw a multicolor circle grow in front of one of the monsters. The circle expanded briefly and stayed at a constant size for a brief interval before appearing to "pop" and disappear. Next, a similar bubble appeared in front of the other monster. The expansion and disappearing of the bubbles was accompanied by "blowing" and "popping" sound effects. After both bubbles disappeared, students tapped the monster they thought blew the biggest bubble. After correct responses, the monsters would laugh and dance to music, while incorrect responses caused them to shake their heads and say "No, no!."

Approximate number discrimination. In the number discrimination game, students were asked to tap the monster that "sneezed" the most "germs" (see Fig. 3B). As in the other games, they saw two monsters on the screen—one on the left and the other on the right. At the start of the trial, the first monster scrunched up its face and made a sneezing sound as a large sneeze-cloud with a certain number of "germs" appeared under its nose. After less than a second, the cloud of germs corresponding to the first monster disappeared and the second monster took its turn to sneeze a cloud of germs. The germs consisted of a discrete number of bounded shapes with different patterns and textures across trials. On some of the trials, the shapes were of equal size, and on other trials the size of the germs varied. On half of the trials, the total area of the germ shapes was equal between monsters (viz., the total area was controlled across numbers), and in the other half of trials the size of each germ and the total area varied (viz., the total area did not consistently correlate with the number of germs). These controls ensured that both individual germ size and total germ area were poor predictors of number in our stimuli. As in the other two games, students received auditory and visual feedback for the responses in the form of the monsters laughing and dancing for correct responses or shaking their heads (i.e., for incorrect responses).

Results

Our most important measures were Arithmetic Ability (Batería III Woodcock-Muñoz), ANS Ability (paper/pencil ANS task), SES (as reported by ANEP),

Repeater Status (determined by students' birthdates), and IQ (Raven's Progressive Matrices). We also include the brief geometry and vocabulary tests as pilot control measures. We first investigated pretest differences in Arithmetic Ability, ANS Ability, and IQ as a function of SES, Grade, and Repeater Status (i.e., Repeater vs. Non-Repeater). In our analysis of pretest abilities, we grouped both Game Intervention and BAU Control children together because, at this point, no children had received an intervention.

IQ and Repeater Status by SES Quintile

Independent from our interests in math intervention, factors like IQ, grade-repeating, and SES may have interesting interdependencies of their own. We first asked whether children's IQ might contribute to determining which children did and did not repeat a grade as a function of SES. In Fig. 6 each dot represents a child's age-normalized percentile score on the Raven's Progressive Matrices (IQ) assessment, arranged by SES Quintile and jittered along the *x*-axis to reveal all children. Filled-in and open dots correspond to Repeaters and Non-Repeaters, respectively. The filled and open gray dots indicate means for each group, with bars representing standard error. Because Raven's scores are age normalized to each student's birthdate, the scores in Fig. 6 can be compared across Grades 2 and 3, and across Repeaters and Non-Repeaters, that is, even though Repeaters were older than their classmates they are not favored on this measure.

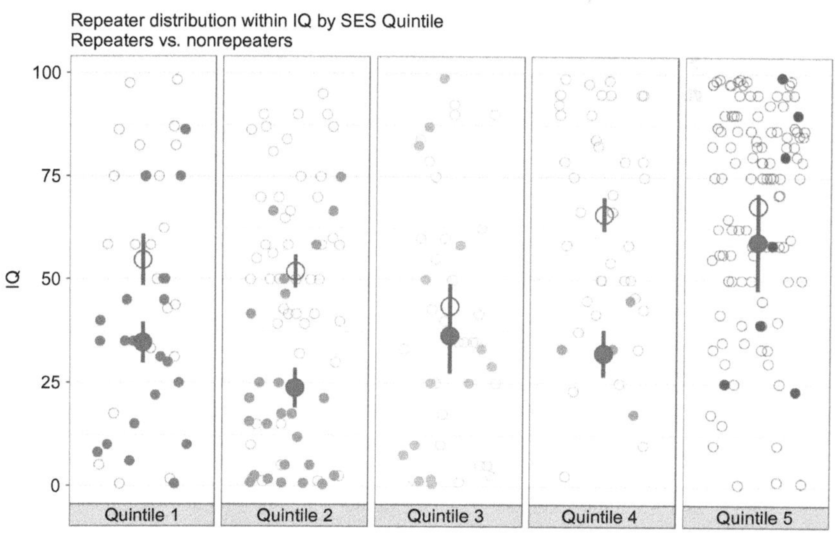

FIG. 6 Repeater Status compared with IQ within each SES Quintile. *Filled-in dots* indicate students who repeated a grade; *open circles* indicate students who never repeated a grade. *Dark gray dots* indicate group means for Repeaters *(filled)* and Non-Repeaters *(open)*, with bars indicating standard error of the mean.

The first thing we notice is that every Quintile contains the full range of IQ scores (i.e., we see scores from approximately 0–100 in each Quintile). Second, there is a small effect of SES on IQ in both Repeaters and Non-Repeaters, with higher SES corresponding to somewhat higher IQ scores on average, though this trend is perhaps minor compared to the broad similarity in spread that we see in this measure (0–100). This difference in measured IQ as a function of SES replicates patterns previously demonstrated in the literature (Turkheimer et al., 2003).

One of the most dramatic patterns in Fig. 6 is that the proportion of Repeaters is much higher in the lower quintiles compared to the higher quintiles. The effect of SES on the percentage of students who were repeating a grade is so dramatic that the difference between SES Quintile 1 (disadvantaged) and SES Quintile 5 (advantaged) was nearly eightfold in our sample: 41.8% of students in SES Quintile 1 had or were repeating a grade while only 5.4% of students in SES Quintile 5 had repeated or were repeating a grade. This trend is noticeable in Table 4, but in Fig. 6 we can see the pattern at the level of the child by looking at the ratio of filled to open circles as a function of SES.

Finally, in Fig. 6, we see that the mean IQ tends to be lower for Repeaters compared to Non-Repeaters. This result is important for considering how to effectively help these struggling students. Further, it is noteworthy that Repeaters tend to cluster at the bottom of the IQ scale in Quintiles 1 and 2, whereas the mean IQ of Repeaters was similar to Non-Repeaters in two of the three upper Quintiles (and note that in the one upper Quintile where this was not true, Quintile 4, there were only four Repeater children and so a larger sample may be required before concluding that this Quintile would show a significant difference). Interestingly, while lower IQ is linked with repeating a grade in SES Quintiles 1 and 2, it is less linked in the more advantaged schools (SES Quintiles 3 and 5).

Thus while it is true that many factors may impact performance on IQ tests like the one we have used here, our data support the inference that SES serves as a protective factor against repeating a grade. One possible explanation is that lower IQ students who might otherwise repeat a grade may receive more support in higher SES schools compared to students in disadvantaged schools and therefore be able to follow the standard trajectory. Conversely, students in low SES schools appear to need a higher IQ to avoid having to repeat a grade. This interpretation would be consistent with recent work showing that SES may moderate the impact of environmental factors that affect school performance. More specifically, higher SES mitigates the potential negative effects of factors such as low IQ (Tucker-Drob & Bates, 2016; Turkheimer et al., 2003).

Pre-Intervention Arithmetic by Grade and Repeater Status

Next, we considered Pre-Intervention Arithmetic Ability. Fig. 7 shows the performance for four groups of children across SES and Grade (i.e., Grade 2 Non-Repeaters, Grade 2 Repeaters, Grade 3 Non-Repeaters, and Grade 3

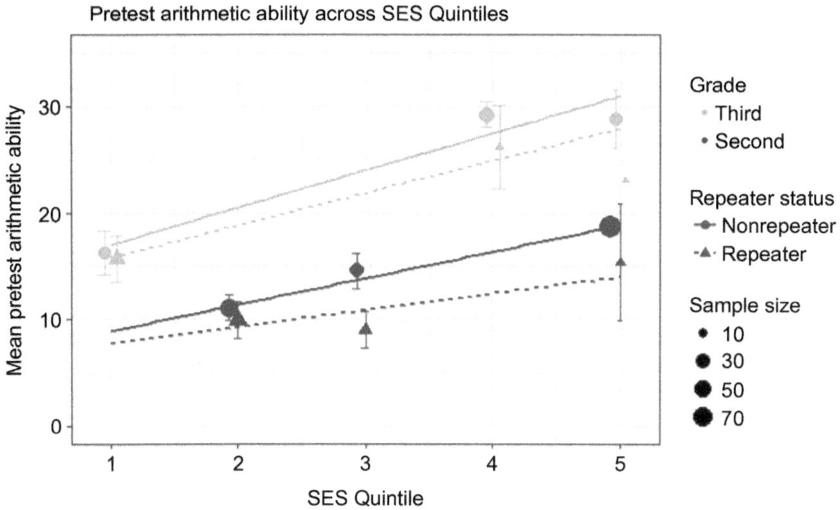

FIG. 7 Pretest Woodcock-Muñoz Arithmetic Ability score by SES quintile separated by Grade (second and third) and Repeater Status (Repeater vs. Non-Repeater). For this figure, four separate planned regressions were run (one for each group). The effect of SES on performance is apparent in each group (all $n \geq 35$), indicated by the upward slope of the lines as SES Quintile increases (all $P < .01$). The *top pair of lines* represents third-graders, while *bottom pair* represents second-graders (reflecting the positive difference that one additional year of school has on speeded arithmetic performance); *dashed lines* through *triangles* and *solid lines* through *circles* correspond to Repeaters and Non-Repeaters, respectively (these lines are regression lines based on the raw data from each child in the group). The size of each triangle and circle is proportional the sample size in that category. *Symbols* within each Quintile have been slightly jittered horizontally to prevent overlap.

Repeaters) in our assessment of Arithmetic Ability. If SES affects Arithmetic Ability, we should see a positive trend across SES with increasing Arithmetic Ability scores as SES Quintile increases. In Fig. 7, we see this pattern of higher scores for children in higher SES schools across Grade and Repeater Status as revealed in a regression of Arithmetic Ability by SES [$F(1,314) = 38.81$, $P < .001$, $R^2 = .11$)]. This reveals that SES impacts arithmetic ability and that children in higher SES Quintiles do better than their peers in less advantaged schools, as found in other studies (Goldin et al., 2014; Klibanoff et al., 2006; Odic et al., 2016; Valle-Lisboa et al., 2016; Wilkinson & Pickett, 2010; Zosh et al., 2018).

Considering Repeater Status in Fig. 7, note that despite Repeaters being a year older than their Non-Repeater classmates they nevertheless performed below the level of the Non-Repeater children on the pretest of Arithmetic Ability as revealed by planned *t*-tests within each Grade (Second Graders: $t(72.53) = 3.47$, $P < .001$; Third Graders $t(49.68) = 3.35$, $P < .01$). This highlights the importance of considering Repeater Status as a factor influencing students' performance.

ANS and Arithmetic Ability

Next, we considered the relationship between Arithmetic Ability and ANS Ability in our sample. There is a continuing debate concerning whether ANS abilities are related to school mathematical abilities (Bugden & Ansari, 2016; Clayton, Gilmore, & Inglis, 2015; DeWind & Brannon, 2012; Fazio, Bailey, Thompson, & Siegler, 2014; Gilmore et al., 2013; Odic et al., 2016; Xenidou-Dervou, Molenaar, Ansari, van der Schoot, & van Lieshout, 2017). We used the preintervention scores from all children to determine whether Arithmetic Ability is correlated with ANS Ability. The top plot in Fig. 8 shows the simple correlation

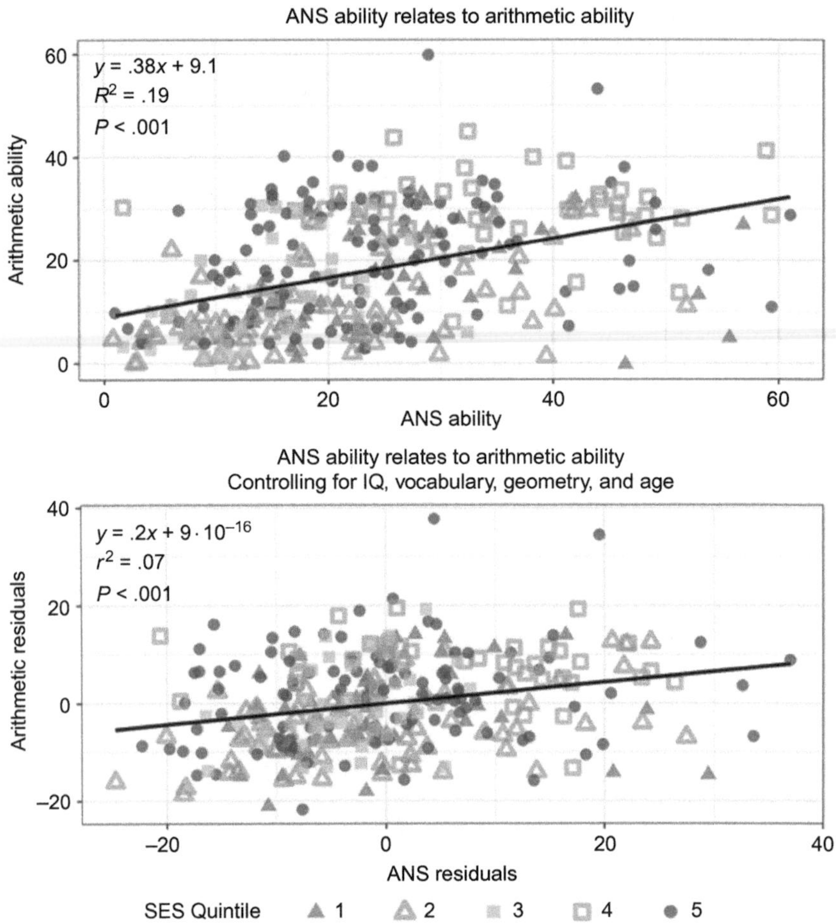

FIG. 8 Correlation and partial correlation (controlling for IQ, Vocabulary, Geometry Ability, and Age) between ANS ability and arithmetic ability. The top figure shows the simple correlation between the measures; the bottom figure shows the partial correlation. Each *symbol* corresponds to a child, and the *shapes* indicate the child's SES Quintile. Both regressions are significant and each SES Quintile contributes to these effects.

between these measures while the bottom shows the partial correlation of Arithmetic Ability and ANS Ability controlling for IQ, Vocabulary, Geometry Ability, and Age. The colors correspond to the SES Quintiles of students' schools. The first thing we notice is that the relationship is significantly positive in both cases ($P < .001$) suggesting a relationship between ANS and Arithmetic Ability, even when controlling for other factors (correlation: $F(1{,}304) = 70.07$, $P < .001$, $R^2 = .19$; partial correlation: $F(1{,}304) = 21.23$, $P < .001$, $r^2 = .065$). An additional age-normalized regression (not shown) between ANS Ability and Arithmetic Ability was also significant (this regression is one way of creating age-normalized standard scores for these tasks): $F(1{,}316) = 67$, $P < .001$, $r^2 = .17$.

Second, we see that the colored dots are evenly spread throughout the trend, indicating that no single SES Quintile is driving the effect. Indeed, individual regressions were also performed for each SES Quintile separately and the regression slopes were all positive suggesting that all SES Quintiles contribute to this effect. Therefore contrary to some claims in the literature (Bugden & Ansari, 2016; Clayton et al., 2015; Gilmore et al., 2013; Xenidou-Dervou et al., 2017) and consistent with others (DeWind & Brannon, 2012; Fazio et al., 2014; Odic et al., 2016) our data show a link between ANS Ability and Arithmetic Ability (across SES Quintiles and controlling for many relevant measures).

Engagement With the Intervention

Thus far we have seen noteworthy effects of SES and Repeater-Status on several measures in our sample. We also found that ANS Ability correlates with Arithmetic Ability while controlling for many factors. Before turning to consider how performance changed from pre- to posttest, we first report children's engagement with the intervention games. Recall that children in the Game Intervention group ($n = 197$) had access to three discrimination games on their tablets for 5 weeks and they were free to play them as much or as little as they liked while at home. Fig. 9 shows children's total number of games played over the course of the 5 weeks. The total number of games played suggests that children liked the games and did volunteer to play them. Because each discrimination game lasted approximately 3 min per play, this intervention can be considered a brief amount of exposure to the intervention games over the course of the 5 weeks.

Pre- to Postintervention Improvement

We turn now to considering how performance changed from pre- to posttest. Fig. 10 shows the percentage change from pretest to posttest performance (viz., performance 5 weeks later), collapsing across SES Quintiles, Grade, and Repeater-Status, for both Game Intervention and BAU Control children on each of our measures: Arithmetic Ability, ANS Ability, Vocabulary, Geometry, IQ. We saw significant gains for all groups of children on all tasks as shown by planned *t*-tests for each bar in Fig. 10 (all $P\text{s} < .05$ except

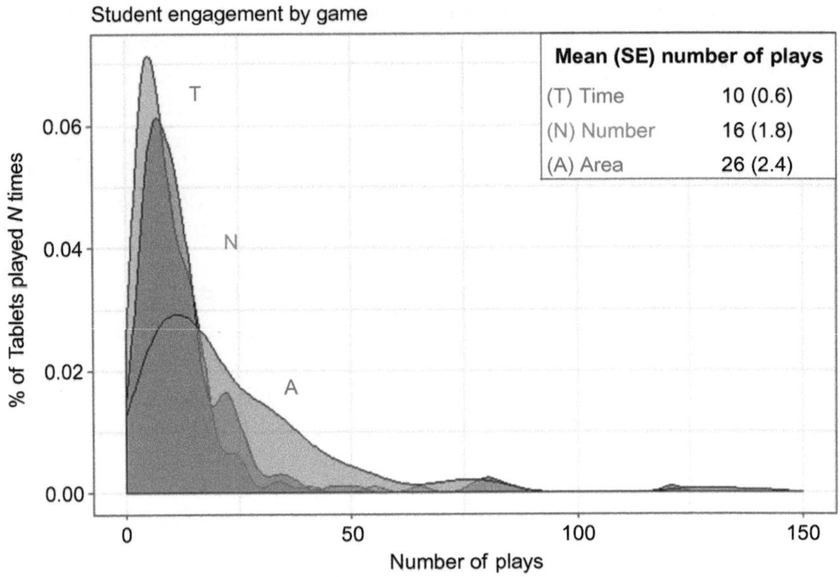

FIG. 9 Distribution of number of plays per child per game over the course of the 5-week intervention for all children in the Game Intervention group. The *graphs* are density distributions and show the proportion of children for each number of plays. Most children played each game around 10 times while the more active players pulled the mean number of plays higher for some games.

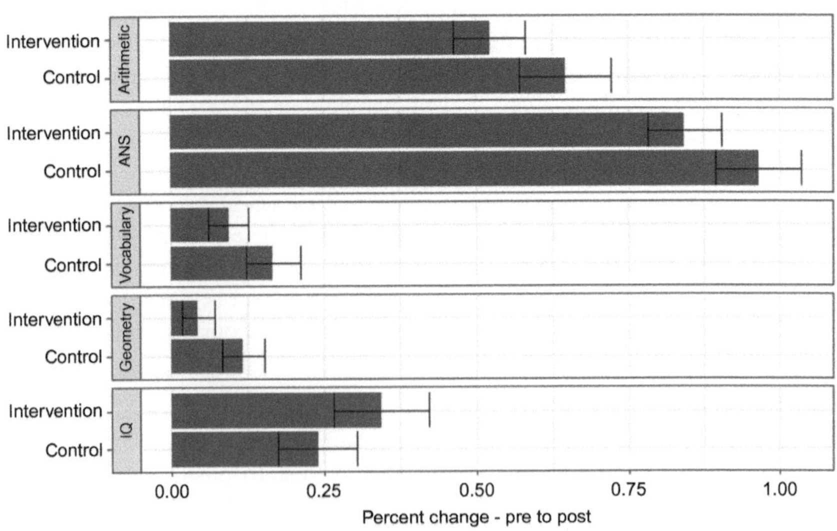

FIG. 10 Percent improvement on each assessment for both BAU Control and Game Intervention children. Both groups showed significant improvements from pre- to posttest on each of our measures (all $P < .05$ except Geometry, Intervention group $P = .09$), with the greatest improvement seen in the math assessments of Arithmetic Ability and ANS ability.

Geometry Ability in the Intervention group: $P = .09$). It is also noteworthy that we saw the greatest gains for our two math assessments: Arithmetic Ability and ANS Ability. Considering possible differences between the gains seen for BAU Control children and those seen for Game Intervention children, Fig. 10 does not reveal noteworthy improvements of the Game Intervention children above and beyond the BAU Control children. This is perhaps because of the large overall improvements we saw in all children. We next look at improvement as a function of SES Quintile and Repeater-Status as these factors may modulate who does and who does not improve.

We computed z-transformed change scores for both Arithmetic Ability and ANS Ability by taking each child's change in number of problems answered correctly from pre- to posttest and dividing by the SD of that child's classroom scores on the pretest:

$$\left(z_{diff} = \frac{Post_{child} - Pre_{child}}{SD(Pre_{child})} \right)$$

Recall that both our Arithmetic Ability assessment (Batería III Woodcock-Muñoz) and ANS Ability assessment were timed assessments which measured the number of problems correctly solved during 3 min. Dividing the difference from pre- to posttest by the SD of the pretest scores for each child's classroom is what makes this a z-normalized change score, and it has the consequence that each child's change score is roughly normalized across our factors of interest. That is, a z-normalized change score from a Grade 2 child in the Game Intervention group of the 5th SES Quintile will be comparable to the z-normalized change score from a Grade 3 child in the BAU Control group of the 1st SES Quintile. In both cases, the z-normalized change score indicates how much the child improved relative to the variability in the pretest scores of their immediate peers.

Arithmetic ability. In Fig. 11, the left side shows the z-normalized change scores in Arithmetic Ability for Non-Repeater children and the right side shows the z-normalized change scores for Repeater children. The first pattern to notice is that all of the mean change scores are above 0 (the level of no change), indicating that all groups of children improved from pretest to posttest. This duplicates the pattern of bars in Fig. 10. The change scores across groups seems to hover around a value of 1. This means that children, in general, improved about 1 SD above the mean of their classroom group from pretest to posttest.

Next, on the left side we see that for no SES Quintile did the Game Intervention children improve significantly more than the BAU Control children, with the exception of trends in the 1st and 3rd SES Quintiles. Being conservative, we can say that this intervention did not show noteworthy improvement over the test-retest improvement in the matched BAU Control classrooms for children who did not repeat a grade.

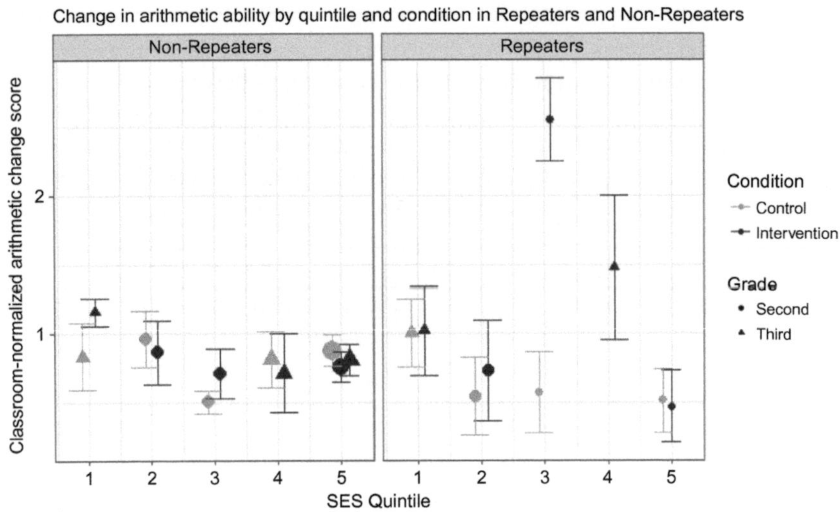

FIG. 11 Change in arithmetic ability by quintile and condition (BAU control vs. game intervention) for Repeaters (right) and Non-Repeaters (left).

Turning now to children who were or had already repeated a grade (right side), we see two noteworthy standouts for *z*-normalized change scores: the 3rd SES Quintile Game Intervention group and the 4th SES Quintile Game Intervention group. Planned *t*-tests compared the Game Intervention children to the BAU Control children in each SES Quintile. The test for the 3rd SES Quintile Repeaters (Game Intervention, $M=2.55$, $SD=0.600$; BAU Control $M=0.575$, $SD=0.588$) revealed that the Game Intervention children showed improvement significantly greater than the BAU Control children [$t(5.998)=-4.693$, $P<.01$]. The children in the 4th SES Quintile Game Intervention group are showing higher *z*-normalized change scores than most other groups; however, we did not have a child in the 4th SES Quintile BAU Control group who was a Repeater, so we cannot carry out the same *t*-test on this group. These groups suggest that the game intervention was more effective than simply test-retest improvement in at least some SES Quintiles for children who had or were repeating a grade and that there may be trends for lower SES children who did not repeat a grade (e.g., SES Quintiles 1 and perhaps 3), but these trends were insufficient in our sample to overcome the already large improvements seen in the BAU Control children and must be taken as merely suggestive.

As one indication that these improvements may generalize—given greater training or more effective yoking of pre- and posttesting to minimize the large test–retest improvements we saw here in the BAU children—we looked at whether the groups who showed noteworthy gains were different from their peers (e.g., examining whether their gains were an epiphenomenal result of their having extremely low pretest scores, which made large gains easier to

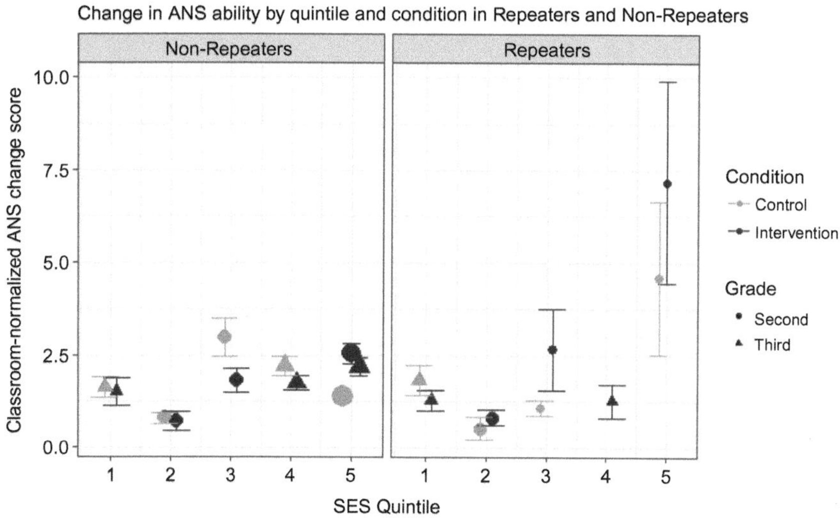

FIG. 12 Change in ANS ability by SES quintile and condition (BAU control vs. game intervention) for repeaters (right) and non-repeaters (left).

achieve). In both pretest ANS Ability and pretest Arithmetic Ability, children who improved more after the Game Intervention were no different than their peers—suggesting that these improvements did not merely result from aberrantly low pretest scores for these children.

ANS ability. Next, in Fig. 12, we consider the *z*-normalized change scores for performance on the ANS Ability assessment. Here again, we computed the change in number of correct answers from pre- to posttest and divided by the SD of scores within each child's classroom. In Fig. 12 we see these scores separated by intervention condition and Repeater Status. The first pattern to notice is that all of the mean change scores are above 0 (the level of no change), indicating that all groups of children improved from pretest to posttest. This again duplicates the pattern of bars in Fig. 10. The change scores across groups seem to hover around a value of 1.75. This means that children, in general, improved about 1.75 SDs above the mean of their classroom group from pretest to posttest.

On the left side we see that Game Intervention children in SES Quintile 5 improved significantly more than their peers in the BAU Control group as revealed by a planned *t*-test [$t(91.212) = -4.553$), $P < .001$]. However, children in SES Quintiles 3 and 4 showed the opposite trend with BAU Control children tending to show more improvement then the Game Intervention children [Q3: $t(20.019) = 1.890$, $P = .07$; Q4: $t(35.900) = 1.445$, $P = .16$]. Thus for children who were not repeating a grade, on the left side of Fig. 12, we cannot say that there was any systematic benefit of the Game Intervention above and beyond did the BAU Control children. All children improved on this assessment.

Considering now the children who had or were repeating a grade, on the right side of Fig. 12, we see a trend where Game Intervention children improved more than the BAU Control children in SES Quintile 3 ($p = .25$). However, none of these trends attained significance, perhaps due to the large variability from these small sample sizes.

Summary of Pre- to Postintervention Improvement

Across all measures, there is noteworthy improvement from pretest to posttest for both the Game Intervention and BAU Control children, most likely due to test-retest improvement as well as the positive effects of having a special visitor come to the classroom and engage children in special tests—note that the tester for the posttest (J.L.), who was blind to the experimental condition for each classroom, was a particularly special visitor as he is male (atypical in education settings in Uruguay), an American, and a Spanish speaker with an American-Iberian accent. This may have contributed to better focus and overall performance on posttest versus pretest (which was administered by female Uruguayan graduate students (D.L. & D.F.).

While this overall improvement made it difficult to see improvement in our Game Intervention children above and beyond the improvement in our BAU Control children, and in spite of the small sample sizes in some of these groups, we did see some significant results and trends suggesting that there may be a positive influence of the game-based intervention training. But for certain, the strongest results in the current sample concern the effects of SES and Repeater-Status on Arithmetic Ability, along with the differential links between IQ and these factors, and the relationship between Arithmetic Ability and ANS Ability in our sample. Continued work to tailor interventions to the community is necessary.

CONCLUSIONS AND FUTURE DIRECTIONS

Researchers and educators have a shared interest in understanding the factors that influence students' educational outcomes, and our work is consistent with previous findings that the socioeconomic status of a child's community is related to their performance in the classroom. Even before our intervention started, students in lower SES schools scored lower on measures of math achievement compared to students at more advantaged schools. We also saw that lower SES students are more likely to repeat a grade compared to higher SES students, and that having a lower IQ has a disproportionately negative impact on the likelihood of repeating a grade for lower SES children. One reason for this may be that lower IQ students at higher SES schools are better supported, either within the classroom or at home, compared to similarly scoring students at lower SES schools. An important future direction may be investigating the specific ways that more affluent schools succeed in supporting students at risk for repeating a grade so that these methods can be implemented at lower SES schools. Certainly availability of resources has a role to play here, but interventions such as the one presented in this study suggest that there are possibilities for reducing the gap between high- and low-achieving students using existing infrastructure.

One method for enriching the educational experience of vulnerable students may be through technology. The Uruguayan educational system provides a unique opportunity to investigate this question, given the country's investment in educational technology in schools that span the SES spectrum. While not conclusive at this stage, our results are consistent with the idea that the tablet can serve as an important vehicle for intervention and that it may be useful for improving math performance for students who are the most disadvantaged. What is more certain is that students seemed to show across-the-board improvement in a number of areas when special adults visited their classrooms and evaluated their performance. To the extent that lower SES students feel different societal expectations in terms of their academic potential, providing special attention to these students may prove an important tool in helping them reach the same levels of achievement as their more affluent peers.

The design of the current study had some limitations which should be addressed in future interventions. A more fine-grained way to measure the impact of the game would have been to look at dosage effects; however, this stage of the project included no standardization of the amount of time that the students interacted with the intervention media. In part, this was by design; the project is a collaboration with the teachers who agreed to use the game in their classrooms. By giving them the freedom to implement the game in the ways they saw fit, we stood to learn from their feedback about how best to integrate the intervention game with the normal classroom curricula. Future deployments of this intervention may include lesson planning guides based on suggestions from teachers in the current study.

While the effects of SES and Repeater Status are central to our interpretation of this study's results, we were unable to balance our sample of students by age across SES Quintiles because classrooms in the study were included based on their willingness to participate. For the most part, only one grade level was represented within each SES Quintile, so it made comparisons across grade levels difficult given the large effect of Grade on our outcome measures (viz., Arithmetic Ability score). Similarly, we would have liked to have been able to examine in more detail the effect of repeating a grade, but the number of Repeater students was impossible to control in the sample. Future studies might focus on schools with high levels of repetition to examine the effects of the intervention in these groups. Having a more complete sample would allow us to identify which students are most helped by our intervention; however, this would depend on the availability of classrooms willing to participate in the study.

This study's main goal was practical: to create a tool for teachers to use in the classroom that could improve students' math performance. However, these results also speak to important scientific questions about the relationship between underlying cognitive abilities and classroom math performance. One of the strongest results in this study was the evidence for a relationship between Arithmetic Ability and ANS Ability. We found that ANS Ability related to Arithmetic Ability across all SES Quintiles controlling for IQ, Vocabulary, Geometry Ability, and Age. The inclusion of these control measures was important because there is a continuing debate concerning whether ANS abilities are related to school mathematical abilities when controlled for other factors (Bugden & Ansari, 2016; DeWind & Brannon, 2012; Fazio et al., 2014; Gilmore et al., 2013; Odic et al., 2016; Xenidou-Dervou et al., 2017). Contrary to some claims in the literature (Bugden & Ansari, 2016; Clayton et al., 2015; Gilmore et al., 2013; Xenidou-Dervou et al., 2017) and consistent with others (Halberda et al., 2012; Odic et al., 2016) our data show a consistent link between ANS Ability and Arithmetic Ability across SES Quintiles.

Considering the importance of intervention research, to the extent that magnitude training can improve formal math ability, this suggests a causal link between intuitive number sense and children's formal understanding of math. Our data are consistent with other findings that have shown such a link (DeWind & Brannon, 2012; Odic et al., 2016; Odic, Libertus, et al., 2013), but further work is needed to explore more fine-grained questions about training type and specific formal skills. The intervention in this study trained children on three different magnitude tasks (area, time, and approximate number). A future study might look at the correlation between amount of student interaction with each of these games and amount of improvement in various achievement domains (e.g., addition/subtraction, geometry, time estimation). For example, if the partial correlation between ANS training and formal math ability were significant when accounting for approximate area training, this would suggest a privileged connection between the ANS system and symbolic math, as found by Lourenco et al. (2012) for adults. This would align with previous work that has shown a privileged relationship between ANS and symbolic math compared to symbolic math and time discrimination (Odic et al., 2016). It may also be the case that different magnitude training tasks are more strongly linked to performance in various outcome domains; for instance, we might observe privileged relationships between area training and formal geometry as well as between ANS training and formal arithmetic. These are important scientific questions about the foundations of human numerical understanding that could be addressed in future studies.

Understanding underlying mechanisms is key to progress in many areas of science. When we succeed in applying this understanding to improving the lives of people, our work becomes even more thrilling. The current project reveals that there are children who will benefit from interventions to improve math ability, and that these improvements may vary as a function of IQ and SES. It also shows that there is a link between ANS Ability and Arithmetic Ability across SES Quintiles. Nevertheless, our work also highlights the challenges of developing an intervention that will bridge the gap between game-based improvements and gains in math understanding. Through our partnership with teachers and students, we aim to continue exploring how equal access to technology and purpose-built software can improve educational outcomes, and ultimately help to create a more equitable society.

REFERENCES

ANEP [Administración Nacional de Educación Primaria]. (2012). *Relevamiento de contexto socio-cultural de escuelas de educación primaria [Survey of primary school's socio-cultural context]*. Montevideo: ANEP. Obtenido de http://www.anep.edu.uy/anep/index.php/codicen-publicaciones/category/117-publicacionesdivision-de-investigacion-y-estadistica-educativa?download=1132:relevamiento-de-contextosociocultural-2010.

Bugden, S., & Ansari, D. (2016). Probing the nature of deficits in the 'Approximate Number System' in children with presistent developmental dyscalculia. *Developmental Science*, *19*(5), 817–833.

Clayton, S., Gilmore, C., & Inglis, M. (2015). Dot comparison stimuli are not all alike: the effect of different visual controls on ANS measurement. *Acta Psychologica*, *19*(5), 177–184.

DeWind, N. K., & Brannon, E. M. (2012). Malleability of the approximate number system: effects of feedback and training. *Frontiers in Human Neuroscience*, *6*, 1–10.

Dillon, M. R., Kannan, H., Dean, J. T., Spelke, E. S., & Duflo, E. (2017). Cognitive science in the field: a preschool intervention durably enhances intuitive but not formal mathematics. *Science*, *357*(6346), 47–55.

Fazio, L. K., Bailey, D. H., Thompson, C. A., & Siegler, R. S. (2014). Relations of different types of numerical magnitude representations to each other and to mathematics achievement. *Journal of Experimental Child Psychology*, *123*, 53–72.

Feigenson, L., Dehaene, S., & Spelke, L. (2004). Core systems of number. *Trends in Cognitive Science*, *8*(7), 307–314.

Gilmore, C., Attridge, N., Clayton, S., Cragg, L., Johnson, S., Marlow, N., et al. (2013). Individual differences in inhibitory control, not non-verbal number acuity, correlate with mathematics achievement. *PLoS ONE*, *8*(6).

Goldin, A. P., Hermida, M. J., Shalom, D. E., Costa, M. E., Lopez-Roesenfeld, M., Segretin, M. S., et al. (2014). Far transfer to language and math of a short software-based gaming intervention. *Proceedings of the National Academy of Sciences of the United States of America*, *111*(17), 6443–6448.

Halberda, J., & Feigenson, L. (2008). Developmental change in the acuity of the "number sense": the approximate number system in 3-, 4-, 5-, and 6-year-olds and adults. *Developmental Psychology*, *44*(5), 1457–1465.

Halberda, J., Ly, R., Wilmer, J. B., Naiman, D. Q., & Germine, L. (2012). Number sense across the lifespan as revealed by a massive Internet-based sample. *Proceedings of the National Academy of Sciences of the United States of America*, *109*(28), 11116–11120.

Hart, B., & Risley, T. R. (1975). Incidental teaching of language in the preschool. *Journal of Applied Behavior Analysis*, *8*(4), 411–420.

Hoff, E. (2003). The specificity of environmental influence: socioeconomic status affects early vocabulary development via maternal speech. *Child Development*, *74*(5), 1368–1378.

Hyde, D. C., Khanum, S., & Spelke, E. S. (2014). Brief non-symbolic, approximate number practice enhances exact symbolic arithmetic in children. *Cognition*, *131*(1), 92–107.

INEEd. (2017). *Informe sobre el estado de la educación en Uruguay: 2015–2016*. Montevideo: INEEd.

Izard, V., Sann, C., Spelke, E., & Streri, A. (2009). Newborn infants perceive abstract numbers. *Proceedings of the National Academy of Sciences of the United States of America*, *106*(25), 10382–10385.

Klibanoff, R. S., Levine, S. C., Huttenlocher, J., Vasilyeva, M., & Hedges, L. V. (2006). Preschool children's mathematical knowledge: the effect of teacher "math talk." *Developmental Psychology*, *42*(1), 59–69.

Larson, K., Russ, S. A., Nelson, B. B., Olson, L. M., & Halfon, N. (2015). Cognitive ability at kindergarten entry and socioeconomic status. *Pediatrics*, *135*(2), 440–448.

Levine, S. C., Vasilyeva, M., Lourenco, S. F., Newcombe, N. S., & Huttenlocher, J. (2005). Socioeconomic status modifies the sex difference in spatial skill. *Psychological Science*, *16*(11), 841–845.

Lourenco, S. F., Bonny, J. W., Fernandez, E. P., & Rao, S. (2012). Nonsymbolic number and cumulative area representations contribute shared and unique variance to symbolic math competence. *Proceedings of the National Academy of Sciences of the United States of America*, *109*(46), 18737–18742.

Mailhos, Á., Maiche, A., Langfus, J., De Leon, D., Fitipalde, D., Sánchez, I., & Halberda, J. (2018). Towards a paper-and-pencil ANS assessment tool. (in prep.).

Odic, D., Lisboa, J. V., Eisinger, R., Gonzalez, M. O., Maiche, A., & Halberda, J. (2016). Approximate number and approximate time discrimination each correlate with school math abilities in young children. *Acta Psychologica*, *163*, 17–26.

Odic, D., Libertus, M., Feigenson, L., & Halberda, J. (2013). Developmental change in the acuity of approximate number and area representations. *Developmental Psychology*, *49*(6), 1103.

Odic, D., Pietroski, P., Hunter, T., & Lidz, J. (2013). Young children's understanding of "more" and discrimination of number and surface area. *Journal of Experimental Psychology: Learning, Memory, and Cognition*, *39*(2), 451–461.

Pilli, O., & Aksu, M. (2013). The effects of computer-assisted instruction on the achievement, attitudes and retention of fourth grade mathematics students in North Cyprus. *Computers & Education*, *62*, 62–71.

Raven, J., Raven, J. C., & Court, J. H. (1998). *Manual for Raven's progressive matrices and vocabulary scales*. San Antonio, TX: Pearson.

Sirin, S. R. (2005). Socioeconomic status and academic achievement: a meta-analytic review of research. *Review of Educational Research*, *75*(3), 417–453.

Sokolowski, H. M., Fias, W., Ononye, C. B., & Ansari, D. (2017). Are numbers grounded in a general magnitude processing system? A functional neuroimaging meta-analysis. *Neuropsychologia*, *105*, 50–69.

Trucano, M. (2011). *What's next for plan ceibal in Uruguay?* EduTech: A World Bank Blog on ICT use in Education. 10 de June de. Obtenido de: http://blogs.worldbank.org/edutech/planceibal2.

Tucker-Drob, E. M., & Bates, T. C. (2016). Large cross-national differences in gene x socioeconomic status interaction on intelligence. *Psychological Science*, *27*(2), 138–149.

Turkheimer, E., Haley, A., Waldron, M., D'Onofrio, B., & Gottesman, I. I. (2003). Socioeconomic status modifies heritability of IQ in young children. *Psychological Science*, *14*(6), 623–628.

Valle-Lisboa, J., Cabana, Á. E., Mailhos, Á., Luzardo, M., Halberda, J., & Maiche, A. (2016). Cognitive abilities that mediate SES's effect on elementary mathematics learning: the Uruguayan tablet-based intervention. *Prospects: Comparative Journal of Curriculum, Learning, and Assessment*, *46*(2), 301–315.

Verdine, B. N., Irwin, C. M., Golinkoff, R. M., & Hirsh-Pasek, K. (2014). Contributions of executive function and spatial skills to preschool mathematics achievement. *Journal of Experimental Child Psychology*, *126*, 37–51.

Walsh, V. (2003). A theory of magnitude: common cortical metrics of time, space and quantity. *Trends in Cognitive Science*, *7*(11), 483–488.

Wilkinson, R., & Pickett, K. (2010). *The spirit level: Why equality is better for everyone*. London: Penguin.

Woodcock, R. W., McGrew, K. S., & Mather, N. (2001/2007). *Woodcock-Johnson III tests of cognitive abilities*. Rolling Meadows, IL: Riverside.

World Computer Exchange. (2016). *WCE FY16 annual report*. Hull, MA: World Computer Exchange.

Xenidou-Dervou, I., Molenaar, D., Ansari, D., van der Schoot, E., & van Lieshout, C. (2017). Nonsymbolic and symbolic magnitude comparison skills as longitudinal predictors of mathematical achievement. *Learning and Instruction*, *50*, 1–13.

Zhang, M., Trussell, R. P., Gallegos, B., & Asam, R. R. (2015). Using math apps for improving student learning: an exploratory study in an inclusive fourth grade classroom. *TechTrends*, *59*(2), 32–40.

Zosh, J., Verdine, B., Halberda, J., Hirsh-Pasek, K., & Golinkoff, R. (2018). Preschoolers' approximate number system varies by socio-economic status. (in prep.).

Chapter 3

Role of Play and Games in Building Children's Foundational Numerical Knowledge

Geetha B. Ramani, Emily N. Daubert and Nicole R. Scalise
Department of Human Development and Quantitative Methodology, University of Maryland, College Park, MD, United States

INTRODUCTION

Adults' mathematical knowledge is important for daily functioning, career advancement, and life success (National Research Council, 2001). Laying the foundation for learning this vital knowledge begins early. During early childhood, children acquire a range of numeracy skills that are critical for later mathematical achievement. For example, children's numerical understanding from the start of kindergarten through the middle of 1st grade is strongly related to mathematics achievement at the end of 1st grade through 3rd grade (Geary et al., 2018; Jordan, Kaplan, Locuniak, & Ramineni, 2007). In a large longitudinal study, children's numerical knowledge at age 4 predicted their mathematical knowledge at age 15, even after controlling for various factors, such as IQ, socioeconomic stats (SES), and reading comprehension abilities (Watts, Duncan, Siegler, & Davis-Kean, 2014). Moreover, children's numerical knowledge at age 7 was related to their SES at age 42, again after controlling for factors such as IQ, childhood SES, academic motivation, and reading achievement (Ritchie & Bates, 2013).

Given the importance of early numerical knowledge for later achievement, a strong emphasis has been placed on ensuring high-quality mathematical instruction in early childhood classrooms. A key goal of mathematics instruction in these early childhood classrooms is the use of developmentally appropriate practices so that mathematics is purposeful and meaningful for young children (Baroody, 2000). One way to achieve this goal is through play and games. While many mathematics skills need to be explicitly taught,

Mathematical Cognition and Learning, Vol. 5. https://doi.org/10.1016/B978-0-12-815952-1.00003-7

playful activities and games in early education curricula can supplement this instruction by providing children with opportunities to practice their early mathematics skills in engaging and meaningful ways.

In this chapter, we will argue for the benefits of using age-appropriate play and games in early childhood classrooms to promote children's numerical knowledge. We will focus on one area of numerical knowledge, specifically numerical magnitude knowledge, by presenting theory and research that underscores the importance of this area In the following section, we present specific examples of games that have been found to promote early numerical magnitude knowledge. Our focus is on both traditional games as well as math-related computer and tablet games to improve numerical knowledge in early childhood. Overall, we will argue for the importance of focusing on theoretically driven design of games to promote young children's mathematical knowledge.

UNDERSTANDING OF NUMERICAL MAGNITUDES

Early childhood is critical for establishing the foundational numerical knowledge that will serve as the basis for children's future mathematics achievement. One key aspect of early mathematical knowledge that has roots in infancy is children's understanding of numerical magnitudes or the relative quantities represented by numbers in their various forms. Numerical magnitude knowledge allows us to determine which of two bushes has more berries and to know whether $10 or $25 per hour is a higher payment for our efforts. Numerical magnitude information is also embedded in much of mathematics. For example, understanding numerical magnitudes can help students recognize implausible arithmetic solutions, such as why 10 plus 2 is unlikely to be equal to 102. Individual differences in numerical magnitude knowledge, assessed with tasks such as comparing sets of objects, identifying which of two numerals is more, and estimating the position of a number on a bounded number line, predict mathematics achievement among children and adults (Schneider et al., 2016). People's numerical knowledge grows throughout the lifespan, typically in conjunction with formal education (Siegler, 2016). The present section reviews an integrated theory of numerical magnitude development and highlights the characteristics of numerical magnitude knowledge in early childhood.

Integrated Theory of Numerical Development

From infancy through adulthood, humans are aware of and can discriminate between numerical magnitudes. The integrated theory of numerical development asserts that numerical knowledge across the lifespan can be characterized as an increasing understanding of numerical magnitude information represented by the mental number line (Siegler, 2016; Siegler, Thompson, & Schneider, 2011). The mental number line is a spatial-numerical association shared by humans and some animals, with most members of Western cultures representing number in a horizontal line moving from left (smaller quantities) to right (larger quantities; Dehaene, 2011). The theory outlines four conceptually distinct but temporally overlapping trends in magnitude understanding, including discriminating between nonsymbolic sets of objects, linking nonsymbolic quantities to small whole numbers, expanding to magnitudes of larger whole numbers, and then expanding to magnitudes of negative and rational numbers (Siegler, 2016).

While each trend centers on a different type of magnitude knowledge, magnitude knowledge is typically assessed with comparison or estimation tasks across trends (Fig. 1). Magnitude comparison tasks involve comparing two or more quantities to determine which is the most, using nonsymbolic quantities (e.g., □□□ vs. □□□□□□) or symbolic numbers (e.g., 3 vs. 5). Depending on the age and skill level of participants, magnitude knowledge is then operationalized as accuracy across trials, average reaction time, or the most difficult ratio that a participant can consistently discriminate. Magnitude estimation tasks involve placing a nonsymbolic quantity or symbolic number in the correct position on a number line with labeled end points.

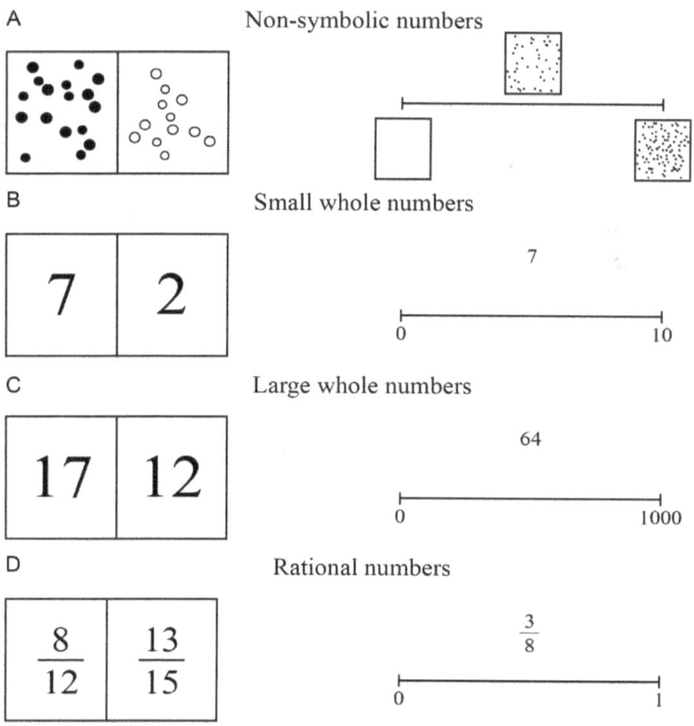

FIG. 1 Magnitude knowledge tasks. This figure illustrates commonly used magnitude knowledge tasks for each of the four types of numerical magnitude understanding (A–D), in magnitude comparison formats (left side) and number line estimation formats (right side).

Consistent with the characteristics of the mental number line, number line estimation tasks use a horizontal line labeled with a smaller quantity on the left endpoint and a larger quantity on the right endpoint. Participants' magnitude knowledge on these types of magnitude estimation tasks is operationalized as the discrepancy between where they positioned the target number and its actual location (percent absolute error), or the degree to which their estimates follow a linear distribution (such that the distance between estimates of 2 and 3 is equivalently spaced to the distance between estimates of 8 and 9).

The first trend of numerical magnitude development involves increasingly precise representations of nonsymbolic magnitude. From birth, humans are sensitive to large differences in nonsymbolic numerical magnitude (e.g., 8 objects vs. 16 objects), often attributed to the evolved approximate number sense (ANS; Feigenson, Dehaene, & Spelke, 2004; Izard, Sann, Spelke, & Streri, 2009). As early as the first few days of life, infants possess the ability to discriminate between visual displays of one and three objects (Antell & Keating, 1983; Starkey & Cooper, 1980). However, the ability to discriminate between sets of objects depends on the ratio of the number of objects in each of the sets. Izard et al. (2009) found that newborns showed a behavioral preference (i.e., longer looking times) when a number of objects displayed that matched the number of sounds played simultaneously, as compared to trials where the number of objects and sounds did not match. The newborns showed this preference for ratios of 1:3 (e.g., 4 objects vs. 12 sounds) but did not differ in their looking time behavior for more difficult ratios such as 1:2 (e.g., 4 objects vs. 8 sounds), demonstrating ratio-dependent processing that is found in both children and adults.

By the time infants are 6 months old, they can discriminate between sets that differ for number, space, and duration with 2:1 ratios, and by 9 months, infants can discriminate between sets with 3:2 ratios (Brannon, Lutz, & Cordes, 2006; Cordes & Brannon, 2008; Lipton & Spelke, 2003; vanMarle & Wynn, 2006; Xu & Spelke, 2000). The developmental trend continues, with 3-year-olds being able to discriminate 4:3 ratios, 6-year-olds able to discriminate 6:5 ratios, and adults able to discriminate 8:7 ratios (Bonny & Lourenco, 2013; Halberda & Feigenson, 2008; Holloway & Ansari, 2008).

Studies of children and adults that employ a comparison task (Fig. 1, set a) often characterize their ANS acuity with a Weber fraction, representing the most difficult ratio of stimuli that an individual can accurately discriminate (Halberda & Feigenson, 2008). Individual differences in nonsymbolic magnitude knowledge, measured by Weber fractions or overall task accuracy, predict concurrent and later mathematics achievement (e.g., Batchelor, Inglis, & Gilmore, 2015; Bonny & Lourenco, 2013; Halberda, Ly, Wilmer, Naiman, & Germine, 2012; Libertus, Feigenson, & Halberda, 2011). Although the relation between nonsymbolic magnitude and math achievement is highly contested, several meta-analyses suggest there is a significant, albeit small, average effect across childhood and adulthood ($r \approx 0.20$; Chen & Li, 2014; Fazio, Bailey, Thompson, & Siegler, 2014; Schneider et al., 2016).

The second trend of numerical magnitude development is the process of linking nonsymbolic representations to symbolic representations of small whole numbers (Siegler, 2016). Children begin to recognize and attach meaning to symbolic number words (e.g., one, two, three) and written numerical symbols (e.g., 1, 2, 3) in early childhood. Learning number words is central to young children's understanding of the counting system (Wynn, 1992). Children begin to use number words to recite the count sequence between the ages of 2 and 3 years (Gelman & Gallistel, 1978), although they may not initially adhere to the standard sequence (i.e., using the words "two, four, five" to count three items). Beginning at age 2 or 3 years, children require about a year of experience counting before they begin to relate specific number words to their corresponding numerosities (Wynn, 1992). By the time they are 3 and 4 years old, children typically know the correct count sequence for numbers one through ten (Fuson, 1988; Siegler & Robinson, 1982). By age five, most children have mastered the main counting principles. For example, children's understanding of cardinality improves, which link a symbolic referent (e.g., number word) to a nonsymbolic quantity (e.g., tokens). However, children may make some errors such as believing items must be counted consecutively from one contiguous item to the next (adjacency). Despite these errors, children are highly proficient counters by the time they reach kindergarten (Geary, 2006). Although these counting skills are critical for numerical magnitude understanding, there is also evidence that they are not sufficient for learning about the magnitudes of numbers. Specifically, children can count in a number range nearly a year before they acquire an understanding of the magnitudes of those numbers (Le Le Corre & Carey, 2007).

Some theorists have suggested children's symbolic magnitude representations, such as written numerals and number words, are mapped directly onto their nonsymbolic magnitude representations (e.g., Dehaene, 2001). Indeed, both show ratio-dependent performance in magnitude comparison tasks and overlapping regions of neural activity (Nieder & Dehaene, 2009; Reynvoet & Sasanguie, 2016). However, the integrated theory of numerical development aligns with an alternative theoretical perspective that suggests children may learn the meaning of small numbers (up to 10) by associating them with nonsymbolic representations, including their own fingers (Siegler, 2016; Sullivan & Barner, 2014). In other words, having something external to represent quantity, including fingers, helps children to learning the meaning of numbers. Once children have linked small symbolic numbers to nonsymbolic representations, they may then abstract to gain meaning of larger symbolic numbers that could not be precisely represented by approximate nonsymbolic representations (e.g., 100 and 101). Variability in preschool and kindergarten children's symbolic magnitude knowledge for small numbers (1–10) assessed through symbolic magnitude comparison and number line estimation tasks predicts their math achievement in later elementary school (Booth & Siegler, 2006; Kolkman, Kroesbergen, & Leseman, 2013; Sasanguie, De Smedt, Defever, & Reynvoet, 2012).

After acquiring an understanding of small symbolic numbers, the third stage in magnitude development is an understanding of larger symbolic numbers. Children's initial performance on 0–10 number line estimation tasks follows a logarithmic pattern, where smaller numbers are overly spread across the left side of the number line and larger numbers are condensed near the right side of the number line (Siegler, 2016). By ages 5 and 6 on 0–10 number line estimation tasks, children can accurately produce a linear distribution of their estimates, where estimated magnitudes increase linearly with actual magnitude. However, it is this age range children show the same logarithmic distribution on 0–100 number line estimation tasks (Berteletti, Lucangeli, Piazza, Dehaene, & Zorzi, 2010; Kim & Opfer, 2017; Laski & Siegler, 2007).

The fourth and final stage of numerical magnitude development is an understanding of rational numbers, including decimals, fractions, ratios, and rates (Siegler, 2016). Although rational number knowledge predicts mathematics achievement and is key to more advanced mathematics and scientific work, many children and adults have a limited understanding of rational magnitudes (Siegler, 2016). Several principles of rational numbers contrast starkly with principles of whole numbers: they do not have singular predecessors or successors, they can be represented by infinitely many symbols (e.g., 0.5 = ½ = 4/8), and may decrease with multiplication or increase with division (see also Vamvakoussi, this volume). These departures from the principles of whole numbers are likely part of the reason why students struggle to develop an understanding of rational number magnitudes. There is some evidence that whole number magnitude understanding relates to later knowledge of fraction magnitudes (Bailey, Siegler, & Geary, 2014). In addition, understanding the relative magnitude of rational numbers may be what allows students to be successful in arithmetic with rational numbers; for example, students who practiced comparing and ordering fractions of varying magnitudes saw greater improvement in their conceptual understanding and arithmetic performance than students who practiced creating part-whole representations of fractions (Fuchs et al., 2013). Thus as with nonsymbolic and symbolic whole numbers, a deeper understanding of the magnitude of rational numbers may underlie numerical development and aid in mathematics performance broadly.

In sum, the integrated theory of numerical development describes a progressive broadening of numerical magnitude understanding from infancy to adulthood that is in turn related to math achievement. Each of the four trends of developing magnitude knowledge is a candidate for interventions to promote mathematics learning, although different types of magnitude are more and less relevant for certain age groups.

Numerical Magnitude Understanding in Early Childhood

As the main focus of this chapter is on early childhood, two of the four trends in the integrated theory of numerical development framework are most relevant to this period. In preschool, children are refining their nonsymbolic magnitude knowledge, able to reliably discriminate between sets of objects at a 3:4 ratio around 3 years old and ratios of 5:6 by 6 years old (Halberda & Feigenson, 2008). Children are also mastering the verbal count sequence, recognizing written Arabic numerals, and learning to connect symbolic number representations like number words and numerals to quantities of nonsymbolic objects.

On the basis of the integrated theory of numerical development, nonsymbolic magnitude representations do not causally lead to all symbolic number knowledge; however, young children may be attaching meaning to their newly acquiring number symbols by associating them with known nonsymbolic representations (Siegler, 2016; Sullivan & Barner, 2014; vanMarle et al., 2018). This theoretical orientation implies that refining the nonsymbolic magnitude representation system in preschool is perhaps less critical for later math achievement than helping children link their nonsymbolic and symbolic number representations. Indeed, preschoolers' performance on cardinality tasks strongly predicts later math achievement (Geary & vanMarle, 2016).

Fluency with small symbolic numerical magnitudes may support the development of simple arithmetic skills, as well as later fluency with larger numerical whole and rational number. Young children demonstrate arithmetic understanding on nonverbal addition tasks as early as 2 years of age (Starkey, 1992). By the time they are three, many children can solve simple nonsymbolic problems of small numbers (2+1), and 4-year-olds can solve addition as well as subtraction problems (3–1; Vilette, 2002). Symbolic numerical magnitude knowledge can help children with their symbolic arithmetic performance by signaling implausible answers (e.g., that the sum of 2+4 could not be less than 4). Thus by the time children reach 4 to 5 years of age, they are using a range of numerical knowledge skills including counting, numeral identification, and numerical magnitude knowledge for more advanced mathematics like arithmetic.

The central reason to focus on the development of numerical knowledge in early childhood period is because of the evidence linking these early quantitative skills to later mathematics achievement. For example, 6-year-olds' understanding of symbolic number magnitude predicted their fraction magnitudes and fractions arithmetic—traditionally difficult mathematical concepts—at age 13 (Bailey et al., 2014). These relations held even after controlling for various factors, such as children's executive function, IQ, race, gender, and SES. Using longitudinal data, multiple studies have found similar relations between early magnitude understanding and later fraction magnitude and arithmetic knowledge (Hecht, Close, & Santisi, 2003; Hecht & Vagi, 2010; Jordan et al., 2013; Vukovic et al., 2014).

Strikingly, children's accuracy and linearity of estimates on the number line estimation task is one of the strongest predictors of concurrent and future mathematical achievement (Booth & Siegler, 2006; Geary, 2011; Geary, Hoard, Nugent, & Byrd-Craven, 2008; Laski & Siegler, 2007; Sasanguie et al., 2012; Sella, Tressoldi, Lucangeli, & Zorzi, 2016; Siegler & Booth, 2004) and arithmetic performance (Ashcraft & Moore, 2012; Gunderson, Ramirez, Beilock, & Levine, 2012; LeFevre et al., 2013). These relations are present even after statistically controlling for a range of related variables, including IQ, working memory, nonsymbolic numerical knowledge, reading achievement, race, ethnicity, and parental income (Bailey et al., 2014; Booth & Siegler, 2006, 2008; Cowan & Powell, 2014; Fazio et al., 2014; Geary et al., 2008). Overall, these studies provide strong empirical support for promoting mathematical knowledge in young children, specifically numerical magnitude knowledge.

PLAY AND GAMES IN MATHEMATICS DEVELOPMENT

Numerous theoretical perspectives argue for using play and games as developmentally appropriate methods to build young children's mathematical development. This section will broadly discuss some of these perspectives, as well as the advantages of using play and games, specifically for mathematics learning.

Classic developmental theories each include arguments for the benefits of play and games for children's cognitive growth. For example, Piaget (1962) argued that play can provide children with opportunities to utilize and practice their existing skills. It is through play and interactions with concrete materials in their environment that children can build their knowledge. From a sociocultural perspective, Vygotsky (1976, 1978) also emphasized the value of play for cognitive growth, suggesting that play provides children with experience using cultural tools, such as language and symbols, creating and following specific rules, and moving from use of just concrete objects to thinking in more abstract and symbolic ways. These experiences can promote growth in children's cognitive abilities, such as attention, memory, and problem-solving. Combining Piaget and Vygotsky's theories can help to provide balanced and developmentally appropriate ways for teaching mathematics, if done judiciously (Fuson, 2009). Concrete materials can be used to help children make meanings of cultural symbols, such as numbers. For example, using number lines in classrooms as a tool for children can provide a visual representation of the order and magnitudes of numbers. Then using these materials in playful contexts could help children learn about numbers in developmentally appropriate ways.

Contemporary theorists and researchers have also discussed the importance of play, specifically in the area of mathematics. As previously discussed, young children possess a range of numerical knowledge prior to entering school. This early knowledge has been theorized to develop through early playful experiences that can help build a foundation for later, more complex mathematics (Ginsburg, 2006). The process through which children build mathematical knowledge through everyday experiences is termed "mathematization" (Sarama & Clements, 2009). This can occur through engaging in mathematical-related behavior during play. For example, when playing with collections of items, such as animal figures, children can sort them by their common characteristics, and then realize that the collections provide a way to compare the quantities of the groups, thereby using mathematical reasoning.

A specific type of play that can be useful for promoting the development of children's mathematical knowledge is games, which provide engaging, enjoyable, and structured experiences while helping children to reach targeted learning goals. Specifically, games are socially interactive and capitalize on children's interest (Hassinger-Das et al., 2017). Games provide an ideal context for playful learning, which involves learning approaches where children are actively involved, engaged with meaningful materials, and interacting with others (Hirsh-Pasek, Golinkoff, & Eyer, 2003; Resnick, 1999). Specifically, for math, there are numerous advantages for using games as teaching tools. According to Davies (1995), a few of these advantages are that games provide meaningful contexts to apply and use mathematical skills. Games also allow children with varying abilities in mathematics to interact and potentially learn from one another. Playing games can also increase children's motivation and build a positive attitude toward mathematics. Thus there are numerous advantages for using play and games as a means for promoting children's numerical knowledge.

Improving Children's Numerical Magnitude Knowledge Through Games and Play

There is strong theoretical grounding for using play and games for promoting children's development in the area of mathematics. As an educational tool, however, it is important for games to match the educational objectives and areas that are being targeted in mathematics lessons (Aldridge & Badham, 1993). In this section, we describe empirical work on play and games that have been found to improve children's early numerical magnitude knowledge. We will argue for the importance of using games that align with this specific area of numerical knowledge.

Playing Traditional Games to Promote Numerical Knowledge

Board games have been used for promoting young children's numerical magnitude knowledge. Siegler and Booth (2004) hypothesized that board games designed with linearly arranged, consecutively numbered spaces of equal sizes can provide multiple cues to both the order of numbers and the numbers' magnitudes. Theoretically, such a game provides a physical representation of the hypothesized mental number line upon which numerical magnitude is built (Dehaene, 2011). One example of such a game is *Chutes and Ladders*. During their play, children experience the visual cues to the order of numbers, auditory cues when saying the numbers on the spaces, and kinesthetic cues when moving the game pieces. Such redundant, multisensory cues are helpful for learning numerical relations even in infancy (Jordan, Suanda, & Brannon, 2008).

Building upon this theoretical analysis, a series of studies tested the benefits of playing simple, linear number board games. The studies focused primarily on preschool children from lower-income backgrounds because the numerical understanding of children from low-income households typically lags behind that of children from middle- and upper-income households by up to 8 months (Ramani & Siegler, 2011; Starkey, Klein, & Wakeley, 2004). Therefore young children from lower-income backgrounds may especially benefit from targeted magnitude interventions. In the studies, preschool children from Head Start classrooms played one-on-one with an experimenter either a linear numerical board game with squares numbered from 1 to 10 or an identical version of the game except the squares varied in colors (Fig. 2). Children in these studies played one of the two games for four 15–20 min sessions. Children who played the number board game considerably improved their performance on a numerical line estimation task of the same number range, whereas children who played the color version of the game did not (Siegler & Ramani, 2008).

In a subsequent study, children who played the linear numerical board game improved their performance on the number line estimation task, as well as their counting skills from 1 to 10, their numeral identification of the numerals 1–10, and their comparison of the magnitudes of numerals. Further, these improvements were stable over a 9-week period of not having played the board game (Ramani & Siegler, 2008). Simple linear board games can also be scaled up for use in early childhood classrooms. After a brief training, assistant teachers from the Head Start classrooms supervised a small group of children from the classroom while they played the game. Children who played the number board game improved their numerical knowledge. The assistant teachers also adapted the amount and type of feedback they provided during the game. For example, they used less instruction and modeling on how to complete the turns in the last game playing session than in the first game playing session (Ramani, Siegler, & Hitti, 2012). Thus playing the linear number board game can have broad and stable benefits for children's numerical knowledge in classrooms.

The design of games is important when considering its potential benefits for young children's numerical knowledge. For example, according to the representational mapping hypothesis (Siegler & Ramani, 2009), the more transparent the mapping between materials and the desired internal representation,

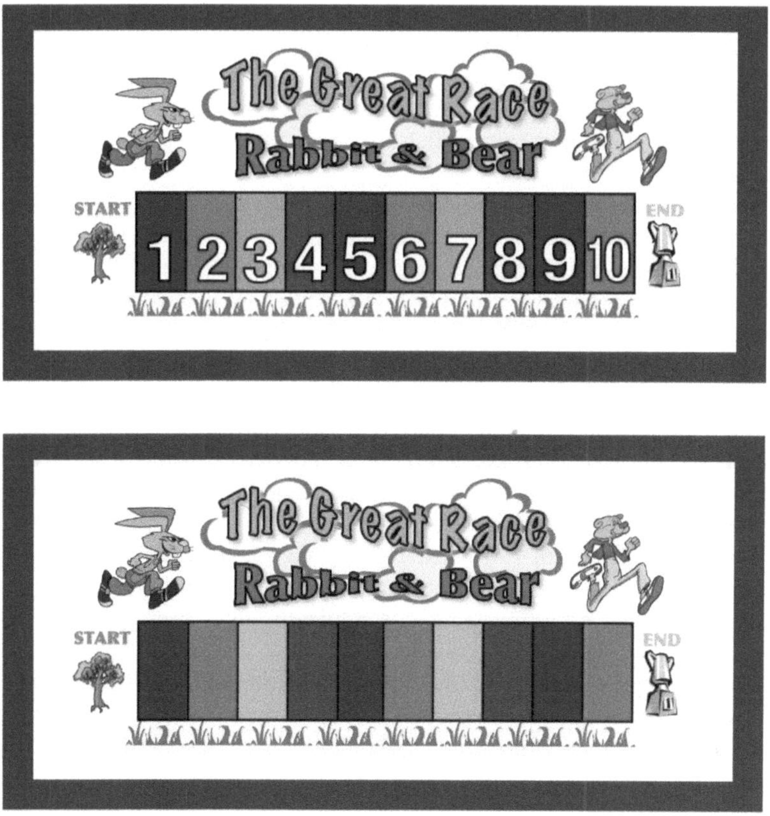

FIG. 2 Linear number board game and linear color board game (Ramani & Siegler, 2008; Siegler & Ramani, 2008).

the greater the learning of these representations. To test this hypothesis, Siegler and Ramani assigned children to play either the linear board game or a circular board game with numbers in the same 1–10 range. It was predicted that the linear game would be more effective than the circular one at increasing numerical magnitude knowledge because the linear board is easier to translate into the mental number line. As predicted, playing the linear board game for roughly an hour increased low-income preschoolers' numerical magnitude knowledge, whereas playing the circular board game did not. However, playing both games improved children's numeral identification skills because of the similar practice both groups received in this area. A third group of children who had practice with counting and numeral naming tasks, but not in a game context, did not show improvements on any measure of numerical knowledge. The latter results suggest even though the children engaged with these number activities for the same length of time as children who played the board game, there was no change in their numerical knowledge suggesting the game context was important. In further support of the importance of the design of the linear game, preschoolers who played the linear board game learned more from subsequent training on simple arithmetic problems than children in the other two conditions. These results suggest that playing the linear board game helped to produce a linear representation of numbers, which makes it easier to solve the arithmetic problems.

Several other empirical studies on board games have found similar improvements in children's numerical knowledge. For example, the benefits of playing number board games on children's numerical knowledge have been replicated with preschool-age children in Scotland (Whyte & Bull, 2008) and Sweden (Elofsson, Gustafson, Samuelsson, & Träff, 2016). Further, board games have been used with Kindergarten children. Specifically, they played a board game with the numbers 0–100 arranged in a 10 x 10 matrix for four sessions with an experimenter (Laski & Siegler, 2014). It was hypothesized that having children count the numbers in the spaces would lead to greater encoding of the numerals. They found that only children who played using the counting-on procedures (e.g., being on the space numbered 5 and counting-on by saying 6, 7, 8 while moving the game piece) improved their performance on a number line estimation task, whereas others who played the same game by counting from one while moving their game piece did not show any significant improvements. Overall, these studies suggest that playing board games can promote children's numerical knowledge and specific aspects of the game design can influence these benefits.

In addition to board games, there are other games and playful materials that can provide children with numerical information and practice to promote their numerical magnitude knowledge. For example, traditional numerical playing cards provide simultaneous exposure to symbolic and nonsymbolic magnitude representations, in the form of Arabic numerals and sets of objects (e.g., hearts, diamonds). As shown through board games, exposing children to materials with multiple redundant cues to numbers and their magnitudes can help build their numerical knowledge (Ramani & Siegler, 2008; Siegler & Booth, 2004). Providing children with both the symbolic and nonsymbolic magnitude representations will allow them to use either type of information when playing with the cards.

To capitalize on the dual representation of numerical information on the design of cards, Scalise, Daubert, and Ramani (2017) randomly assigned preschool-age children from Head Start classrooms to play one of two games with numerical cards that ranged from 1 to 10. Specifically, children played either a numerical magnitude comparison (commonly known as "War") or a numerical matching card game (commonly known as "Memory"). The children played for four 15–20 min sessions over a 3-week period. The different games using the same cards allowed for testing how specific card games that vary in the type and amount of numerical knowledge they require and use to complete the game can influence children's learning. As predicted, the magnitude comparison card game led to significant improvements in children's symbolic magnitude comparison skills, whereas both games led to improvements in children's counting and numeral identification skills. These results suggest a brief intervention using games with materials that combine nonsymbolic and symbolic magnitude representations can support low-income preschoolers' numerical magnitude knowledge.

Similar to cards, other number-based games can include materials with nonsymbolic and symbolic numerical information. For example, Brankaer, Ghesquière, and De Smedt (2015) randomly assigned 8-year-olds with mild intellectual disabilities to play one of two games with dominoes for eight 15-min sessions. One game included dominoes with the Arabic numeral on one side and the corresponding number in dots on the other side, whereas the other game included dominoes with colors and pictures. Both games involved playing a matching game with the dominoes. They found that although there were no differences in accuracy, children who played the numerical domino game became faster on a symbolic magnitude comparison task compared to children who played the color domino game.

Together, these studies show that including multiple representations of the magnitude of numbers in game materials are important features of promoting young children's numerical knowledge. Materials that use redundant cues, such as the spatial information in linear number board games, as well as the symbols and objects in cards and dominoes can help support the development of numerical magnitude knowledge. Further, playful, interactive settings where children take an active role in the game can be an important context for their learning.

Computer and Tablet Games

In addition to board and card games, there are numerous technology-based games that have been shown to be effective at improving young children's numerical knowledge. Preschool- and kindergarten-age children from lower-income backgrounds played an adaptive computer, number-based game that involved performing numerical magnitude comparisons using dots, numbers, or arithmetic problems. Playing a number-based game compared to playing a reading game improved the children's numerical magnitude knowledge (Wilson, Dehaene, Dubois, & Fayol, 2009). A more recent study has replicated these results with similar-age children from middle-income backgrounds (Sella et al., 2016). In another study, Toll and Van Luit (2013) randomly assigned kindergarten-age children to play one of two adaptive computer games for eight 25-min sessions; one game focused on comparing the quantities of numbers, and the other focused on learning procedural and conceptual counting knowledge. Compared to a control group, children in both numerical conditions improved their counting and arithmetic skills after the intervention and 6 months later. Thus playing well-designed and validated computer-based games can benefit young children's numerical knowledge.

Recently, there has been a rapid growth in tablet computer technology. Given the portability and lower cost of tablet computers compared to desktops computers, tablet computers have become more accessible for use in schools and homes. Further, the touchscreen technology of tablets makes their use extremely intuitive, requiring little instruction so they can be used by a wide audience (Dubé & McEwen, 2015).

A recent study tested the benefits of playing tablet computers for improving kindergarten children's numerical knowledge. The games targeted improving either children's domain-specific skills, their numerical magnitude knowledge, or their underlying domain-general skills, specifically working memory. Geary and Hoard's (2005) framework of mathematical development suggests that underlying cognitive systems, including the amount of information one can hold and manipulate in memory, directly supports complex mathematical activities. Working memory predicts children's mathematics performance (Friso-van den Bos, van der Ven, Kroesbergen, & van Luit, 2013) and is likely involved in their process of identifying, encoding, and manipulating numerical magnitude information. Thus promoting children's working memory skills in addition to their specific numerical skills could lead to greater improvements in their mathematical knowledge. For example, working memory plays a substantial role in the performance of numerical magnitude knowledge tasks (Kolkman, Hoijtink, Kroesbergen, & Leseman, 2013) and is important for solving arithmetic problems (Geary, Hoard, Byrd-Craven, & DeSoto, 2004).

In the study, the tablet game that targeted improving children's domain-specific knowledge of numerical magnitude understanding was similar to a traditional board game. Adapted from Laski and Siegler (2014), the game used a 10 x 10 array with the number from 1 to 100 in each square. The other tablet game targeted improving children's domain-general knowledge, specifically their working memory. Children were shown a series of the same characters in various colors which were presented upside-down or right-side up. Children had to touch one of the corresponding buttons in the bottom corners of the screen to indicate the orientation of the character. They then had to recall the sequence of colors of the characters. Playing both tablet games improved children's numerical magnitude knowledge compared to children who instead received the ongoing classroom instruction. After playing one of the two tablet games for 10 sessions that each lasted about 10–15 min, children's number line estimation performance improved suggesting that both domain-specific and domain-general interventions can facilitate mathematical learning (Ramani, Jaeggi, Daubert, & Buschkuehl, 2017).

Other studies are also beginning to show the benefits of touchscreen technology for children's mathematics skills. Specifically, children played one of three tablet games for 6 sessions of approximately 10 min over a period of

3 weeks. In the number line estimation game, children were given a number (both symbolic and nonsymbolic quantities) and were asked to place their vehicle on the corresponding place on a number line. In the magnitude comparison game, children were shown two quantities (both symbolic and nonsymbolic) and were asked to tap the larger quantity. The active control game was a memory matching game. Children received auditory and visual feedback in all three games. They found that playing the number line estimation game improved children's number line performance with the numbers 1 to 10; however, playing both the number line game and the magnitude comparison game improved performance on an arithmetic task (Maertens, De Smedt, Sasanguie, Elen, & Reynvoet, 2016). Overall, the results demonstrate the potential of using a rapidly growing technology to promote young children's numerical knowledge.

Preschool Programs Using Games and Play

One benefit of using games in early childhood classrooms is that they can be integrated with instruction and other games to build children's numerical knowledge. For example, Passolunghi and Costa (2016) examined the benefits of having children play a board game as well as seven other number-related games targeted at improving different categories of numerical knowledge. These games included a finger counting game, a game requiring children to identify which character had more of an item, and a game involving connecting quantities with a numeral. Other children were randomly assigned to play games targeted toward improving children's working memory. Children played the games for 1 h twice a week for 5 weeks. They found that number games improved children's numerical knowledge, whereas the working memory games improved children's working memory, as well as their numerical knowledge. Another such intervention targeted the understanding of whole numbers, number operation, and number relations of kindergarten-age children from lower-income backgrounds (Dyson, Jordan, & Glutting, 2013). The 8-week intervention provided children with 30-min small group sessions 3 days a week. The program included 11 number games in the intervention, such as board game, counting games, and numerical comparison games. Children who played the number games made larger gains on the number sense measures than children who did not play the games. Importantly, these improvements were still found 6 weeks after the intervention.

Similar to traditional games and computer games, tablet-based games can also be used as a part of a larger curriculum. Schacter and Jo (2016) evaluated the benefits of *Math Shelf*, a preschool mathematics curriculum for tablets with over 100 games that target foundational numerical knowledge, such as subitizing and counting, as well as more advanced concepts, such as cardinal principle, place values, and comparing the magnitudes of numbers. Children from predominantly lower-income backgrounds played games appropriate for their skill level for 10 min twice a week for 15 weeks. Compared to children who participated in the standard classroom activities, children who played the *Math Shelf* games performed significantly higher on a variety of measures of numerical knowledge. They also found that students with the lower numerical knowledge at the beginning of the intervention showed the greatest gains from playing the games. Together, these studies suggest that both traditional games and technology-based games can be used with other instructional materials to maximize their benefits.

Several well-known preschool programs targeted at improving mathematics skills have taken this approach by including many playful activities. Many of these programs integrate informal learning activities and play with age-appropriate, explicit classroom instruction. One such curriculum is *Building Blocks* (Clements & Sarama, 2007, 2008), which includes whole classroom math-related activities with small group activities, and computer games. Preschoolers from low-income backgrounds given the *Building Blocks* curriculum made much greater progress than a control group in number, geometry, measurement, and recognition of patterns. In *Pre-K Mathematics* (Klein, Starkey, Clements, Sarama, & Iyer, 2008), children engage in small group activities and games. One critical aspect of the program is that these activities and games focus on using concrete manipulatives. In large randomized control trials, prekindergarten children from lower-income backgrounds showed greater growth in their mathematical knowledge when their teachers used *Pre-K Mathematics* in the classroom as opposed to the standard mathematics curriculum.

CONCLUSIONS AND FUTURE DIRECTIONS

Overall, there is strong theoretical and empirical support for the benefits of playing games on children's numerical knowledge. Activities that target promoting children's numerical magnitude knowledge are advantageous, given the importance of this knowledge for children's later mathematical achievement. Both traditional games and computer games that are designed to focus children's attention on the numerical information in the materials seem most effective in promoting children's numerical magnitude knowledge. This can include focusing their attention through the materials used in the game, instructions on how to play the game, or the practice with numbers children have while playing the game.

In this chapter, we focused on numerical magnitude knowledge, but there are other related foundational areas of numeracy, such as cardinality and counting principles, where children could likely benefit from play-based activities. Future research that continues to identify how various aspects of games influence children's learning of foundational numerical knowledge could be particularly important. This could help to determine which games should be used to target the learning of specific skills or mathematical concepts, or how the games can be used in conjunction with other instructional materials. Future research in this area could also address the common fade-out effects found with early mathematics interventions (Bailey, Duncan, Watts, Clements, & Sarama, 2018). Specifically, longer term follow-up studies are needed to assess the long-term benefits of the math-related games, and if there are fade-out effects, to develop booster interventions that reduce could these effects. Research that continues to investigate the benefits of games will help in the development of age-appropriate practices so that mathematics instruction in early childhood classrooms is purposeful and meaningful for young children.

REFERENCES

Aldridge, S., & Badham, V. (1993). Beyond just a game. *Pamphlet Number, 21, Primary Mathematics Association.*

Antell, S. E., & Keating, D. P. (1983). Perception of numerical invariance in neonates. *Child Development, 54*, 695–701.

Ashcraft, M. H., & Moore, A. M. (2012). Cognitive processes of numerical estimation in children. *Journal of Experimental Child Psychology, 111*, 246–267.

Bailey, D. H., Duncan, G. J., Watts, T., Clements, D., & Sarama, J. (2018). Risky business: correlation and causation in longitudinal studies of skill development. *American Psychologist, 73*(1), 81–94.

Bailey, D. H., Siegler, R. S., & Geary, D. C. (2014). Early predictors of middle school fraction knowledge. *Developmental Science, 17*(5), 775–785. https://doi.org/10.1111/desc.12155.

Baroody, A. J. (2000). Does mathematics instruction for three-to five-year-olds really make sense? *Young Children, 55*(4), 61–67.

Batchelor, S., Inglis, M., & Gilmore, C. (2015). Spontaneous focusing on numerosity and the arithmetic advantage. *Learning and Instruction, 40*, 79–88. https://doi.org/10.1016/j.learninstruc.2015.09.005.

Berteletti, I., Lucangeli, D., Piazza, M., Dehaene, S., & Zorzi, M. (2010). Numerical estimation in preschoolers. *Developmental Psychology, 46*, 545–551. https://doi.org/10.1037/a0017887.

Bonny, J. W., & Lourenco, S. F. (2013). The approximate number system and its relation to early math achievement: evidence from the preschool years. *Journal of Experimental Child Psychology, 114*, 375–388.

Booth, J. L., & Siegler, R. S. (2006). Developmental and individual differences in pure numerical estimation. *Developmental Psychology, 42*, 189–201. https://doi.org/10.1037/0012-1649.41.6.189.

Booth, J. L., & Siegler, R. S. (2008). Numerical magnitude representations influence arithmetic learning. *Child Development, 79*, 1016–1031.

Brankaer, C., Ghesquière, P., & De Smedt, B. (2015). The effect of a numerical domino game on numerical magnitude processing in children with mild intellectual disabilities. *Mind, Brain, and Education, 9*, 29–39. https://doi.org/10.1111/mbe.12065.

Brannon, E. M., Lutz, D., & Cordes, S. (2006). The development of area discrimination and its implications for number representation in infancy. *Developmental Science, 9*(6), F59–F64. https://doi.org/10.1111/j.1467-7687.2006.00530.x.

Chen, Q., & Li, J. (2014). Association between individual differences in non-symbolic number acuity and math performance: a meta-analysis. *Acta Psychologica, 148*, 163–172. https://doi.org/10.1016/j.actpsy.2014.01.016.

Clements, D. H., & Sarama, J. (2007). Effects of a preschool mathematics curriculum: summative research on the building blocks project. *Journal for Research in Mathematics Education, 38*(2), 136–163.

Clements, D. H., & Sarama, J. (2008). Experimental evaluation of the effects of a research-based preschool mathematics curriculum. *American Educational Research Journal, 45*(2), 443–494.s.

Cordes, S., & Brannon, E. M. (2008). Quantitative competencies in infancy. *Developmental Science, 11*(6), 803–808. https://doi.org/10.1111/j.1467-7687.2008.00770.x.

Cowan, R., & Powell, D. (2014). The contributions of domain-general and numerical factors to third-grade arithmetic skills and mathematical learning disability. *Journal of Educational Psychology, 106*(1), 214–229. https://doi.org/10.1037/a0034097.

Davies, B. (1995). The role of games in mathematics. *Square One, 5*(2), 7–17.

Dehaene, S. (2001). Précis of the number sense. *Mind & Language, 16*(1), 16–36. https://doi.org/10.1111/1468-0017.00154.

Dehaene, S. (2011). *The number sense: How the mind creates mathematics* [rev. and updated ed.]. New York: Oxford University.

Dubé, A. K., & McEwen, R. N. (2015). Do gestures matter? The implications of using touchscreen devices in mathematics instruction. *Learning and Instruction, 40*, 89–98. https://doi.org/10.1016/j.learninstruc.2015.09.002.

Dyson, N. I., Jordan, N. C., & Glutting, J. (2013). A number sense intervention for low-income kindergartners at risk for mathematics difficulties. *Journal of Learning Disabilities, 46*(2), 166–181. https://doi.org/10.1177/0022219411410233.

Elofsson, J., Gustafson, S., Samuelsson, J., & Träff, U. (2016). Playing number board games supports 5-year-old children's early mathematical development. *The Journal of Mathematical Behavior, 43*, 134–147. https://doi.org/10.1016/j.jmathb.2016.07.003.

Fazio, L. K., Bailey, D. H., Thompson, C. A., & Siegler, R. S. (2014). Relations of different types of numerical magnitude representations to each other and to mathematics achievement. *Journal of Experimental Child Psychology, 123*, 53–72. https://doi.org/10.1016/j.jecp.2014.01.013.

Feigenson, L., Dehaene, S., & Spelke, E. (2004). Core systems of number. *Trends in Cognitive Sciences, 8*, 307–314. https://doi.org/10.1016/j.tics.2004.05.002.

Friso-van den Bos, I., van der Ven, S. H. G., Kroesbergen, E. H., & van Luit, J. E. H. (2013). Working memory and mathematics in primary school children: a meta-analysis. *Educational Research Review, 10*, 29–44. https://doi.org/10.1016/j.edurev.2013.05.003.

Fuchs, L. S., Schumacher, R. F., Long, J., Namkung, J., Hamlett, C. L., et al. (2013). Improving at-risk learners' understanding of fractions. *Journal of Educational Psychology, 105*, 683–700. https://doi.org/10.1037/a0032446.

Fuson, K. C. (1988). Children's counting and concepts of number. *Developmental Psychology, 6*(4), 395–397. https://doi.org/10.1111/j.2044-835X.1988.tb01118.x.

Fuson, K. C. (2009). Avoiding misinterpretations of Piaget and Vygotsky: mathematical teaching without learning, learning without teaching, or helpful learning-path teaching? *Cognitive Development, 24*(4), 343–361.

Geary, D. C. (2006). Development of mathematical understanding. In D. Kuhn, R. S. Siegler, W. Damon, & R. M. Lerner (Eds.), *Handbook of child psychology. Cognition, perception, and language* (Vol. 2, pp. 777–810). Hoboken, NJ: John Wiley.

Geary, D. C. (2011). Cognitive predictors of individual differences in achievement growth in mathematics: a five-year longitudinal study. *Developmental Psychology, 47*, 1539–1552.

Geary, D. C., & Hoard, M. K. (2005). Learning disabilities in arithmetic and mathematics: theoretical and empirical perspectives. In J. I. D. Campbell (Ed.), *Handbook of mathematical cognition* (pp. 253–267). New York, NY: Psychology Press.

Geary, D. C., Hoard, M. K., Byrd-Craven, J., & DeSoto, M. C. (2004). Strategy choices in simple and complex addition: contributions of working memory and counting knowledge for children with mathematical disability. *Journal of Experimental Child Psychology, 88*(2), 121–151. https://doi.org/10.1016/j.jecp.2004.03.002.

Geary, D. C., Hoard, M. K., Nugent, L., & Byrd-Craven, J. (2008). Development of number-line representations in children with mathematical learning disability. *Developmental Neuropsychology, 33*, 277e299. https://doi.org/10.1080/87565640801982361.

Geary, D. C., & vanMarle, K. (2016). Young children's core symbolic and nonsymbolic quantitative knowledge in the prediction of later mathematics achievement. *Developmental Psychology, 52*(12), 2130–2144. https://doi.org/10.1037/dev0000214.

Geary, D. C., vanMarle, K., Chu, F., Rouder, J., Hoard, M. K., & Nugent, L. (2018). Early conceptual understanding of cardinality predicts superior school-entry number system knowledge. *Psychological Science, 29*(2), 181–205. https://doi.org/10.1177/0956797617729817.

Gelman, R., & Gallistel, C. R. (1978). *The child's understanding of number.* Cambridge, MA: Harvard University Press.

Ginsburg, H. P. (2006). Mathematical play and playful mathematics: a guide for early education. In D. Singer, R. M. Golinkoff, & K. Hirsh-Pasek (Eds.), *Play=Learning: how play motivates and enhances children's cognitive and social-emotional growth* (pp. 145–165). New York, NY: Oxford University Press.

Gunderson, E. A., Ramirez, G., Beilock, S. L., & Levine, S. C. (2012). The relation between spatial skills and early number knowledge: the role of the linear number line. *Developmental Psychology, 48*(5), 1229–1241. https://doi.org/10.1037/a0027433.

Halberda, J., & Feigenson, L. (2008). Developmental change in the acuity of the "number sense": the approximate number system in 3-, 4-, 5-, and 6-year-olds and adults. *Developmental Psychology, 44*, 1457–1465.

Halberda, J., Ly, R., Wilmer, J. B., Naiman, D. Q., & Germine, L. (2012). Number sense across the lifespan as revealed by a massive Internet-based sample. *Proceedings of the National Academy of Sciences of the United States of America, 109*, 11116–11120.

Hassinger-Das, B., Toub, T. S., Zosh, J. M., Michnick, J., Golinkoff, R. M., & Hirsh-Pasek, K. (2017). More than just fun: a place for games in playful learning/Más que diversión: el lugar de los juegos reglados en el aprendizaje lúdico. *Infancia y Aprendizaje, 40*(2), 191–218. https://doi.org/10.1080/02103702.2017.1292684.

Hecht, S. A., Close, L., & Santisi, M. (2003). Sources of individual differences in fraction skills. *Journal of Experimental Child Psychology, 86*, 277–302.

Hecht, S. A., & Vagi, K. J. (2010). Sources of group and individual differences in emerging fraction skills. *Journal of Educational Psychology, 102*, 843–859.

Hirsh-Pasek, K., Golinkoff, R. M., & Eyer, D. (2003). *Einstein never used flashcards: How our children really learn- and why they need to play more and memorize less.* Emmaus, PA: Rodale.

Holloway, I. D., & Ansari, D. (2008). Domain-specific and domain-general changes in children's development of number comparison. *Developmental Science, 11*(5), 644–649. https://doi.org/10.1111/j.1467-7687.2008.00712.x.

Izard, V., Sann, C., Spelke, E. S., & Streri, A. (2009). Newborn infants perceive abstract numbers. *Proceedings of the National Academy of Sciences of the United States of America, 106*, 10382–10385. https://doi.org/10.1073/pnas.0812142106.

Jordan, K., Suanda, S., & Brannon, E. (2008). Intersensory redundancy accelerates preverbal numerical competence. *Cognition, 108*, 210–221.

Jordan, N. C., Hansen, N., Fuchs, L. S., Siegler, R. S., Gersten, R., & Micklos, D. (2013). Developmental predictors of fraction concepts and procedures. *Journal of Experimental Child Psychology, 116*, 45–58. https://doi.org/10.1016/j.jecp.2013.02.001.

Jordan, N. C., Kaplan, D., Locuniak, M. N., & Ramineni, C. (2007). Predicting first-grade math achievement from developmental number sense trajectories. *Learning Disabilities Research and Practice*, *22*(1), 36–46. https://doi.org/10.1111/j.1540-5826.2007.00229.x.

Kim, D., & Opfer, J. E. (2017). A unified framework for bounded and unbounded numerical estimation. *Developmental Psychology*, *53*(6), 1088.

Klein, A., Starkey, P., Clements, D., Sarama, J., & Iyer, R. (2008). Effects of a pre-kindergarten mathematics intervention: a randomized experiment. *Journal of Research on Educational Effectiveness*, *1*(3), 155–178. https://doi.org/10.1080/19345740802114533.

Kolkman, M. E., Hoijtink, H. J. A., Kroesbergen, E. H., & Leseman, P. P. M. (2013). The role of executive functions in numerical magnitude skills. *Learning and Individual Differences*, *24*, 145–151.

Kolkman, M. E., Kroesbergen, E. H., & Leseman, P. P. M. (2013). Early numerical development and the role of non-symbolic and symbolic skills. *Learning and Instruction*, *25*, 95–103.

Laski, E. V., & Siegler, R. S. (2007). Is 27 a big number? Correlational and causal connection among numerical categorization, number-line estimation and numerical magnitude comparison. *Child Development*, *78*, 1723–1744. https://doi.org/10.1111/j.1467-8624.2007.01087.x.

Laski, E. V., & Siegler, R. S. (2014). Learning from number board games: you learn what you encode. *Developmental Psychology*, *50*(3), 853–864. https://doi.org/10.1037/a0034321.

Le Corre, M., & Carey, S. (2007). One, two, three, four, nothing more: an investigation of the conceptual sources of the verbal counting principles. *Cognition*, *105*(2), 395–438. https://doi.org/10.1016/j.cognition.2006.10.005.

LeFevre, J. A., Jimenez Lira, C., Sowinski, C., Cankaya, O., Kamawar, D., & Skwarchuk, S. L. (2013). Charting the role of the number line in mathematical development. *Frontiers in Psychology*, *4*, 641. https://doi.org/10.3389/fpsyg.2013.00641.

Libertus, M. E., Feigenson, L., & Halberda, J. (2011). Preschool acuity of the approximate number system correlates with school math ability. *Developmental Science*, *14*, 1292–1300. https://doi.org/10.1111/j.1467-7687.

Lipton, J. S., & Spelke, E. S. (2003). Origins of number sense: large-number discrimination in human infants. *Psychological Science*, *14*, 396–401.

Maertens, B., De Smedt, B., Sasanguie, D., Elen, J., & Reynvoet, B. (2016). Enhancing arithmetic in pre-schoolers with comparison or number line estimation training: does it matter? *Learning and Instruction*, *46*, 1–11. https://doi.org/10.1016/j.learninstruc.2016.08.004.

National Research Council. (2001). Adding it up: helping children learn mathematics. In Mathematics Learning Study Committee, J. Kilpatrick, J. Swafford, & B. Findell (Eds.), *Center for education, division of behavioral and social sciences and education.* Washington, DC: National Academy Press.

Nieder, A., & Dehaene, S. (2009). Representation of number in the brain. *Annual Review of Neuroscience*, *32*, 185–208. https://doi.org/10.1146/annurev.neuro.051508.135550.

Passolunghi, M. C., & Costa, H. M. (2016). Working memory and early numeracy training in preschool children. *Child Neuropsychology*, *22*(1), 81–98. https://doi.org/10.1080/09297049.2014.971726.

Piaget, J. (1962). *Play, dreams and imitation in childhood.* New York: Norton.

Ramani, G. B., Jaeggi, S. M., Daubert, E., & Buschkuehl, M. (2017). Domain-general and domain-specific training to improve kindergarten children's mathematics. *Journal of Numerical Cognition*, *3*(2), 468–495. https://doi.org/10.5964/jnc.v3i2.31.

Ramani, G. B., & Siegler, R. S. (2008). Promoting broad and stable improvements in low-income children's numerical knowledge through playing number board games. *Child Development*, *79*, 375–394.

Ramani, G. B., & Siegler, R. S. (2011). Reducing the gap in numerical knowledge between low- and middle-income preschoolers. *Journal of Applied Developmental Psychology*, *32*, 146–159. https://doi.org/10.1016/j.appdev.2011.02.005.

Ramani, G. B., Siegler, R. S., & Hitti, A. (2012). Taking it to the classroom: number board games as a small group learning activity. *Journal of Educational Psychology*, *104*(3), 661–672.

Resnick, L. B. (1999). Making America smarter. *Education Week*, 38–40.

Reynvoet, B., & Sasanguie, D. (2016). The symbol grounding problem revisited: a thorough evaluation of the ANS mapping account and the proposal of an alternative account based on symbol-symbol associations. *Frontiers in Psychology*, *7*(1581), 1–11. https://doi.org/10.3389/fpsyg.2016.01581.

Ritchie, S. J., & Bates, T. C. (2013). Enduring links from childhood mathematics and reading achievement to adult socioeconomic status. *Psychological Science*, *24*(7), 1301–1308. https://doi.org/10.1177/0956797612466268.

Sarama, J., & Clements, D. H. (2009). *Early childhood mathematics education research: learning trajectories for young children*. New York, NY: Routledge.

Sasanguie, D., De Smedt, B., Defever, E., & Reynvoet, B. (2012). Association between basic numerical abilities and mathematics achievement. *British Journal of Developmental Psychology*, *30*, 344–357.

Scalise, N., Daubert, N. A., & Ramani, G. B. (2017). Narrowing the early mathematics gap: a play-based intervention to promote head start preschoolers' number skills. *Journal of Numerical Cognition*, *3*(3), 559–581. https://doi.org/10.5964/jnc.v3i3.72.

Schacter, J., & Jo, B. (2016). Improving low-income preschoolers mathematics achievement with Math Shelf, a preschool tablet computer curriculum. *Computers in Human Behavior*, *55*, 223–229.

Schneider, M., Beeres, K., Coban, L., Merz, S., Schmidt, S. S., Stricker, J., et al. (2016). Associations of non-symbolic and symbolic numerical magnitude processing with mathematical competence: a meta-analysis. *Developmental Science*, *20*, e12372. https://doi.org/10.1111/desc.12372.

Sella, F., Tressoldi, P., Lucangeli, D., & Zorzi, M. (2016). Training numerical skills with the adaptive videogame "the number race": a randomized controlled trial on preschoolers. *Trends in Neuroscience and Education*, *5*, 20–29. https://doi.org/10.1016/j.tine.2016.02.002.

Siegler, R. S. (2016). Magnitude knowledge: the common core of numerical development. *Developmental Science*, *19*, 341–361. https://doi.org/10.1111/desc.12395.

Siegler, R. S., & Booth, J. (2004). Development of numerical estimation in young children. *Child Development*, *75*, 428–444.

Siegler, R. S., & Ramani, G. B. (2008). Playing board games promotes low-income children's numerical development. *Developmental Science*, *11*, 655–661. Special Issue on Mathematical Cognition.

Siegler, R. S., & Ramani, G. B. (2009). Playing linear number board games – but not circular ones – improves low-income preschoolers' numerical understanding. *Journal of Educational Psychology*, *101*, 545–560.

Siegler, R. S., & Robinson, M. (1982). The development of numerical understandings. *Advances in Child Development and Behavior*, *16*, 241–312.

Siegler, R. S., Thompson, C. A., & Schneider, M. (2011). An integrated theory of whole number and fractions development. *Cognitive Psychology*, *62*, 273–296.

Starkey, P. (1992). The early development of numerical reasoning. *Cognition*, *43*(2), 93–126.

Starkey, P., & Cooper, R. G. (1980). Perception of numbers by human infants. *Science*, *210*(4473), 1033–1035.

Starkey, P., Klein, A., & Wakeley, A. (2004). Enhancing young children's mathematical knowledge through a pre-kindergarten mathematics intervention. *Early Childhood Research Quarterly, 19*, 99–120.

Sullivan, J., & Barner, D. (2014). Inference and association in children's early numeral estimation. *Child Development, 85*, 1740–1755.

Toll, S. W., & Van Luit, J. E. (2013). Accelerating the early numeracy development of kindergartners with limited working memory skills through remedial education. *Research in Developmental Disabilities, 34*, 745–755. https://doi.org/10.1016/j.ridd.2012.09.003.

vanMarle, K., Chu, F. W., Mou, Y., Seok, J. H., Rouder, J., & Geary, D. C. (2018). Attaching meaning to the number words: contributions of the object tracking and approximate number systems. *Developmental Science, 21*(1), 1–17.

vanMarle, K., & Wynn, K. (2006). Six-month-old infants use analog magnitudes to represent duration. *Developmental Science, 9*, F41–F49.

Vilette, B. (2002). Do young children grasp the inverse relationship between addition and subtraction?: evidence against early arithmetic. *Cognitive Development, 17*(3), 1365–1383.

Vukovic, R. K., Fuchs, L. S., Geary, D. C., Jordan, N. C., Gersten, R., & Siegler, R. S. (2014). Sources of individual differences in children's understanding of fractions. *Child Development, 85*(4), 1461–1476.

Vygotsky, L. S. (1976). Play and its role in the mental development of the child. In J. S. Bruner, A. Jolly, & K. Sylva (Eds.), *Play* (pp. 537–559). New York: Harper and Row.

Vygotsky, L. S. (1978). *Mind and society: The development of higher psychological processes.* Cambridge, MA: Harvard University.

Watts, T. W., Duncan, G. J., Siegler, R. S., & Davis-Kean, P. E. (2014). What's past is prologue: relations between early mathematics knowledge and high school achievement. *Educational Researcher, 43*(7), 352–360. https://doi.org/10.3102/0013189X14553660.

Whyte, J. C., & Bull, R. (2008). Number games, magnitude representation, and basic number skills in preschoolers. *Developmental Psychology, 44*(2), 588. https://doi.org/10.1037/0012-1649.44.2.588.

Wilson, A. J., Dehaene, S., Dubois, O., & Fayol, M. (2009). Effects of an adaptive game intervention on accessing number sense in low-socioeconomic-status kindergarten children. *Mind, Brain, and Education, 3*, 224–234. https://doi.org/10.1111/j.1751-228X.2009.01075.x.

Wynn, K. (1992). Addition and subtraction by human infants. *Nature, 358*(6389), 749–750.

Xu, F., & Spelke, E. S. (2000). Large number discrimination in 6-month-old infants. *Cognition, 74*, B1–B11. https://doi.org/10.1016/S0010-0277(99)00066-9.

Chapter 4

Understanding the Link Between the Approximate Number System and Math Abilities

Melissa E. Libertus
Department of Psychology and Learning Research and Development Center, University of Pittsburgh, Pittsburgh, PA, United States

INTRODUCTION

Imagine you are at the grocery store to get a few items that you forgot for dinner. As you reach the checkout, you scan the lanes trying to figure out which will be the fastest. Most likely you will look at the other shoppers' carts and compare which lane has the fewest shoppers with the fewest items without exactly counting how many items each shopper is purchasing. You can complete this comparison process quickly and effortlessly and within seconds decide where you will line up.

Quick estimation and comparison processes such as the one described in the previous scenario rely on our approximate number system (ANS). The ANS allows us to form approximate representations of quantities without the need to count or understand symbolic labels for quantities (Dehaene, 1997). Since these representations are inherently imprecise and this imprecision increases as the quantities increase, our ability to discriminate between quantities is determined by their ratio. For example, it is easier to determine that a shopper with 15 items in his cart has more items than a shopper with only 10 items compared to determining that a shopper with 45 items has more than a shopper with 40 items. While ratio-dependent discrimination is present in children and adults at all ages (Halberda & Feigenson, 2008; Xu & Spelke, 2000), individuals differ in the precision of their ANS representations (Halberda, Ly, Willmer, Naiman, & Germine, 2012; Halberda, Mazzocco, & Feigenson, 2008; Libertus & Brannon, 2010).

Even though the ANS can be useful under certain circumstances such as quickly comparing which checkout lane will move the fastest, it does not afford

Mathematical Cognition and Learning, Vol. 5. https://doi.org/10.1016/B978-0-12-815952-1.00004-9

precise number representation that may be needed to figure out how much our items will cost together. As educated humans, we have acquired symbolic representations such as number words and Arabic numerals to represent exact quantities (Fuson, 1988; Gelman & Gallistel, 1978; Wynn, 1992). Importantly, these exact representations allow us to perform precise arithmetic and even learn more complex math such as trigonometry, algebra, and calculus.

Despite the fact that the ANS only offers imprecise numerical representations, it seems to be linked to people's symbolic math abilities. Three recent meta-analyses (Chen & Li, 2014; Fazio, Bailey, Thompson, & Siegler, 2014; Schneider et al., 2016) have shown that across a large number of correlational studies, there is a small but significant association such that people with more precise ANS representations tend to perform better on a variety of math assessments. This association appears to be stronger in children than adults (Fazio et al., 2014; Inglis, Attridge, Batchelor, & Gilmore, 2011) and for math assessments that rely more heavily on the processing of quantities instead of the application of rote memorized procedures (Schneider et al., 2016).

While these meta-analyses support the existence of a link between the ANS and math, they are based on correlational findings, which leave open the possibility that other factors may explain the observed associations. For example, some researchers have argued that tasks that are designed to measure the ANS and math both require visuospatial processing and that positive correlations may be explained by this shared need (Allik & Tuulmets, 1991; Dakin, Tibber, Greenwood, Kingdom, & Morgan, 2011; Durgin, 1995; Gebuis, Cohen Kadosh, & Gevers, 2016; Ginsburg & Nicholls, 1988; Leibovich, Katzin, Harel, & Henik, 2016; Tibber et al., 2013; Tibber, Greenwood, & Dakin, 2012; Tokita & Ishiguchi, 2010; Vos, van Oeffelen, Tibosch, & Allik, 1988; Zhou, Wei, Zhang, Cui, & Chen, 2015). Other researchers have argued that the shared reliance on executive functions, especially inhibitory control, may explain the observed correlations (Fuhs & McNeil, 2013; Gilmore et al., 2013) although the evidence for this explanation is mixed (Keller & Libertus, 2015).

TRAINING STUDIES

Training studies are one way to circumvent the limitations of correlational studies and to examine whether the ANS and math are causally linked. If participants are randomly assigned to different conditions, some of which may train the ANS and others may not, any subsequent improvements in math ability in participants who completed ANS-related training conditions are most likely due to the reliance on the ANS during training.

Training Studies Using "The Number Race"

Early training studies often combined several different number processing tasks that tapped into the ANS and other basic number processing skills. For example, "The Number Race" is an adaptive computer game that was initially designed to improve math abilities in children with dyscalculia (Wilson, Dehaene, et al., 2006). It adapts to each child's individual performance to minimize failure and maintain an appropriate level of difficulty. Children perform number comparisons and simple arithmetic using a mix of both nonsymbolic and symbolic numerical information. Speed is emphasized by asking children to perform the comparisons quicker than a virtual opponent. Additionally, choosing the larger quantity is incentivized by having children advance their game token on a board by the number of items in the chosen set. At the same time, this step is intended to help children train the connection between numbers and space.

In the first evaluation of "The Number Race," a sample of nine dyscalculic children played the game 4 days a week for 5 weeks (Wilson, Revkin, Cohen, Cohen, & Dehaene, 2006). Each training session lasted about 30 min. Comparisons between pre- and posttraining performance revealed significant increases in subitizing speed (i.e., the speed in which children enumerated small quantities) and subtraction accuracy after training. However, the lack of a control group limits the conclusions that can be drawn from this study.

Three follow-up studies were designed to remedy this shortcoming. In the first follow-up study, Wilson, Dehaene, Dubois, and Fayol (2009) used a cross-over design to train 4- to 6-year-olds from low socioeconomic status households with "The Number Race" and a commercially available reading software as a control training. Half of the children completed six sessions of "The Number Race" first and then four sessions of the reading software, the other half first completed four sessions of the reading software and then six sessions of "The Number Race." Each session lasted about 20 min. Unlike in the earlier study, Wilson and colleagues only found significant improvements in symbolic number comparison (i.e., identifying the larger of two Arabic numerals) following training on "The Number Race." No transfer to addition skills was observed.

In the second follow-up study, Räsänen, Salminen, Wilson, Aunio, and Dehaene (2009) divided a sample of 6-year-olds into three different groups: Half of the children were assigned to the passive control group, that is, they only completed the pre- and postassessments and participated in regular classroom instruction in the interim. Half of the other children completed daily 10- to 15-min sessions of "The Number Race" for a total of 3 weeks. The remaining children completed "Graphogame-Math," another computerized number game that emphasizes the correspondence between small quantities and their labels. For example, children hear a number word such as "three" and have to select the correct match from 2 to 5 visually presented options. Choices include dot patterns, number symbols, and in the end symbolic addition (e.g., "1 + 2") and subtraction problems (e.g., "4 – 1"). Children in both training conditions showed significant improvements in symbolic number comparisons as compared with the children in the passive control condition. However, similar to the findings by Wilson et al. (2009), no transfer to improvements in addition and subtraction skills were observed.

In the third follow-up study, Sella, Tressoldi, Lucangeli, and Zorzi (2016) used "The Number Race" in a randomized controlled trial with 5-year-old Italian preschoolers. Half of the children completed the Italian version of "The Number Race," the other half played a computerized game to foster drawing skills. Both groups completed 10 weeks of training, two sessions per week for about 20 min each. Children in "The Number Race" training showed significantly more precise mapping of numbers onto a number line (ranging from 1 to 10 and from 1 to 20) as well as better mental calculation skills after training compared to the children in the control group.

While these previous studies focused on preschool-aged children and those that just started formal education, Kuhn and Holling (2014) designed a computerized number sense training similar to "The Number Race" for 3rd and 4th graders. This training consisted of two number line tasks and a number comparison task. In the number line tasks children had to indicate the position of a single number, the result of a computation, or multiple numbers on number lines that ranged from 0 to 20 or 0 to 100. In the number comparison task, children had to select the largest number out of two or four whereby numbers were shown as nonsymbolic arrays of shapes, numerals, or symbolic calculations (e.g., "8 + 6"). Children completed two out of the three tasks each day for 10 min each for a total of 15 days (3 weeks of school). Children in the number sense training showed significantly greater gains on a standardized math assessment compared to a passive control group but not compared to a working memory training group. However, children in the number sense training group improved mainly in arithmetic, while children in the working memory training group improved in word problem solving. These findings corroborate that a computerized training targeting basic number processing such as mapping numbers onto space and comparing numbers may improve some aspects of children's math abilities. These findings further suggest that such gains can be accomplished in older, typically achieving children (average age 9 years).

To disentangle the role of number comparison versus number line tasks for training-related improvements, Maertens, De Smedt, Sasanguie, Elen, and Reynvoet (2016) designed two separate training games and tested them in a sample of 5-year-old children. In the number comparison training, children saw nonsymbolic and symbolic stimuli as well as a combination of both types of stimuli and had to choose the numerically larger of the two. In the number line training, children saw a line anchored either by nonsymbolic or symbolic stimuli and had to place a number on the line relative to the two anchors. Maertens and colleagues compared these number training conditions to a memory-game control group and a passive control group. Children in the three active training groups completed six sessions of their respective games, each lasting for about 10 min, over a period of 3 weeks. Children in both number training conditions but not the control conditions improved in arithmetic (e.g., symbolic problems such as "6 + 3 = ..." and word problems such as

"Here you see two red balloons and three blue balloons. How many balloons are there together?"). These findings suggest that comparison skills and the mappings of numbers onto space are linked to math abilities.

In sum, findings from training studies using "The Number Race" and similar training games that required symbolic and nonsymbolic number comparisons as well as mappings of numbers onto space are mixed and largely inconsistent. While performance on the trained tasks generally improves, transfer to math more generally is less reliable. However, variations in the samples (ages, socioeconomic status, initial math abilities), duration and timing of the intervention, the nature of the control, as well as the pre- and posttraining assessments make it difficult to draw firm conclusions about the effectiveness of these kinds of training approaches.

Nonsymbolic vs. Symbolic Training

All of the previously reviewed training studies used both nonsymbolic and symbolic training stimuli. Thus it is unclear whether the observed improvements in some of the math-related tasks are due to the training with nonsymbolic or symbolic representations or the need to connect the two. To disentangle whether training on nonsymbolic stimuli differs from training on symbolic stimuli, Obersteiner, Reiss, and Ufer (2013) trained four groups of first graders on slightly modified versions of "The Number Race." In the "Approximate Version" of the game, children performed nonsymbolic number comparisons as well as estimations in which they had to judge which of two symbolic estimates was nearest to the correct answer. In contrast, in the "Exact Version" of the game, children had to judge the exact number of dots in the nonsymbolic stimuli. One group of children was trained on the Approximate and the Exact Versions of the computer game, another group was trained only on the Approximate Version, and a third only on the Exact Version. The fourth group of children was trained on a computer game designed to improve language skills. All students completed 10 training sessions, 30 min each over a period of 4 weeks. Children who completed only the Approximate or the Exact Version, but not a combination of both or the language training, performed higher on a subsequent arithmetic test. While these findings suggest that training involving nonsymbolic numerical stimuli can lead to improvements in math, the "Approximate Version" required children to also process number symbols in the estimation portion of the game. Thus there were again some symbolic representations, which make it impossible to ascertain that this training was purely targeting the ANS.

To avoid these confounds, Honore and Noel (2016) created two parallel training conditions, one that used entirely nonsymbolic number stimuli and another that only used symbolic number stimuli. They compared these training conditions along with a storybook reading control condition in a sample of 5-year-olds. In both nonsymbolic and symbolic number training conditions, children had to complete a comparison task and a number line task using the respective stimuli. In the comparison task, children were shown three arrays of dots or three Arabic numerals and three animals in various sizes. Children were instructed to give the smallest number to the smallest animal, the medium number to the medium-sized animal, and the largest number to the largest animal. In the number line task children were shown a line bounded by two arrays of dots or two numerals (1 and 10 or 1 and 20) and asked to indicate where a third number presented either nonsymbolically or symbolically would be with respect to the two anchors. All children completed ten 30-min sessions of their assigned training condition, 15 min per task during each session. While the nonsymbolic training condition yielded significant improvements in nonsymbolic number comparison and nonsymbolic number line performance, only the symbolic training yielded improvements in arithmetic.

The findings from these two studies thus suggest that the link between symbolic number representations and math may be stronger than the link between the ANS and math. However, they show that the ANS is malleable and leave open the possibility that improvements in math may occur with more or different forms of nonsymbolic training. These conclusions are in line with those by Dillon, Kannan, Dean, Spelke, and Duflo (2017) who found that nonsymbolic numerical and geometrical training implemented in a large number of preschools in India led to long-term gains in nonsymbolic skills but not school math. Dillon and colleagues suggest that a more effective intervention may need to align the nonsymbolic content more clearly with the school math curriculum and/or that the training should occur at the same time that children receive formal instruction in school math so that children may form connections between nonsymbolic training content and school math.

Training Specific Aspects of Nonsymbolic Number Processing

To more clearly isolate specific aspects of approximate number processing for math, a number of recent training studies have focused on single nonsymbolic tasks during training. Some of these studies used brief training durations to assess their effects on math abilities within the same session, while others looked at the effects of long-term training.

Brief ANS Training

Hyde, Khanum, and Spelke (2014) used brief training on approximate numerical addition, approximate number comparison, line length addition, and brightness comparison to address whether approximate number training was more effective at improving arithmetic performance in first graders than nonnumerical addition or nonnumerical comparison training. In the approximate numerical addition training, children saw two arrays of dots successively moving behind an occluder and they had to indicate whether the sum was more or less numerous than a third comparison array (Fig. 1).

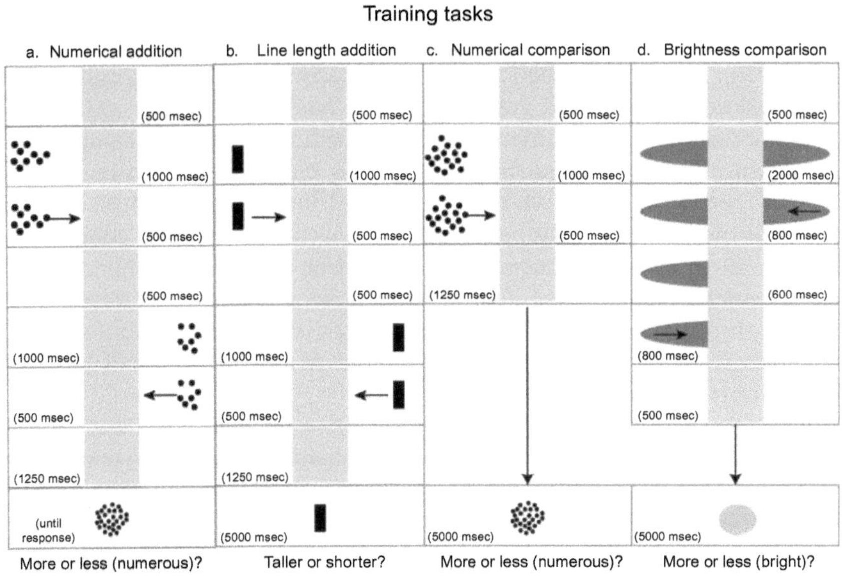

FIG. 1 Schematic depiction of training tasks used by Hyde et al. (2014). Stimulus events are organized horizontally from start (top) to finish (bottom). The numbers indicate the duration of presentation. The *horizontal arrows* indicate stimulus movement. The *vertical arrows* indicate the following event. Reprinted with permission.

In the approximate numerical comparison training, children saw one array of dots move behind an occluder and they had to indicate whether it was more or less numerous than a comparison array. In the line length addition training, children saw two line segments successively move behind an occluder and they had to indicate whether the sum was taller or shorter than a third line segment. Finally, in the brightness comparison training, children saw an oval of a certain brightness disappear behind an occluder and they had to indicate whether it was more or less bright than a comparison circle. Hyde and colleagues found that approximate numerical addition and approximate number comparison training but not line length addition or brightness comparison improved first graders' symbolic addition performance immediately after the training. This transfer was specific to math as they did not find subsequent improvements in children's reading performance. In a follow-up study, Khanum, Hanif, Spelke, Berteletti, and Hyde (2016) replicated the finding that approximate numerical addition but not line length addition improves symbolic addition in Pakistani first graders.

Instead of comparing approximate number training to training conditions using other types of stimuli, Wang, Odic, Halberda, and Feigenson (2016) made use of the fact that children perform significantly better on nonsymbolic number comparisons when they start out with the easiest comparisons and gradually receive more difficult comparisons than when they begin with the hardest comparisons and receive gradually easier ones (Odic, Hock, & Halberda, 2014). In this study, Wang et al. (2016) asked 5-year-old children to first complete a nonsymbolic number comparison task starting either with the easiest comparisons and gradually getting harder or vice versa. Immediately following these two different training conditions, children completed a math assessment tapping into counting, number comparison, calculation, and numeral literacy. Children who first completed the easiest number comparisons performed significantly better on the nonsymbolic number comparison task as well as on a subsequent math assessment compared to children who first completed the hardest number comparisons and these effects were medium sized. Importantly, these improvements did not extend to a vocabulary assessment supporting the idea that engagement of the ANS leads to changes in math, but not vocabulary.

While these brief training studies suggest that engagement of the ANS can lead to immediate improvements in math performance, the underlying mechanisms for these improvements are unclear. Hyde et al. (2014) suggest that "symbolic arithmetic draws on at least some overlapping cognitive and/or neural structures used to represent approximate number" (p. 104). Activating these overlapping structures during the brief training session may serve as a "warm-up" before performing math. However, this account cannot fully explain the results of Wang et al. (2016) because in their study both training conditions engaged the ANS. Instead, these results could be explained by training-related differences in "children's internal sense of their ability to reason about quantities," which may have led those children who completed the easiest comparisons first "to think of themselves as mathematically competent, thereby leading to more perseverance in subsequent symbolic math problems" (Wang et al., 2016, p. 94). Regardless, it is likely that these brief training interventions do not exert strong, long-lasting effects on children's math abilities even though this assertion remains to be experimentally validated. Long-term training studies are more likely to yield these benefits.

Long-Term ANS Training

A set of recent ANS training studies sought to determine if long-term training of the ANS can lead to changes in adults' math abilities. To this end, Park and Brannon (2013) trained a group of adults on a nonsymbolic arithmetic task for ten sessions every day, each lasting about 25 min. Similar to the training task used in Hyde et al.'s (2014) approximate addition training, participants saw an array of dots move behind an occluder and a second array was added (addition) or a subset moved out from behind the occluder (subtraction). Participants then either saw a third array and had to indicate whether the sum or difference of the first two arrays was more or less than the third array or they saw two arrays and had to indicate which contained the sum or difference of the first two arrays. Park and Brannon (2013) found that this nonsymbolic arithmetic training led to significant improvements in a symbolic arithmetic assessment compared to a passive test-retest control and an active general world-knowledge training. A follow-up study (Park & Brannon, 2014) replicated the improvements in symbolic arithmetic following nonsymbolic arithmetic training while failing to find such improvements following a nonsymbolic number comparison training similar to the approximate number comparison training used by Hyde et al. (2014). The authors argue that the manipulation of nonsymbolic number representations may be a critical component of this training, which relates to the manipulations necessary to successfully solve symbolic arithmetic.

Recent efforts have expanded on these initial studies to examine whether young children benefit in similar ways from long-term ANS training. Park, Bermudez, Roberts, and Brannon (2016) trained 3- to 5-year-old children on a nonsymbolic addition and subtraction computer game similar to the one used with adults (Park & Brannon, 2013, 2014). Children in the ANS training group completed ten sessions, 12 min each, over a period of 2–3 weeks while children in the control group completed a computerized memory game for the same period of time. They found significantly greater improvement on a standardized math assessment for those children who completed the nonsymbolic arithmetic training compared to those who completed the memory training.

Building on these findings, Libertus, Odic, Feigenson, and Halberda (2018) showed that 5 weeks of nonsymbolic number comparison training in young children also leads to greater improvements on a standardized math assessment compared to a phoneme awareness training. In both training conditions, children were trained every other day for 15 min for a total of 5 weeks. Interestingly, even though the nonsymbolic number comparison training relied entirely on visual stimuli (as did all other training studies reviewed before), children in the number comparison training group also performed significantly better on an auditory approximate number comparison task after training. In this task, they heard two sequences of rapidly presented tones, one high pitched and one low pitched, and had to indicate which of the two sequences contained more tones. This training-related transfer suggests that the visual nonsymbolic number comparison training does not merely improve children's ability to inhibit irrelevant perceptual cues that are specific to the visual stimuli (e.g., total surface area, density, or convex hull), but instead sharpens their ANS representations that are independent of the format in which stimuli are presented (i.e., dot arrays in the visual modality or tone sequences in the auditory modality).

MECHANISMS BEHIND LONG-TERM ANS TRAINING IMPROVEMENTS

While these long-term training studies suggest that ANS training can lead to improvements in math performance in both adults and children, the underlying mechanisms for these improvements are unclear. Hyde, Berteletti, and Mou (2016) suggested two possible mechanisms: the Overlapping

Operations Hypothesis and the Representational Overlap Hypothesis. According to the Overlapping Operations Hypothesis, the shared reliance on similar mental operations (e.g., addition and subtraction) may explain why training on nonsymbolic number formats may lead to improvements in math abilities (see Park & Brannon, 2014, for a similar explanation). This account fits well with training studies that used nonsymbolic arithmetic and showed subsequent improvements in symbolic arithmetic (Hyde et al., 2014; Park & Brannon, 2013, 2014), but it cannot explain why nonsymbolic number comparison training yields improvements in symbolic arithmetic (Hyde et al., 2014) or math abilities more broadly (Libertus et al., 2018; Wang et al., 2016). Instead, these findings may be better explained by the Representational Overlap Hypothesis, which purports that symbolic number representations rely to some extent on ANS representations. However, this account cannot explain why nonsymbolic arithmetic but not nonsymbolic number comparison training in adults yields improvements in symbolic arithmetic (Park & Brannon, 2014).

While these previous hypotheses focused on training-related changes to mental processes closely related to the training content, another possible explanation is that long-term ANS training leads to greater engagement with, interest in, or attention to math-related information outside of the training context, which in turn may improve math abilities. For example, children who complete ANS training repeatedly over the course of several weeks may become more interested in their math lessons at school or engage in more math-related activities such as number games outside of school. They might also pay more attention when other people in their environment talk about numbers and math such as a parent talking about the costs of groceries while shopping.

Preliminary support for the viability of this explanation comes from correlational studies examining the role of math-related input for children's math abilities. Direct observations typically quantify math input in terms of the frequency of math talk, that is, math-related utterances such as "we need three spoons of sugar" or "there are more than ten people on the bus." Previous studies examining the amount of parental math talk that young children are exposed to have found positive associations with children's math ability (Casey et al., 2018; Elliott, Braham, & Libertus, 2017; Gunderson & Levine, 2011; Levine, Suriyakham, Rowe, Huttenlocher, & Gunderson, 2010; Susperreguy & Davis-Kean, 2016; see also Levine et al., this volume). The more parents talked about numbers and math with their children during everyday activities or play in the lab, the greater the children's counting skills and performance on standardized math assessments. Similarly, the frequency of parent-reported math activities in the home has also been linked to children's math achievement (Anders et al., 2012; Blevins-Knabe & Musun-Miller, 1996; Kleemans, Peeters, Segers, & Verhoeven, 2012; Missall, Hojnoski, Caskie, & Repasky, 2015; Skwarchuk, 2009). For example, home activities, such as counting, relate to children's understanding of numerical symbols (LeFevre et al., 2009) and activities such as playing board games and puzzles predict children's performance on nonsymbolic math tasks and children's developing fluency with basic number skills (Skwarchuk, Sowinski, & LeFevre, 2014; Vandermaas-Peeler, Nelson, Bumpass, & Sassine, 2009; see also Ramani et al., this volume).

While variability in parents' math-related input has been observed in a number of different studies, it is unclear why parents differ in the provision of math-related input. On the one hand, it seems that if parents perceive early math learning to be important, they report greater engagement in math activities such as playing board games with numbers or measuring ingredients while cooking than if they perceive math to be less important (Blevins-Knabe, Berghout Austin, Musun, Eddy, & Jones, 2000; Missall et al., 2015; Musun-Miller & Blevins-Knabe, 1998; Sonnenschein et al., 2012). On the other hand, parents who report feeling more comfortable teaching math or report having better math skills report greater engagement in math-related activities with their children (Blevins-Knabe et al., 2000; Missall et al., 2015).

One of the shortcomings of these previous studies is that they used parental report to measure parents' math skills and math-related input. To test whether similar relations between parents' cognitive abilities, their provision of math-related input, and children's math abilities could be observed when parents and children were tested directly we completed a series of studies. First, we found that parents with more precise ANS representations also had children with more precise ANS representations (Braham & Libertus, 2017; Navarro, Braham, & Libertus, in press). Second, parents' ANS precision was a significant predictor of children's ability to solve math word problems, but not written calculation or speeded mental arithmetic problems (Braham & Libertus, 2017). Thus we hypothesized that parents with more precise ANS representations may talk more about math with their children during everyday activities which may provide the children with more learning opportunities to apply math in everyday contexts.

In a follow-up study, we provided evidence that parents' ANS indeed relates to their provision of math-related input (Elliott et al., 2017). In this study, mothers were observed while they played with their children in the lab for 10 min. We quantified the amount of math talk that occurred during these 10 min by counting how often mothers used a number word. We found that mothers' ANS precision as well as their self-reported math ability were significant predictors of how often they used number words greater than ten. In turn, the more frequently mothers talked about numbers greater than ten, the better their children tended to perform on a standardized math assessment. Thus individual differences in ANS precision can be linked to attention to and engagement with math-related information in everyday contexts, which in the case of greater parental conversation about these topics, leads to more opportunities for their children to learn math.

Obviously, more research is needed to determine whether children may prompt their parents to provide content or whether parents initiate the math-related input themselves. Regardless, linking the math input findings to the previous ANS training studies raises the intriguing possibility that ANS training may heighten people's interest in or attention to numbers and other math-related information outside of the training which in turn may increase their opportunities to improve their math abilities. For example, engaging in ANS training may lead children to seek out more math-related activities alone or with their parents and this greater math-related input may lead to improvements in math ability.

We recently showed that experimentally manipulating children's exposure to math talk indeed results in greater subsequent focusing on number in an unrelated context (Braham, Libertus, & McCrink, in press). In this study, we asked preschool-aged children and their parents to play together in a children's museum exhibit using either a numerical prompt or a nonnumerical prompt. Before and after playing with their parents, children completed assessments to measure their tendency to spontaneously focus on number when imitating the actions of an experimenter. After playing with their parents, children whose parents received the numerical prompt showed greater spontaneous focus on number compared to children whose parents received the nonnumerical prompt. These findings suggest that when parents provide their children with math-related input, it increases children's later spontaneous attention to numerical information, which in turn has been shown to predict children's math abilities (Hannula, Lepola, & Lehtinen, 2010; Hannula-Sormunen, Lehtinen, & Räsänen, 2015).

CONCLUSIONS AND FUTURE DIRECTIONS

In sum, previous correlational and training studies provide compelling support for the link between the ANS and math. However, attempts to explain the mechanisms behind this link via reliance on shared mental operations or shared representations have not been successful at integrating all of the findings across studies. One alternative explanation raised here is that ANS training may lead to greater engagement in math-related activities or heightened attention to math-related learning opportunities, which in turn improve math abilities.

The initial findings presented in this chapter at least suggest that this idea is tenable, but further work is needed to directly test this hypothesis. For example, future long-term ANS training studies could assess participants' interest in math-related activities and information before, during, and after the training to test whether ANS training leads to increased interest. Relatedly, future studies could vary whether children are receiving concurrent instruction in math while completing ANS training and what math concepts children are learning. These studies would provide important information for potential applications of ANS training in the classroom but also elucidate which aspects of math are most closely linked to the ANS.

REFERENCES

Allik, J., & Tuulmets, T. (1991). Occupancy model of perceived numerosity. *Perception & Psychophysics*, *49*(4), 303–314.

Anders, Y., Rossbach, H.-G., Weinert, S., Ebert, S., Kuger, S., Lehrl, S., et al. (2012). Home and preschool learning environments and their relations to the development of early numeracy skills. *Early Childhood Research Quarterly*, *27*, 231–244.

Blevins-Knabe, B., Berghout Austin, A., Musun, L., Eddy, A., & Jones, R. M. (2000). Family home care providers' and parents' beliefs and practices concerning mathematics with young children. *Early Child Development and Care*, *165*(1), 41–58.

Blevins-Knabe, B., & Musun-Miller, L. (1996). Number use at home by children and their parents and its relationship to early mathematical performance. *Early Development and Parenting*, *5*(1), 35–45.

Braham, E. J., & Libertus, M. E. (2017). Intergenerational associations in numerical approximation and mathematical abilities. *Developmental Science*, *20*(5), e12436.

Braham, E. J., Libertus, M. E., & McCrink, K. (in press). Improving young children's spontaneous focus on number through guided parent-child interactions at a children's museum. Developmental Psychology.

Casey, B. M., Lombardi, C. M., Thomson, D., Nguyen, H. N., Paz, M., Theriault, C. A., & Dearing, E. (2018). Maternal support of children's early numerical concept learning predicts preschool and first-grade math achievement. *Child Development*, *89*(1), 156–173.

Chen, Q., & Li, J. (2014). Association between individual differences in non-symbolic number acuity and math performance: a meta-analysis. *Acta Psychologica*, *148*, 163–172.

Dakin, S. C., Tibber, M. S., Greenwood, J. A., Kingdom, F. A., & Morgan, M. J. (2011). A common visual metric for approximate number and density. *Proceedings of the National Academy of Sciences of the United States of America*, *108*(49), 19552–19557. https://doi.org/10.1073/pnas.1113195108.

Dehaene, S. (1997). *The number sense: How the mind creates mathematics*. New York: Oxford University Press.

Dillon, M. R., Kannan, H., Dean, J. T., Spelke, E., & Duflo, E. (2017). Cognitive science in the field: a preschool intervention durably enhances intuitive but not formal mathematics. *Science*, *357*, 47–55.

Durgin, F. H. (1995). Texture density adaptation and the perceived numerosity and distribution of texture. *Journal of Experimental Psychology: Human Perception and Performance*, *21*(1), 149–169.

Elliott, L. E., Braham, E. J., & Libertus, M. E. (2017). Understanding sources of individual variability in parents' number talk with young children. *Journal of Experimental Child Psychology*, *159*, 1–15.

Fazio, L. K., Bailey, D. H., Thompson, C. A., & Siegler, R. S. (2014). Relations of different types of numerical magnitude representations to each other and to mathematics achievement. *Journal of Experimental Child Psychology*, *123*, 53–72.

Fuhs, M. W., & McNeil, N. M. (2013). ANS acuity and mathematics ability in preschoolers from low-income homes: contributions of inhibitory control. *Developmental Science*, *16*(1), 136–148. https://doi.org/10.1111/desc.12013.

Fuson, K. C. (1988). *Children's counting and concepts of number*. New York: Springer-Verlag.

Gebuis, T., Cohen Kadosh, R., & Gevers, W. (2016). Sensory-integration system rather than approximate number system underlies numerosity processing: a critical review. *Acta Psychologica*, *171*, 17–35.

Gelman, R., & Gallistel, C. R. (1978). *The child's understanding of number*. Cambridge: Harvard University Press.

Gilmore, C., Attridge, N., Clayton, S., Cragg, L., Johnson, S., Marlow, N., et al. (2013). Individual difference in inhibitory control, not non-verbal number acuity, correlate with mathematics achievement. *PLoS ONE*, *8*(6), e67374.

Ginsburg, N., & Nicholls, A. (1988). Perceived numerosity as a function of item size. *Perceptual and Motor Skills*, *67*, 656–658.

Gunderson, E. A., & Levine, S. C. (2011). Some types of parent number talk count more than others: relations between parents' input and children's cardinal-number knowledge. *Developmental Science*, *14*(5). https://doi.org/10.1111/j.1467-7687.2011.01050.x 21884318.

Halberda, J., & Feigenson, L. (2008). Developmental change in the acuity of the 'number sense': the approximate number system in 3-, 4-, 5-, and 6-year-olds and adults. *Developmental Psychology*, *44*(5), 1457–1465.

Halberda, J., Ly, R., Willmer, J., Naiman, D., & Germine, L. (2012). Number sense across the lifespan as revealed by a massive internet-based sample. *Proceedings of the National Academy of Sciences of the United States of America*, *109*(28), 11116–11120. https://doi.org/10.1073/pnas.1200196109.

Halberda, J., Mazzocco, M. M., & Feigenson, L. (2008). Individual differences in non-verbal number acuity correlate with maths achievement. *Nature*, *455*(7213), 665–668.

Hannula, M. M., Lepola, J., & Lehtinen, E. (2010). Spontaneous focusing on numerosity as a domain-specific predictor of arithmetical skills. *Journal of Experimental Child Psychology*, *107*(4), 394–406. https://doi.org/10.1016/j.jecp.2010.06.004.

Hannula-Sormunen, M. M., Lehtinen, E., & Räsänen, P. (2015). Preschool children's spontaneous focusing on numerosity, subitizing, and counting skills as predictors of their mathematical performance seven years later at school. *Mathematical Thinking and Learning*, *17*(2–3), 155–177.

Honore, N., & Noel, M. P. (2016). Improving preschoolers' arithmetic through number magnitude training: the impact of non-symbolic and symbolic training. *PLoS ONE*, *11*(11), e0166685.

Hyde, D. C., Berteletti, I., & Mou, Y. (2016). Approximate numerical abilities and mathematics: insight from correlational and experimental training studies. In M. Cappelletti & W. Fias (Eds.), *Vol. 227. Progress in brain research* (pp. 335–351). Amsterdam: Elsevier.

Hyde, D. C., Khanum, S., & Spelke, E. S. (2014). Brief non-symbolic, approximate number practice enhances subsequent exact symbolic arithmetic in children. *Cognition*, *131*, 92–107.

Inglis, M., Attridge, N., Batchelor, S., & Gilmore, C. (2011). Non-verbal number acuity correlates with symbolic mathematics achievement: but only in children. *Psychonomic Bulletin & Review*. https://doi.org/10.3758/s13423-011-0154-1.

Keller, L., & Libertus, M. E. (2015). Inhibitory control may not explain the link between approximation and math abilities in kindergarteners from middle class families. *Frontiers in Psychology*, *6*, 685. https://doi.org/10.3389/fpsyg.2015.00685.

Khanum, S., Hanif, R., Spelke, E., Berteletti, I., & Hyde, D. C. (2016). Effects of non-symbolic approximate number practice on symbolic numerical abilities in Pakistani children. *PLoS ONE*, *11*(10), e0164436.

Kleemans, T., Peeters, M., Segers, E., & Verhoeven, L. (2012). Child and home predictors of early numeracy skills in kindergarten. *Early Childhood Research Quarterly*, *27*, 471–477.

Kuhn, J.-T., & Holling, H. (2014). Number sense or working memory? The effect of two computer-based trainings on mathematical skills in elementary school. *Advances in Cognitive Psychology*, *10*(2), 59–67.

LeFevre, J. A., Skwarchuk, S. L., Smith-Chant, B. L., Fast, L., Kamawar, D., & Bisanz, J. (2009). Home numeracy experiences and children's math performance in the early school years. *Canadian Journal of Behavioural Science*, *41*(2), 55–66.

Leibovich, T., Katzin, N., Harel, M., & Henik, A. (2016). From 'sense of number' to 'sense of magnitude'—the role of continuous magnitudes in numerical cognition. *Behavioral and Brain Sciences*, *40*, 1–62.

Levine, S. C., Suriyakham, L. W., Rowe, M. L., Huttenlocher, J., & Gunderson, E. A. (2010). What counts in the development of young children's number knowledge? *Developmental Psychology*, *46*(5), 1309–1319. https://doi.org/10.1037/a0019671 20822240.

Libertus, M. E., & Brannon, E. M. (2010). Stable individual differences in number discrimination in infancy. *Developmental Science*, *13*(6), 900–906. https://doi.org/10.1111/j.1467-7687.2009.00948.x.

Libertus, M. E., Odic, D., Feigenson, L., & Halberda, J. (2018). *Visual training of approximate number sense improves auditory number sense and school math ability*. [in prep].

Maertens, B., De Smedt, B., Sasanguie, D., Elen, J., & Reynvoet, B. (2016). Enhancing arithmetic in pre-schoolers with comparison or number line estimation training: does it matter? *Learning and Instruction*, *46*, 1–11.

Missall, K., Hojnoski, R. L., Caskie, G. I., & Repasky, P. (2015). Home numeracy environments of preschoolers: examining relations among mathematical activities, parent mathematical beliefs, and early mathematical skills. *Early Education and Development*, *26*(3), 356–376.

Musun-Miller, L., & Blevins-Knabe, B. (1998). Adults' beliefs about children and mathematics: how important is it and how do children learn about it. *Early Development and Parenting*, *7*(4), 191–202.

Navarro, M., Braham, E. J., & Libertus, M. E. (in press). Intergenerational associations of the approximate number system in toddlers and their parents. British Journal of Developmental Psychology.

Obersteiner, A., Reiss, K., & Ufer, S. (2013). How training on exact or approximate mental representations of number can enhance first-grade students' basic number processing and arithmetic skills. *Learning and Instruction*, *23*, 125–135.

Odic, D., Hock, H., & Halberda, J. (2014). Hysteresis affects approximate number discrimination in young children. *Journal of Experimental Psychology: General*, *143*(1), 255–265.

Park, J., Bermudez, V., Roberts, R. C., & Brannon, E. M. (2016). Non-symbolic approximate arithmetic training improves math performance in preschoolers. *Journal of Experimental Child Psychology*, *152*, 278–293.

Park, J., & Brannon, E. M. (2013). Training the approximate number system improves math proficiency. *Psychological Science*, *24*(10), 2013–2019.

Park, J., & Brannon, E. M. (2014). Improving arithmetic performance with number sense training: an investigation of underlying mechanism. *Cognition*, *133*(1), 188–200.

Räsänen, P., Salminen, J., Wilson, A. J., Aunio, P., & Dehaene, S. (2009). Computer-assisted intervention for children with low numeracy skills. *Cognitive Development*, *24*, 450–472.

Schneider, M., Beeres, K., Coban, L., Merz, S., Schmidt, S. S., Stricker, J., et al. (2016). Associations of non-symbolic and symbolic numerical magnitude processing with mathematical competence: a meta-analysis. *Developmental Science*, *20*(3), e12372. https://doi.org/10.1111/desc.12372.

Sella, F., Tressoldi, P., Lucangeli, D., & Zorzi, M. (2016). Training numerical skills with the adaptive videogame "the number race": a randomized controlled trial on preschoolers. *Trends in Neuroscience and Education*, *5*, 20–29.

Skwarchuk, S. L. (2009). How do parents support preschoolers' numeracy learning experiences at home? *Early Childhood Education Journal*, *37*, 189–197.

Skwarchuk, S. L., Sowinski, C., & LeFevre, J. A. (2014). Formal and informal home learning activities in relation to children's early numeracy and literacy skills: the development of a home numeracy model. *Journal of Experimental Child Psychology*, *121*, 63–84.

Sonnenschein, S., Galindo, C., Metzger, S. R., Thompson, J. A., Huang, H. C., & Lewis, H. (2012). Parents' beliefs about children's math development and children's participation in math activities. *Child Development Research. 2012*, Article ID 851657, https://doi.org/10.1155/2012/851657.

Susperreguy, M. I., & Davis-Kean, P. E. (2016). Maternal math talk in the home and math skills in preschool children. *Early Education and Development*, *27*(6), 841–857.

Tibber, M. S., Greenwood, J. A., & Dakin, S. C. (2012). Number and density discrimination rely on a common metric: similar psychophysical effects of size, contrast, and divided attention. *Journal of Vision*, *12*(6), 8. https://doi.org/10.1167/12.6.8.

Tibber, M. S., Manasseh, G. S. L., Clarke, R. C., Gagin, G., Swanbeck, S. N., Butterworth, B., et al. (2013). Sensitivity to numerosity is not a unique visuospatial psychophysical predictor of mathematical ability. *Vision Research*, *89*, 1–9.

Tokita, M., & Ishiguchi, A. (2010). Effects of element features on discrimination of relative numerosity: comparison of search symmetry and asymmetry pairs. *Psychological Research*, *74*(1), 99–109. https://doi.org/10.1007/s00426-008-0183-1.

Vandermaas-Peeler, M., Nelson, J., Bumpass, C., & Sassine, B. (2009). Numeracy-related exchanges in joint storybook reading and play. *International Journal of Early Years Education*, *17*(1). https://doi.org/10.1080/09669760802699910.

Vos, P. G., van Oeffelen, M. P., Tibosch, H. J., & Allik, J. (1988). Interactions between area and numerosity. *Psychological Research*, *50*(3), 148–154.

Wang, J., Odic, D., Halberda, J., & Feigenson, L. (2016). Changing the precision of preschoolers' approximate number system representations changes their symbolic math performance. *Journal of Experimental Child Psychology*, *147*, 82–99.

Wilson, A. J., Dehaene, S., Dubois, O., & Fayol, M. (2009). Effects of an adaptive game intervention on accessing number sense in low-socioeconomic-status kindergarten children. *Mind, Brain and Education*, *3*(4), 224–234.

Wilson, A. J., Dehaene, S., Pinel, P., Revkin, S. K., Cohen, L., & Cohen, D. (2006). Principles underlying the design of "the number race", an adaptive computer game for remediation of dyscalculia. *Behavioral and Brain Functions*, *2*, 19. https://doi.org/10.1186/1744-9081-2-19.

Wilson, A. J., Revkin, S. K., Cohen, D., Cohen, L., & Dehaene, S. (2006). An open trial assessment of the number race, an adaptive computer game for remediation of dyscalculia. *Behavioral and Brain Functions*, *2*(1), 20.

Wynn, K. (1992). Children's acquisition of the number words and the counting system. *Cognitive Psychology*, *24*, 220–251.

Xu, F., & Spelke, E. S. (2000). Large number discrimination in 6-month-old infants. *Cognition*, *74*(1), B1–B11.

Zhou, X., Wei, W., Zhang, Y., Cui, J., & Chen, C. (2015). Visual perception can account for the close relation between numerosity processing and computational fluency. *Frontiers in Psychology*, *6*, 1364.

Chapter 5

Mathematical Development in the Early Home Environment

Susan C. Levine[*,†,‡], **Dominic J. Gibson**[*] **and Talia Berkowitz**[*]
[*]*Department of Psychology, University of Chicago, Chicago, IL, United States*
[†]*Department of Comparative Human Development, University of Chicago, Chicago, IL, United States*
[‡]*Committee on Education, University of Chicago, Chicago, IL, United States*

INTRODUCTION

Parent support of children's early math learning tends to take a back seat to other aspects of children's development—notably early language and literacy skills. Indeed, parents of preschoolers report that they emphasize language and literacy more than mathematics in the home environment, and they also report that they believe this should be the case in preschool settings (e.g., Cannon & Ginsburg, 2008; LeFevre et al., 2009). While parent support of children's early math development should not supplant their support of language/literacy development, we argue that early math development is important and consequential, and that family involvement is integral to children's development in this domain. We base this argument on a few key findings. First, the math support parents provide to young children, notably the math language and activities they engage in with their children, is related to children's foundational math skills as they enter school (e.g., Gunderson & Levine, 2011; Levine, Ratliff, Cannon, & Huttenlocher, 2012; Levine, Suriyakham, Rowe, Huttenlocher, & Gunderson, 2010; Pruden, Levine, & Huttenlocher, 2012). Second, the level of children's math knowledge at kindergarten entry predicts their long-term academic success not only in math, but also in reading (e.g., Claessens & Engel, 2013; Duncan et al., 2007), although the processes that link early skills to later skills are not fully understood (Bailey, this volume; Bailey, Duncan, Watts, Clements, & Sarama, 2018).

At a societal level, these findings are gaining attention because of projections for steep growth in STEM-related (science, technology, engineering, and mathematics) jobs in the coming decades and the central role of mathematics

Mathematical Cognition and Learning, Vol. 5. https://doi.org/10.1016/B978-0-12-815952-1.00005-0

in broad aspects of STEM achievement (e.g., U.S. Department of Education, 2016). Issues of equity are also paramount, because many children from lower socioeconomic backgrounds enter school behind in math, which places them at a disadvantage for the STEM employment opportunities that are expected to grow during their lifetime. Beyond considerations of STEM workforce needs, and associated equity issues, mathematics knowledge is helpful to making good decisions, both related to one's personal life (e.g., health, finances, etc.) and to being an informed citizen who can understand data that bear on a wide range of societal issues (e.g., Reyna, Nelson, Han, & Dieckmann, 2009).

We argue that providing a home environment that supports early math development is an attainable goal for everyone, because adults engage in mathematical thinking every day—when they shop, cook, measure, make plans, think about time, plan routes, assemble, and build things. Thus there are ample opportunities for parents and caregivers to engage children in mathematical thinking. In this chapter, we primarily examine the home math environment through the lens of early "math talk," both numerical and spatial, not only because of its relation to early math skill, but because of evidence that this kind of talk is malleable in terms of quantity and quality (e.g., Gibson, Gunderson, & Levine, under review; Vandermaas-Peeler, Boomgarden, Finn, & Pittard, 2012). In addressing this important kind of early math support, we not only consider the math talk parents provide but also the cospeech gestures that commonly accompany math talk, asking if this makes a difference to children's mathematical learning and if so, why. In addition, we consider how the attitudes that parents have about math might influence their math talk. We know from prior research that math anxious individuals tend to avoid math themselves, and we have preliminary evidence that math anxious parents not only engage in less math talk with their young children (Eason, Nelson, Dearing, & Levine, 2017) but that their math interactions tend to be less effective (Herts et al., 2017; Maloney, Ramirez, Gunderson, Levine, & Beilock, 2015).

In considering the relation of math talk in the home environment to young children's math development, we organize the chapter around the following the three questions:

(1) To what extent is parent math talk in the early home environment related to children's math outcomes, and is there evidence that this relation is causal?
(2) What is the role of gesture in supporting math learning, and why might this be another important facet of children's math input?
(3) Are parent math attitudes and beliefs related to math in the home environment and by extension, to children's math achievement?

PARENT MATH LANGUAGE AND ASSOCIATED OUTCOMES

Number Talk

When parents are asked about early math supports, they are highly likely to think about counting and other number-related activities (e.g., Anderson, 1997;

Cannon & Ginsburg, 2008). Indeed, both questionnaire and observational studies of the early home environment reveal that number-related activities and talk are a common form of parent-child math engagement. Although these methods have differing strengths and weaknesses, and provide different kinds of data, the results that have emerged are largely convergent, supporting a relation between parents' number-relevant talk and activities, and children's number knowledge.

Questionnaire Studies

While parent questionnaires have the advantage of capturing general levels and types of number engagement, they are limited by parents' memories of what actually occurred, and may be subject to demand characteristics, particularly if the respondents know the research is focused on early math. In addition, because questionnaires require the recall of past events, they cannot provide detailed information about the quantity and quality of number talk or accompanying co-speech gestures that may be important to children's number learning. Despite these limitations, questionnaire studies reveal a positive correlation between parents' reported engagement in home numeracy activities and children's number knowledge (e.g., Blevins-Knabe & Musun-Miller, 1998; LeFevre et al., 2009; LeFevre, Clarke, & Stringer, 2002; Manolitsis, Georgiou, & Tziraki, 2013; Niklas & Schneider, 2014; Skwarchuk, Sowinski, & LeFevre, 2014).

Early questionnaire studies mainly focused on parent reports of formal math activities that involved directly teaching children to count, to read and write numerals, or to do simple calculations. These studies largely showed a positive relation between the frequency of these activities and children's math skills (e.g., Blevins-Knabe & Musun-Miller, 1998; LeFevre et al., 2002). More recently, LeFevre and colleagues drew a distinction between direct math activities where the purpose was to teach about number (e.g., counting, numeral recognition, simple arithmetic) and indirect math activities, where there is the potential to learn about number but this is incidental (LeFevre et al., 2009). They asked how these different categories of activities, direct and indirect, related to kindergarten-2nd graders' number knowledge (number, addition, and subtraction) and math fluency (latency to correct responses on addition calculations). In a parent questionnaire study, they found that, controlling for a range of background factors that related to children's math scores, the frequency of indirect math experiences was significantly related to both math knowledge and math fluency, whereas the frequency of direct math experiences was significantly related only to math fluency.

In a related study, Skwarchuk et al. (2014) found that parent reports about the frequency of kindergarteners' direct numeracy activities were related to children's symbolic number skills whereas their reports about the frequency of indirect math activities were related to children's nonsymbolic arithmetic. In contrast to these findings, a questionnaire study carried out with the parents of 3- to 8-year-olds found that a general home math factor (but not a direct math activities factor, an indirect math activities factor, or a spatial activities factor) was significantly related to children's math skills, as reported by parents (accounting for 8% of the variance), controlling for parents' approximate number sense (ANS) acuity, their math anxiety, and household income (Purpura, Napoli, Wehrspann, & Gold, 2017).

Considered together, these studies provide some evidence for differential relations of direct and indirect math activities to children's math skills, but there are inconsistencies across studies, perhaps due to different age groups and measures. It is also possible that the distinction between indirect and direct math activities is not tapping the key factors that are important to children's math skills. Instead, it may be that the math content that is taught, and the ways it is being taught during direct and indirect activities, rather than the formality or directness of the math activities per se, is driving the relations to child math outcomes. Observing parent-child math interactions during direct and indirect math activities opens the door to probing this issue more deeply.

Observational Studies

Like studies utilizing questionnaires, observational studies have both advantages and disadvantages. On the plus side, they allow researchers to measure actual parent supports of young children's math rather than relying on parent reports of such behaviors. Further, they provide detailed information about what was actually said and what accompanying supports were used, such as the use of gestures and manipulatives, thus providing more detailed information about the qualitative as well as quantitative aspects of math interactions. However, by their very nature, observational studies are time limited, capturing only a small slice of life. Thus there is a risk that the math that occurs during the observations might not be representative of the math support the child actually receives. Moreover, as is the case in questionnaire studies, if parents are told that the researchers are interested in math, there are likely demand characteristics that lead to more math talk than is typical. Yet, even with these limitations, observational studies, like questionnaire studies, show that parents' number talk is related to children's number knowledge, and that these relations hold controlling for many variables that could provide alternative explanations for these relations, such as parent education, overall parent talk, child math knowledge, and family SES (e.g., Casey et al., 2018; Levine et al., 2010; Ramani, Rowe, Eason, & Leech, 2015).

Observational Studies in the Lab

In observational studies, researchers commonly videotape parent-child dyads playing with a standard set of toys for a fixed amount of time and code the number talk provided by the parent. In one study of this kind, Casey et al. (2018) utilized videotapes of 140 mothers interacting with their 36-month-old children who participated in the NICHD Study of Early Child and Youth Development from the Boston site. They coded maternal number talk during interactions with two of the boxes in the 3-Boxes task (NICHD Early Child Care Research Network, 1999). These two boxes contained a cash register and dress up clothes (Box 2) and Duplo blocks of various sizes (Box 3), both of which afforded the opportunity for number talk. Mothers were simply asked to have their child play with the toys in each of the boxes, reducing the chance that demand characteristics could influence parent math talk. The researchers coded parent labeling of numerals, labeling the cardinal value of sets, and engaging in one-to-one counting. Controlling for variables including child gender, child IQ, maternal education, income to needs ratio, as well as other variables that might be associated with number talk, results showed that labeling the cardinal value of sets, but not the other categories of number talk, predicted children's performance on the Woodcock-Johnson Applied Problems subtest when children were 4.5 years old and in 1st grade.

Using a similar approach, Ramani et al. (2015) examined parent number talk with 3- to 5-year-old children during a 15-min observation session where dyads were asked to play with a book, puzzle, and board game. All parent number talk was coded and categorized into the following categories: "foundational" number talk (i.e., counting, numerical identification) and "more advanced" number talk (i.e., cardinality, ordinal relations, and arithmetic). Parent talk about foundational number concepts was not related to children's math skills, either foundational (e.g., counting, numeral identification) or more advanced (counting principles, enumeration and cardinality, number line estimation, and magnitude comparison), possibly because many children had already learned the foundational concepts and input focused on these concepts was not helpful to children's learning of more advanced number concepts. In contrast, parent use of more advanced number talk was related to children's advanced number knowledge, controlling for child age.

Naturalistic Home Observations

In other observational studies, researchers have observed parents and children engaging in naturalistic interactions and have then coded number talk that occurred, typically videotaping but occasionally audiotaping the interactions. We took this approach in a longitudinal language project examining the relation of parent language input to children's language and cognitive development. The families were recruited to represent the demographics of the greater Chicago area with the requirement that all participating families reported that English was the language spoken in the home. In this study, we visited parent-child dyads in their homes every 4 months, beginning when children were 14 months of age and continuing until they were 30 months of age. During each visit, the researcher videotaped the parent-child dyads for approximately 90 min, instructing the parents to go about their normal activities. At certain visits, particular aspects of children's language and cognitive skills were assessed, including their math knowledge.

Levine et al. (2010) found substantial variation in the amount of number talk parents engaged in during five 90-min sessions with their 14- to 30-month-old children (Mean parent number tokens (use of numbers between 1 and 10) = 90.8 (sd = 61.3); range: 4–257). In other words, there was about a 60-fold difference between the children who heard the fewest number words and those who heard the most number words. Extrapolating, this range amounts to a substantial input gap between a child hearing as few as 1500 number words to as many as 100,000 number words in a year, assuming 8-h waking days. The majority of these parent number words involved labeling the cardinal value of sets ((e.g., "You have three bears!") (50%) or counting (32%)—either counting objects (three trucks: "One (points to first truck), two (point to second truck, three") or reciting the count list without present objects, with the former being more common. Further, parent number words were most frequent for smaller numbers. As is shown by Fig. 1, number tokens fall off steeply from 1 to 10, with 76.0% of all parent number tokens consisting of 1–3, a pattern that has been reported by others (Ramscar, Dye, Popick, & O'Donnell-McCarthy, 2011).

As is the case for many aspects of language input to young children, number talk was associated with socioeconomic status, as assessed by a composite of education and income ($r = .30$, $P < .05$). Moreover, recent analyses show that the SES difference was significant for cardinal number talk ($r = .49$, $P < .01$) but not for counting ($r = .06$, $P = .75$). We also asked whether parents engage in more number talk with boys than girls but did not find significant differences.

The main question Levine and colleagues asked (Levine et al., 2010) was whether amount of parent number talk as assessed by their number word tokens while interacting with their child during five sessions from 14 to 30 months, predicted children's number knowledge. Using multiple regression, they found that variation in parent number talk predicted children's cardinal number knowledge at 46 months of age as assessed on the Point-to-X task

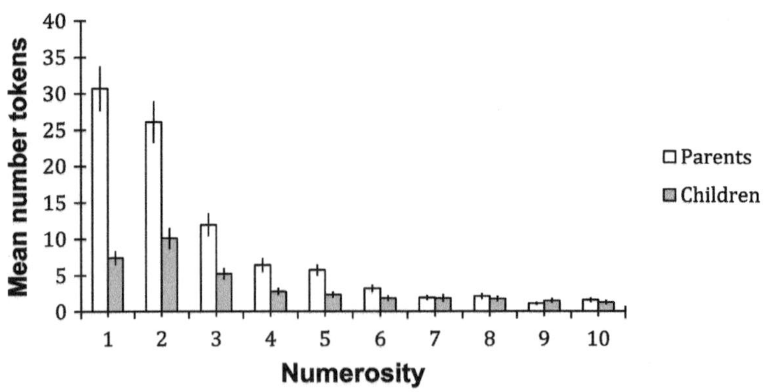

FIG. 1 Cumulative production of each numerosity 1–10, by parents and children. *(From Levine, Huttenlocher, Taylor, & Langrock, 1999; reproduced with the permission of the publisher.)*

(where children see two groups of objects and are asked to point to the group with *x* objects), and that this relation held controlling for family socioeconomic status, child number tokens, and parent other word tokens (see Fig. 2).

In a follow-up study, Gunderson and Levine (2011) coded qualitative aspects of the same parent-child interactions coded by Levine et al. (2010). In a multiple regression analysis they found that parents' counting or labeling the cardinal value of visible sets when children were 14–30 months of age was significantly related to children's cardinal number knowledge at 46 months whereas this was not the case for counting and cardinal utterances that did not refer to visible sets (e.g., practice reciting the count sequence). Further, despite the fact that larger number words were relatively rare, parents' counting or labeling visible sets from 4 to 10 was significantly related to children's understanding of the cardinal meanings of number words at 46 months of age whereas parents' counting or labeling set sizes in the subitizable range (1–3) was not. In Table 1 we provide two examples that illustrate the difference between a number interaction that uses number words to refer to present objects (Example 1) versus one that does not (Example 2). Although the first exchange does not contain as many number words as the second, the child has more opportunity to link the count sequence to the cardinality of the set in this interaction than in the second.

Other studies have examined parent number talk during daily routines such as meal time. Susperreguy and Davis-Kean (2016) coded 4-h of talk that occurred during breakfast time and dinner time, because parent-child conversations are particularly rich during these times (e.g., Tabors, Beals, & Weizman, 2001). Parent number talk was highly variable across the 40 families in the study, ranging from 4 to 195 instances. As in the Levine et al. (2010) study, the most common type of parent math talk involved labeling cardinal values of sets (42%). Counting, ordinal numbers, and naming digits each accounted

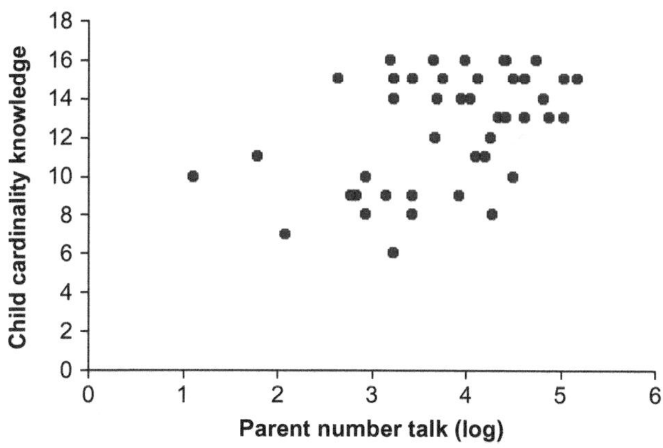

FIG. 2 Scatterplot displaying the relation between parent cumulative number word token (log) when child age was between 14 and 30 months and child cardinal number knowledge at 46 months ($n = 44$). *(From Levine et al., 2010; reproduced with the permission of the publisher.)*

TABLE 1 Examples of Types of Math Talk Parents Engage in with their Young Children

Example 1: 30 months old boy; Father and son looking at photo album
F: I think that was his fourth birthday so there's four candles, see? One, two, three, four. (father points to each candle in turn).
Example 2: 26 month old girl; Mother repeatedly tossing child in air
M: One, two, three, whee!
C: Again!
M: Again?
M: How many more times?
C: Four?
M: Four? Ok.
M: One, two, three!
C: Again! ....(continues until the child has been thrown in the air four times, each time with the count "One, two, three" accompanying each throw, sometimes with a whee!).

for an additional 10% of parent number talk. Further, they found that parent number talk was related to children's math knowledge, as assessed by their scores on the Test of Early Mathematics Ability (TEMA; Ginsburg & Baroody, 2003), administered 1 year after the naturalistic observations took place.

Experimental Studies

As summarized previously, questionnaire and observational studies provide evidence that parent number talk and children's number knowledge are positively related. However, these studies do not provide evidence of a causal relation even when they control for other variables that could account for this relation. One could argue, for example, that children who have more math knowledge or are more interested in math may elicit or initiate number talk with their parents, although Anderson (1997) reports that parents are typically the ones to incorporate math into activities in the home. It is also possible that parents with higher math ability not only talk more about number, but also have children who are predisposed to learn math with greater ease (e.g., Elliott, Braham, & Libertus, 2017).

A growing set of experimental studies are providing evidence of a causal link between parent number talk and activities and children's number knowledge. These studies show that it is possible to enhance the number talk parents provide to young children and that doing so leads to gains in children's number knowledge. Vandermaas-Peeler and colleagues conducted two studies in which parents were instructed to support their children's number learning either in the context of playing a board game (Vandermaas-Peeler, Ferretti, & Loving, 2012) or making a recipe (Vandermaas-Peeler, Boomgarden, et al., 2012). In both of these studies, the parents in the numeracy awareness intervention group provided more number talk and more advanced number talk than parents who were not given number talk instructions (i.e., the control group). For example, in the cooking context the parents in the intervention group were more likely to engage the child in addition and subtraction, moving beyond the kinds of counting activities that they were prompted to engage in. In both numeracy awareness interventions groups, children were asked more numeracy questions and likely because of this more frequently provided numerical information. In the cooking study, children's learning was assessed by administering the TEMA-2 immediately following the activity. The intervention group did not score higher than the control group on this test, likely because it takes more input than was provided in a single activity to develop the math concepts that are tapped on this test, such as cardinal number knowledge. Nonetheless, both the cooking and the board game interventions resulted in significant enhancements in the quality and quantity of parent number support, suggesting that such interventions could result in changes in children's math knowledge over time.

Experiments in the Lab

Experimental studies have also shown that certain activities and kinds of talk lead to gains in children's number knowledge (see also Libertus, this volume; Ramani et al., this volume). For example, researchers have shown that when 4-year-olds from low socioeconomic backgrounds played a linear number board game, they showed gains in number line representations and on comparing number magnitudes, relative to control groups. Those who played the linear board game also showed greater gains than control children in simple arithmetic, after they were provided with a lesson on this topic (e.g., Ramani & Siegler, 2008; Siegler & Ramani, 2009). In terms of the number talk provided, Laski and Siegler (2014) found that a critical component of the linear board game intervention is that it requires counting on, for example, when the child spun a "two" and was on the fifth space, they counted "six, seven" rather than "one, two."

Mix, Sandhofer, Moore, and Russell (2012) conducted a training study, manipulating the context in which children heard new number words. Once a week for 6 weeks, children heard experimenters either only count sets, only label the cardinal values of sets, label the cardinal values of sets and then immediately count each object in the set, or alternate between counting and labeling cardinal values at each session. At the end of the training period, only those children who had received the combined counting and cardinal labeling training showed significant improvements in their understanding of the cardinal principle. This finding indicates that hearing an adult both count and label the cardinal value of sets helps children learn a foundational math concept and does so more effectively than providing children with counting or set size labels alone or even alternating between these two kinds of input. Unfortunately, this kind of input, counting and labeling set sizes, occurs relatively rarely in naturalistic interactions (Casey et al., 2018; Levine et al., unpublished data).

The results of two other training studies suggest that young children learn more about number when object sets are not distracting. That is, perceptual richness of familiar objects may make them overly attractive and focus children's attention on the objects themselves rather than on their role as members of a set that is being counted (Petersen & McNeil, 2013). Consistent with this finding, young children's understanding of cardinality improved when they were trained with pictures of objects but not when they were trained with the objects themselves (Petersen et al., 2014). Together, these studies demonstrate that small differences in both the type of input and the manipulatives involved in various activities can lead to differences in children's learning and performance.

Experiments in the Field

Building on findings showing that families are more likely to embrace their role in supporting children's language and literacy than to embrace their role in supporting young children's math (e.g., Cannon & Ginsburg, 2008; Skwarchuk, 2009), several research teams have examined whether interventions that involve sharing number books with young children at home can increase their number knowledge (Gibson et al., under review; Purpura et al., 2017).

In a parent-delivered book intervention study, Gibson et al. assessed children's understanding of cardinal number using the Give-N task (Wynn, 1990) and randomized them into groups that received one of three book types: number books that focused on 1–3, number books that focused on 4–6, or control adjective books. Parents were asked to read the books to their child as many times a week as possible over the course of a month. At the end of the month, children were once again tested on their understanding of number words. As predicted, children who received the adjective book did not make significant gains in their number knowledge over the month long period. However, children in the number book conditions progressed about one knower level, on average. For instance, some children understood the quantities represented by 'one' and 'two' at the beginning of the month, and understood 'one,' 'two,' and 'three' at the end of the month. Further, children who only understood the number 1, or the numbers 1 and 2 at pretest learned from the small (1–3) but not the large (4–6) number books, whereas children who also understood the numbers 3 and 4 at pretest learned from both the small and large number books. These findings suggest that at the beginning of number word learning, children benefit from having the input tailored to their level of knowledge whereas once children understand three or four number words, they can benefit from exposure to a wider range of number input. This difference suggests that the knowledge children are working on when they are just beginning to learn the cardinal meaning of number words—learning the meaning of the next number word—is qualitatively different from the knowledge they are working on when they are three- and four-knowers—learning the cardinal principle. Even though parents tend to tailor their number input to young children's number knowledge (Saxe, Guberman, & Gearhart, 1987), some parents appear to do this in a more effective way than others, opening the door for taking this factor into account in the design of interventions.

Spatial Talk

While there is no doubt that numerical skills are central to children's mathematical development, there is evidence that spatial skills contribute to this development. There is accumulating evidence from developmental studies that spatial abilities are correlated with mathematical abilities (e.g., Casey et al., 2018; Gunderson, Ramirez, Beilock & Levine, 2012; Li & Geary, 2013, 2017; Mix et al., 2016, 2017), as well as evidence from studies of adults that spatial abilities are broadly predictive of STEM achievement and career paths, even controlling for verbal and mathematical abilities (e.g., Wai, Lubinski, & Benbow, 2009). There is some evidence for a causal relation, as several studies have shown that training a spatial skill—for example, mental rotation—leads to gains in numerical aspects of math—for example solving equivalence problems (e.g., Cheng & Mix, 2014), although not all studies find this relation (e.g., Hawes, Moss, Caswell, & Poliszczuk, 2015).

As was the case for parent number talk, observational studies have revealed marked variation in parent spatial talk to young children. In our longitudinal language study, we coded and analyzed three categories of spatial words that parents and children use to describe objects, which we dubbed "what spatial words," including nine observation sessions that occurred every 4 months between child ages 14 and 46 months. The categories we coded were talk about shapes (e.g., circle, square triangle), sizes (e.g., tall, short, wide, narrow), and spatial features (curved, straight, corner, etc.). Although "where spatial words" (e.g., in, on, under, in front of, above, etc.) are frequently used to talk about locations, we did not include these words because their use was very highly correlated with overall parent talk, and thus it is difficult to examine how these types of words relate to children's spatial thinking. Over nine 90-min observations (total 13.5 h), parents averaged 167 spatial words referring to these categories, with a range of 5–525 words. Assuming 8-h days, we estimated that a child at the high end of input would hear over 113,000 "what spatial words" in a year and a child at the low end would only hear only about 1100 of these words, more than a 100-fold difference.

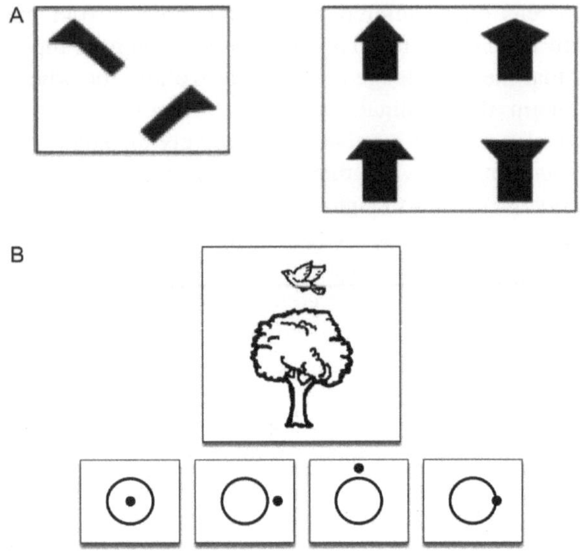

FIG. 3 (A) Sample item from the Spatial Transformation task (Levine et al., 1999; Pruden et al., 2012). Children were asked to point to the shape the pieces would make if they were put together. (B) Sample item from the Spatial Analogies test. Children were asked to select the one picture that "goes best" with the target picture. *(From Huttenlocher & Levine's (1990) Primary Test of Cognitive Skills, Monterey, CA: CTB/McGraw-Hill Companies, Inc. Reproduced with permission of McGraw-Hill Companies, Inc.)*

Moreover, as for parents' number talk, parents use of spatial words when talking to their children was related to children's spatial tokens, controlling for parent other-word tokens. Further, parent spatial language was related to children's spatial skills on nonverbal spatial tests given at 54 months age, including a mental rotation test and a spatial analogies test (see Fig. 3), relations that were mediated by children's spatial language production (Pruden et al., 2012).

Other studies also have provided evidence that spatial language is related to young children's spatial thinking. For example, Verdine et al. (2014) assessed 3-year-old children's spatial thinking using a spatial assembly task. Parents of the participating children were asked to check off which of 14 spatial words they used with their children. Findings showed that parents' report of their use of particular words—*between, below, above, and near*—predicted children's performance on the spatial assembly task. Moreover, the relations between the spatial words parents checked off and children's performance on particular assembly items were quite specific. For example, two of the spatial words, *above* and *below,* were significantly related to children's vertical location scores, but not to their other spatial skills. Verdine et al. hypothesized that this was because knowledge of these words was particularly helpful in determining the correct level of a piece in the construction of an object, perhaps by providing children with another way to encode the spatial relationship present in the model or by focusing their attention on the dimension and actions that were relevant to achieving a match to the model.

The relation of spatial language to spatial thinking has led researchers to ask whether parent spatial talk to young children is more likely to occur in certain contexts than others. One study found more parent spatial talk to their 3- to 5-year-olds during block play than during non-construction activities (Ferrara, Hirsh-Pasek, Newcombe, Golinkoff, & Lam, 2011). Moreover, parent and child spatial language was more frequent when parent-child dyads engaged in guided block play that involved following step-by-step instructions to assemble a structure than during either free play with blocks or play with an already assembled structure.

In the longitudinal language study described earlier, Levine and colleagues examined the frequency of parent-child spatial activities such as block play and puzzle play with preschool boys and girls. Between 26 and 46 months of age, about half of the dyads (51%) engaged in puzzle play, with no sex difference in the number of boys and girls who played with puzzles at least once over the six sessions that were coded (Levine et al., 2012). Nonetheless, parents tended to play with more complex puzzles with boys than girls, and parent spatial language was more frequent with more complex puzzles, resulting in boys hearing more parent spatial language than girls during puzzle play. Further, in this same database, parent-child block play was more common for boys than girls between 26 and 46 months of age, although not between 14 and 22 months (Petersen & Levine, in preparation). Given the Ferrara et al. (2011) findings, greater engagement of boys than girls in block play could result in them hearing more parent spatial language than girls. In fact, Pruden and Levine (2017) found that boys in our longitudinal language study use more spatial language than girls during naturalistic interactions with their parents between 14 and 46 months of age, although this was not the case for other aspects of language. This sex difference was fully mediated by a difference in parent spatial language use. This finding raises the possibility that differences in parent spatial talk to boys and girls could contribute to the male advantage on certain spatial tasks (e.g., Linn & Petersen, 1985; Moore & Johnson, 2008; Quinn & Liben, 2008; Uttal et al., 2012) perhaps by calling their attention to spatial information in the world and/or by reducing cognitive load while they are solving spatial problems.

Summary: Math Talk

Considered together, the take-home message from studies of parents' math talk with their children is that both their number and spatial talk are related to children's talk about these domains and the development of children's knowledge and skills in these domains. Parents' early number talk is related to children's cardinal number knowledge, controlling for a variety of other alternative factors that could account for this relation. Further, parents' spatial talk is related to children's spatial thinking as tapped by nonverbal tasks such as those assessing mental transformation skill and block construction. Importantly, experimental studies show that it is possible to enhance parents' math conversations with young children, and that this results in enhanced child performance on outcome measures, providing evidence of a causal link between the parents' math input and children's math learning. A remaining question is whether these interventions lead to enduring achievement gains and/or greater interest in mathematics. There is some indication that this may be the case for number talk, based on the few studies with longer term follow-up tests, but even these only follow-up a few years later. For example, Geary and colleagues found that children's early acquisition of the cardinal principle is associated with better school entry number knowledge, controlling for other factors (Geary et al., 2018).

GESTURE: AN ADDITIONAL SUPPORT FOR CHILDREN'S MATH LEARNING

One of the most common features of early math talk between parents and young children is the frequent use of gestures (Fuson, 1988; Goldin-Meadow, Levine, & Jacobs, 2014; Suriyakham, 2007). For instance, pointing and finger-counting are some of the most frequent strategies used by children during counting and arithmetic activities (Fuson, 1982, 1988; Geary & Burlingham-Dubree, 1989; Gelman & Gallistel, 1978; Graham, 1999; Saxe, 1977; Siegler & Shrager, 1984). As children get older, they continue to rely on their hands when they solve more complicated mathematical problems such as when explaining their solutions to mathematical equivalence problems (e.g., "$3+5+7=8+7$"; Perry, Church, & Goldin-Meadow, 1988). Likewise, parents and teachers frequently gesture when counting sets or talking about arithmetic (Flevares & Perry, 2001; Goldin-Meadow, Kim, & Singer, 1999; Suriyakham, 2007).

Accordingly, many theories suggest that number gestures play a role in the development of verbal number knowledge and counting principles (e.g., Butterworth, 1999, 2005; Fuson, 1982, 1988; Gelman & Gallistel, 1978; Gracia-Bafalluy & Noël, 2008). However, despite the ubiquity of number gestures, there is surprisingly little agreement concerning the precise role (or roles) that gestures play in number development. This partly stems from the fact that unlike in the case of parent number talk there is not yet a substantial body of research linking parents' number gestures to children's number knowledge and academic math performance more broadly. Nevertheless, research into how children and parents gesture about number and what function children's own gestures play in their developing understanding of number and mathematics can provide clues into how input involving number gestures may benefit children. Therefore in this section we focus on how parents and children use number gestures as well as highlight some initial evidence that input involving gestures may have a positive impact on children's mathematical development.

Counting Gestures

Many of the proposed functions of number gestures revolve around children's comprehension and implementation of the counting principles (Alibali & DiRusso, 1999; Di Luca & Pesenti, 2008; Fuson, 1988; Gelman & Gallistel, 1978; Graham, 1999; Potter & Levy, 1968; Saxe, 1977; Saxe & Kaplan, 1981). Counting gestures, such as pointing to objects while counting and raising fingers while reciting the count list, are used by children as young as the age of two (Fuson, 1982; Gelman & Gallistel, 1978). As children age and improve in their ability to accurately count, their use of counting gestures appears to increase. For instance, Suriyakham (2007) found that 30-month-olds produce more counting utterances without pointing gestures than with counting gestures, but by 38 months children count with pointing gestures as frequently as they count without them. Similarly, Saxe (1977) found that three-year-olds either do not gesture or fail to coordinate their points and verbal counting when tasked with reproducing sets containing 4–9 items. By the age of four, children almost always pointed while counting in this context and did so correctly on the majority of trials.

It has been hypothesized that these pointing gestures play a functional role in the development and implementation of the counting principles. For instance, children may use counting gestures to keep track of the number words while reciting the count list and to help maintain the stable order of the numbers in the count list (Fuson, 1982; Wiese, 2003). Pointing while counting also appears to be particularly useful for understanding and implementing the one-to-one correspondence principle by helping children map a single number word to each item being counted (Alibali & DiRusso, 1999). In fact, 3- to 4-year-old children who watched an experimenter count believed a count was incorrect if the experimenter pointed to one object twice even if the experimenter used the correct order and number of number words when counting (Briars & Siegler, 1984). Moreover, Saxe (1977) observed that 4-year-olds almost always pointed during successful implementation of one-to-one correspondence during counting.

Researchers have also highlighted the connection between number processing and finger representation as evidence that children's fingers play a role in the development of counting (Domahs, Moeller, Huber, Willmes, & Nuerk, 2010; Gerstmann, 1940; Moeller et al., 2012). Notably, children's ability to represent their fingers mentally and discriminate between their fingers is correlated with math ability (Fayol, Barrouillet, & Marinthe, 1998; Noël, 2005). Number gestures have also been linked to numerical cognition by studies demonstrating common neuroanatomical substrates of finger representation and number processing (Pesenti, Thioux, Seron, & Volder, 2000; Piazza, Mechelli, Butterworth, & Price, 2002; Pinel, Piazza, Le Bihan, & Dehaene, 2004).

Cardinal Number Gestures

Cardinal number gestures (e.g., holding up five fingers to indicate 5) represent another significant portion of children's early number input and developing ability to communicate about numbers. Although the specific combinations of fingers that make up canonical number gestures often differ between cultures, the use of conventionalized cardinal number gestures is common across cultures (Bender & Beller, 2012). Cardinal number gestures are particularly interesting since they can be used in combination with or in place of number words to label specific set sizes. However, they also are iconic representations of number in that they represent quantities through one-to-one correspondence between fingers in a gesture and items in a set. There is some evidence to suggest that learning iconic representations may be easier than learning arbitrary symbol-meaning mappings, particularly around the age of two when children begin to learn about numbers (Namy, Campbell, & Tomasello, 2004). Accordingly, researchers have argued that number gestures and other iconic number systems are easier for children to learn than the purely symbolic systems, like number words (Wiese, 2003) and may therefore serve as a bridge between nonsymbolic and symbolic representations of number (Gunderson, Spaepen, Gibson, Goldin-Meadow, & Levine, 2015).

Some have argued that children become proficient with number gestures only after becoming proficient with number words and therefore that number gestures cannot support the acquisition of number words. For instance, in one study, Nicoladis, Pika, and Marentette (2010) found that 4- to 5-year-olds (but not 2- to 3-year-olds) were more accurate when labeling sets using verbal number labels (number words) than cardinal number gestures, with the greatest difference in accuracy coming on larger sets (6, 7, 8, and 9). They concluded that children's knowledge of number words exceeds their knowledge of number gestures, and thus that there was no evidence to suggest children use their knowledge of number gestures to learn number words. Crollen, Seron, and Noël (2011) interpreted these findings to mean that number gestures do not precede the acquisition of number words and thus that the acquisition of number words is not rooted in children's acquisition of cardinal number gestures.

Importantly, Nicoladis et al.'s (2010) finding that 4- to 5-year-olds have great proficiency with number words than number gestures may be due to children at this age already knowing the cardinal principle. Having knowledge of the cardinal principle would mean that these children were likely well practiced using number words and at or near ceiling on tasks involving number words. Consistent with this interpretation, we found that subset knowers (children who had not yet learned the cardinal principle) were significantly more accurate when labeling the number of items in a set using gesture compared to speech (Gunderson et al., 2015). This effect appeared to be strongest for numbers immediately above children's knower level (i.e., numbers which children had not yet mastered in speech). In contrast, cardinal principle knowers showed either no difference or a slight advantage for labeling sets in speech compared to gesture depending on the set size. Subset knowers also showed an advantage when labeling larger sets (above four) using number gestures. Specifically, subset knowers made larger cardinal gestures when labeling larger sets sizes (i.e., 10) than when labeling smaller set sizes (i.e., 5). In contrast, subset knowers' verbal responses did not increase with the size of the set.

Another interesting facet of Gunderson et al.'s (2015) findings is children's simultaneous use of verbal labels and gestures, which often did not match. In such cases, children's gestures were much more accurate than their speech. Previous studies outside the domain of early number development have found that such gesture-speech mismatches are predictive of subsequent language (Iverson & Goldin-Meadow, 2005) and conceptual development (Church & Goldin-Meadow, 1986; Perry et al., 1988). Therefore in a follow-up training study, Gibson et al. (under review) looked at whether children who produced such mismatches in the early stages of number development (one- and two-knowers) were more likely to learn a new number word than those who did not make such mismatches. They found that indeed children who mismatched and received rich number input during training (counting, cardinal labeling of the set size, and comparison of adjacent set sizes) were more likely to learn a new number word than both those who did not mismatch and those who received sparse number input during training (just counting). This finding suggests that number gestures in combination with speech are at least predictive of children's subsequent learning and perhaps play a functional role in this learning.

Analyses of parents' and children's use of number gestures at home provides additional evidence that number gestures may be a pathway for positively impacting children's verbal number knowledge. Suriyakham (2007) found that children whose parents frequently gestured when counting and labeling sets said more number words and performed better on early measures of cardinality such as the Point-to-X task than children whose parents gestured less frequently when counting and labeling sets.

Together, these findings show that prior to fully comprehending number words, children are more accurate when labeling both small and large sets using gestures compared to speech (Gunderson et al., 2015). Moreover, how children gesture about numbers in combination with their speech can help reveal whether children are on the cusp of learning a new number word (Gibson et al., under review). Although more experimental research is necessary to establish a causal link between parents' gestures and children's subsequent number knowledge, there is some evidence that these are related (Suriyakham, 2007). These findings open the door to the possibility that parents' number gestures and children's knowledge of cardinal number gestures could play a supportive role in children's acquisition of number words and could be leveraged to promote verbal number knowledge. Again, this is important given the relation between children's early cardinal number knowledge and their later math skills (Geary et al., 2018).

Gesture and Arithmetic

Gestures continue to be revealing and useful as children progress to solving simple arithmetic problems. For instance, Levine, Jordan, and Huttenlocher (1992) asked 4.5- to 7-year-olds to solve basic addition and subtraction problems that were presented in the form of word problems, number fact problems, and nonverbal problems. The latter were presented with sets of disks that were either added to form a larger hidden set or were subtracted to form a smaller hidden set. Children below the age of 5.5 years rarely used their fingers when attempting to solve the arithmetic problems. However, the older children used their fingers in much the same way that they would use manipulatives. In fact, when manipulatives were not available to children, as was the case on word problems and number fact problems, children used their fingers more than when manipulatives were available, as was the case on the nonverbal problems, suggesting that they viewed fingers and manipulatives as substitutable. Specifically, children may use their fingers as a working memory aid during problem solving (Geary, 1990). Furthermore, children were more accurate in solving the verbal arithmetic problems when they used their fingers than when they did not, a finding that mirrors that of Jordan, Kaplan, Ramineni, and Locuniak (2008) for kindergarten age children.

Children from different socioeconomic backgrounds have been shown to differ in their finger use during calculations. For instance, in kindergarten, middle-income children are more likely to use their fingers to perform arithmetic than low-income children, but by first grade, the reverse is true: low-income children are more likely to use their fingers than middle-income children (Jordan, Levine, & Huttenlocher, 1994). Likewise, middle-SES children's use of fingers during arithmetic decreases between kindergarten and second grade while low-SES children's use of fingers increases over this period (Jordan et al., 2008). Importantly, Jordan et al. (2008) found that finger use in kindergarten was a strong positive predictor of calculation abilities in second grade. However, finger use in second grade was a negative predictor of children's calculation abilities. These findings suggest that gestures may serve as a helpful tool for children's introduction to arithmetical operations but may become a crutch as children are required to move to more advanced strategies.

In support of this idea, Siegler and Jenkins (1989) have argued that children's use of their fingers when solving problems evolves over the course of mathematical development. For example, initially children use a "count all" strategy (holding up fingers for each addend of an addition problem and then counting all of the raised fingers) before moving to a "count on" strategy (counting on from the first addend, e.g., holding up two fingers and then counting "three, four, five, and so on" for the second addend). This progression suggests that children's gestures can reveal how their strategies for arithmetic are changing and becoming more advanced and may even help support that developmental change. For instance, Siegler and Shrager (1984) suggested that early arithmetic strategies such as finger and verbal counting might help children form problem-answer associations that enable them to progress to more advanced retrieval-based problem solving.

As children progress to more advanced mathematical concepts, such as mathematical equivalence, gestures continue to play an important role in how children think and communicate about math problems (McNeil et al., 2006). Likewise, adults frequently gesture when explaining difficult math problems to children (Crowder & Newman, 1993; Flevares & Perry, 2001; Neill, 1991; Roth & Welzel, 2001; Zukow-Goldring, Romo, & Duncan, 1994). In a study of one-on-one instruction with 9- and 10-year-old children, teachers communicated 40% of the problem solving strategies they taught through gesture (Goldin-Meadow et al., 1999). Importantly, there is some evidence to suggest that children benefit more from instruction that involves gesture than instruction that does not involve gestures (Church, Ayman-Nolley, & Mahootian, 2004; Valenzeno, Alibali, & Klatzky, 2003). In fact, gestures may be especially effective in getting children to generalize to new types of problems when learning abstract mathematical concepts (Novack, Congdon, Hemani-Lopez, & Goldin-Meadow, 2014).

As an inherently spatial form of communication, co-speech gesture appears to be helpful in supporting children's spatial language and spatial skill as well. Parents vary widely in how often their spatial utterances are accompanied by gestures, ranging from 0% to 44% of their spatial utterances in one study (Cartmill, Pruden, Levine, & Goldin-Meadow, 2010). Furthermore, young children's spatial language is more highly predicted by parents' spatial language that is accompanied by gesture than by parents' spatial language that is unaccompanied by gesture (Cartmill et al., 2010). Additionally, Young and colleagues have shown in an experiment, that spatial language that is accompanied by gesture is more effective in supporting children's puzzle ability than spatial language not accompanied by co-speech gesture or by nonspatial language, either with or without co-speech gesture (Young, Cartmill, Levine, & Goldin-Meadow, 2014).

Summary: Gesture

In sum, young children frequently gesture when communicating about number and other aspects of mathematics (e.g., Jordan et al., 2008; Pruden et al., 2012). Children's gestures can reveal a lot about what they know and how they are thinking about math problems, enabling parents and teachers to target input to children's current stage of development (e.g., Gibson et al., under review; Jordan et al., 2008; Perry et al., 1988). Moreover, children's own gestures may help scaffold their early verbal mathematical knowledge. Similarly, parents and teachers use gesture to aid in communicating abstract mathematical ideas (e.g., Crowder & Newman, 1993; Flevares & Perry, 2001; Neill, 1991) and there is some evidence to suggest that such uses of gesture can help children better understand numbers, arithmetic, and spatial thinking (Novack et al., 2014; Suriyakham, 2007; Young et al., 2014).

PARENTAL MATH ATTITUDES AND BELIEFS: INTERGENERATIONAL FINDINGS

As discussed previously, parents vary greatly in the quantity and quality of math talk they engage in with their children. Even when controlling for factors such as income level and parental education, we see variation in parents' home numeracy practices (e.g., Levine et al., 2010; Susperreguy & Davis-Kean, 2016). It is important to understand the factors that might undermine parents' engagement in high-quality math talk with their children, because little or low-quality talk can influence children's mathematical development (e.g., Hyde, Else-Quest, Alibali, Knuth, & Romberg, 2006; Maloney et al., 2015).

One such factor is parents' math anxiety, or the fear and apprehension around doing math (Hembree, 1990; Richardson & Suinn, 1972). Math anxiety is a global phenomenon, experienced by both children and adults (Foley et al., 2017; Harari, Vukovic, & Bailey, 2013; Ramirez, Gunderson, Levine, & Beilock, 2013), and is associated with worse performance on standard math achievement tests (e.g., Ashcraft, 2002; Hembree, 1990; Ramirez et al., 2013). This underperformance in math may then continue to spiral downward as math anxiety and poor performance feed off of one another in a vicious cycle (Gunderson, Park, Maloney, Beilock, & Levine, 2017).

One reason math anxious individuals may exhibit poorer performance on math tasks is that the cognitive resources (e.g., working memory) that are normally used when solving math problems are depleted by the fears and negative emotions they experience in these contexts (Beilock & Carr, 2005; Beilock, Schaeffer, & Rozek, 2017; Foley et al., 2017; Hopko, Ashcraft, Gute, Ruggiero, & Lewis, 1998). In support of this idea, in both children (Ramirez et al., 2013) and adults (Beilock & Carr, 2005), high (but not low) working memory individuals show performance deficits in high pressure situations because the worries they experience intrude into their working memory and thereby reduce the resources they otherwise could deploy to solve these problems.

Spatial ability, which is related to math achievement, is also vulnerable to the effects of anxiety. Studies with adults have found that anxiety related to spatial tasks, such as a females fear of confirming gender stereotypes about spatial abilities, hinders performance on those tasks (Campbell & Collaer, 2009; Lawton, 1994; McGlone & Aronson, 2006). As with math anxiety, this relation starts early in schooling. Female students with high working memory who experience spatial anxiety exhibit reduced mental rotation ability compared to their high working memory peers who are not anxious about spatial tasks (Ramirez, Gunderson, Levine & Beilock, 2012).

Intergenerational Effects of Math Anxiety

These findings provide a sense of the impact that math and spatial anxiety have on the lives of those who experience these anxieties, but work from our lab has also found an intergenerational impact of math anxiety, with adults' math anxiety leading to lower math achievement for children. In one study, we assessed the math anxiety of 17 first- and second-grade female teachers, and then looked at the math achievement and beliefs about math of the 117 students in their classrooms (Beilock, Gunderson, Ramirez, & Levine, 2010). We found that while there was no relation between students' math achievement and their teachers' level of math anxiety at the beginning of the year, girls' math achievement was negatively correlated with teachers' math anxiety at the end of the year. The relation of teacher math anxiety to girls' math achievement was mediated by girls' gender ability beliefs—female students with math anxious teachers came to have relatively increased acceptance of traditional gender norms in school (i.e., girls are better at reading, boys are better at math), which in turn impacted their math performance.Although the relationship of teacher math anxiety to boys' math achievement was not significant in the Beilock et al. (2010) study, in a much larger study, teacher math anxiety was negatively associated with both girls' and boys' math learning over the school year (Schaeffer et al., under review).

And it is not only teachers' math anxiety that is negatively related to children's math achievement, but also parents' math anxiety (Berkowitz et al., 2015; Casad, Hale, & Wachs, 2015; Maloney et al., 2015). When highly math anxious parents report frequently helping their children with math homework, their children tend to learn less math over the year compared to children of high math anxious parents who help with homework less and to children of low math anxious parents, even when controlling for parent and child math ability (Maloney et al., 2015). In fact, we start to see negative effects of math anxiety on parent-child math interactions even during the preschool years. Using the same longitudinal database as the Levine et al. (2010) study described previously, we examined whether parent math anxiety was related to the quantity of number talk they engaged in with their children (Berkowitz, Gibson, Monahan, & Levine, 2017). Since math anxious individuals tend to avoid math, we reasoned that they might also avoid talking about math topics with their children. We found a relation between parent number talk when children were young (14–30 months old) and parent math anxiety, as assessed using the Short Mathematics Anxiety Rating Scale (Alexander & Martray, 1989) for higher, but not lower SES families. Among the higher SES parents, even the most basic types of math talk (e.g., cardinal labeling of sets) were lower in math anxious parents. One reason we might only have seen the relation of parent math anxiety and early number talk among higher SES families is that there is less math talk in lower SES families overall, regardless of level of math anxiety.

Another recent study in our lab with 50 parent-child dyads from largely middle income backgrounds found preliminary evidence that math anxious parents engage in less number talk with their preschool children than less math anxious parents (Berkowitz, 2018; Eason et al., 2017). Additionally, we showed that differences in the number prompts (e.g., "How many are there?" and "Can you count the plates?") parents produced was a greater predictor of child number talk than parents' number statements. While we did not find conclusive evidence in this sample that the difference in types of parent number talk was related to parents' math anxiety, future work is planned to explore this possibility.

While less work has explored the intergenerational impacts of spatial anxiety, we have found evidence that adult's spatial anxiety may impact children's performance on spatial tasks. Gunderson, Ramirez, Beilock, and Levine (2013) looked at teachers' spatial anxiety and measured the spatial skills of the students in their classrooms. They found that at the end of the year, student spatial skills were negatively related to teachers' spatial anxiety, even after accounting for other variables such as teachers' math anxiety. This may be due to teachers with high levels of spatial anxiety avoiding introducing spatial activities into the classroom, or presenting spatial activities in a less effective manner, though the exact mechanisms that account for this relation are not yet fully understood. Since spatial skills are an important component of math and other STEM areas, further study of the negative impacts of adults' spatial anxiety on children's spatial ability is an important area for future research.

Other Negative Attitudes Toward Math

Looking beyond math anxiety, parental attitudes about math can influence their children's achievement. Children of parents who frequently express their negative attitudes toward math (e.g., "I am not good at math") have both decreased interest and lower achievement in math, even when controlling for parents' education (Hyde et al., 2006). In part this is because parents' negative attitudes toward math may lead them to have low expectations for their children's math achievement or to fail to impart the value of math to their children during everyday interactions (Harackiewicz, Smith, & Priniski, 2016; Rozek, Svoboda, Harackiewicz, Hulleman, & Hyde, 2017).

Furthermore, differences in parents' expectations for children's math achievement predict children's math achievement as early as the age of five (DeFlorio & Beliakoff, 2015). It may be that children pick up on their parents' attitudes and beliefs about math, and when parents model positive attitudes about math in their own behaviors, their children are more likely to adopt and act on those values (Simpkins, Fredricks, & Eccles, 2012). Parents' process praise ("You did a great job!") rather than person praise ("You're so smart") has also been found to relate to children's growth mind-sets, which in turn has been found to relate to their math achievement (Cimpian, Arce, Markman, & Dweck, 2007; Gunderson, Gripshover, et al., 2013; Gunderson et al., 2018; Skwarchuk et al., 2014).

These findings underline the importance of developing interventions that combat the detrimental impact of parents' negative math attitudes on their children's math achievement. In one such effort, we carried out a longitudinal study assessing the effectiveness of a math app, Bedtime Math, in boosting the math achievement of children of math anxious parents (Berkowitz et al., 2015). As mentioned previously, in other studies we observed a negative relationship between the math anxiety of important adults in children's lives (i.e., parents and teachers) and students' performance in mathematics in early elementary school (e.g., Beilock et al., 2010; Maloney et al., 2015). Therefore we hypothesized that helping parents and children engage in high-quality math interactions could help change the nature of this relation between parent math anxiety and student math achievement. The app contained nightly passages, along with five to six math-related questions for parents and children to solve together. As such, the app not only provided parents with an opportunity for meaningful math interactions with their child, but also helped model how math can be incorporated into children's many interests via engaging discussions that could spill over into daily activities and routines. We found that the children of high math anxious parents showed greater math gains over the school year when they were randomized into the math group compared to a reading control group (see Fig. 4). Moreover, the achievement of this group, unlike that of the children of high math anxious parents who were in the control group, was similar to that of children whose parents were low in math anxiety.

Interestingly, the math app interactions were of higher quality for the dyads where the parent was not math anxious than for the dyads with a math anxious parent (Herts et al., 2017). Nonetheless, the children of the math anxious parents benefited most from this intervention, suggesting that the interactions that took place were of sufficient quality to support math gains. Importantly, we find that children of math anxious parents sustain these gains (compared to children of math anxious parents in the control group) through at least the end of 3rd grade, even though app use dramatically decreases after the first year. One potential mechanism for the sustained effect, supported by

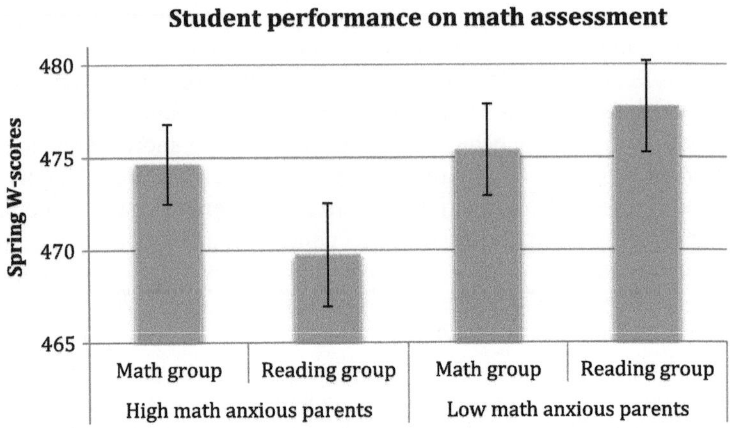

FIG. 4 Students' W-scores on the WJ Applied Problems Subtest in Spring of 1st grade. Students with high math anxious parents who were assigned to the math group performed as well as students of low math anxious parents in both the math and the reading (control) groups. In contrast, students of high math anxious parents in the reading group performed worse than all the other groups on the math assessment.

a mediation analysis, is a change in the math anxious parents' expectations and values for their children's math achievement (Schaeffer, Rozek, Berkowitz, Levine, & Beilock, in press).

Summary: Parental Math Attitudes and Beliefs

To recap, individuals' negative attitudes around math, including their math anxiety, are associated with poorer performance on math-related tasks in both adults and children. Additionally, there is evidence that parents' negative math attitudes and beliefs have adverse implications for their children's performance and achievement in mathematics. The quantity and quality of parents' math interactions with their children appear to be lower when parents are math anxious. However, existing studies indicate that interventions supporting parent-child math interactions can make a meaningful difference in the math achievement of children with math anxious parents.

CONCLUSIONS AND FUTURE DIRECTIONS

Our review highlights several important messages about the early home math environment and children's early math learning. First, it shows that the home math environments young children experience vary widely, and that these variations predict children's early mathematical knowledge. Second, it shows that specific contexts may support particular kinds of math knowledge. For example, indirect math activities such as games tend to support nonsymbolic arithmetic whereas more direct math activities such as numeral recognition and counting tend to support symbolic number knowledge, although these differences may be related to the kinds of math learning experiences provided in these contexts. That is, games that involve counting and learning to recognize numerals may result in the same kinds of learning as more direct math activities. Third, our review shows that it is important to tune home math talk to the level of children's understanding and thus, for parents to be aware of children's knowledge levels. Although Saxe et al. (1987) have shown that parents are sensitive to children's math knowledge levels, their intuitions about the kinds of inputs that are most helpful in enhancing children's knowledge are likely to benefit from evidence-based tips. Fourth, children's gestures can provide information about their readiness to learn a new number concept and particular kinds of cospeech gestures can be an effective way for parents to help children understand challenging math concepts. Fifth, both the math supports that parents provide and children's math outcomes appear to be linked to parents' attitudes about math, notably parents' math anxiety but also other attitudes they have about math. Finally, and most importantly, several studies show that parent math talk is malleable and that it is possible to increase the quantity and quality of this kind of talk through various means (e.g., nudges to support number learning during cooking, games, number books, math apps such as Bedtime Math), which in turn can lead to gains in children's math language and math knowledge.

There are many outstanding basic and applied research questions involving early math development. To highlight just a few, we know little about the long-term effects of enhancing the early home math environment because most intervention studies only test immediate math gains or at most, gains that are present a few years postintervention (see Bailey, this volume). Second, we need to learn more about the kinds of early math experiences that are most effective in supporting not only children's math knowledge, but also their math attitudes. We also need to probe how negative math attitudes, such as math anxiety, get started in the first place and how to prevent these attitudes from taking hold and contributing to the vicious cycle of low math achievement and negative math attitudes (Gunderson et al., 2017).

There are also important questions about how to scale up knowledge about effective home math supports. One important factor in successful scale up efforts is to develop messages and interventions that are not only evidence based but that also resonate with cultural practices of families. For example, a recent set of family discussions suggested that both Latino and Chinese families resonate more toward math supports being talked about as educational activities rather than games (ZenoMath, 2017). Additionally, Marta Civil's work brings the funds of knowledge framework to math, which recognizes and harnesses the strengths that families bring to the task of effectively supporting children's mathematics education, rather than approaching families through the lens of a deficit model. Her work shows that families can effectively support children's math learning and also demonstrates how parents can serve as resources for other parents, and how this can help build a community focused on this important goal (e.g., Civil, 1998; Civil, Bratton, & Quintos, 2005).

Fortunately, work with families and community partners to support children's early math learning is gaining traction and is having positive results. One promising approach is to create communities that embrace the importance of early math and that teach parents how they can support early math (Blumenstock, 2016). These efforts are likely to be more effective when they involve school-home partnerships that enlist parents in supporting curricular goals. Use of bi-generational interventions that involve teaching math to Head Start students and at the same time teaching their families how they can support their children's math learning in the home environment has been found to be an effective way to increase young children's math knowledge (e.g., Starkey & Klein, 2000). Community partnerships can also help with supporting children's math learning and reaching families with messages about how they can support their children's math learning. A study with the National Museum of Math in New York City where researchers and museum professionals codeveloped tip cards that could support parents' math conversations with their children exemplifies work with community partners. In this study, families randomized into the tip card group gave higher ratings of enjoyment and learning at the museum than those in the control group (Herts et al., 2018).

Another approach involves incorporating parent voices into the design of interventions. In our lab, we have held focus groups to help us incorporate parent voices in developing math messages that resonate and in developing effective math interventions. Our effort to develop effective math messages involved convening focus groups with two groups of African American parents. In two successive focus groups, we were able to iteratively refine messages based on what we learned from parents. We learned that parents preferred messages with specific information that could guide their math supports, messages about including math in everyday activities (but not the message that they did not need to make time for math because of this), and messages that math is more than numbers, but again wanted specific information about other important aspects of math. Additionally, parents liked the comparison of math to reading, which highlighted that parent support is important for both domains and that learning in these domains go hand in hand (Levine, Bradley, & Mix, under review).

Technology offers potentially powerful ways to deliver high-quality math activities and math tips to families. This can be done through apps, websites, and texts that can remind families to engage their children in math activities and help them do this effectively. With the increasing spread of smart devices, this opportunity is increasing and becoming ubiquitous. However, it is important to note that not all apps are created equal, and despite the 1000s of "educational" apps that are available, researchers are only starting to develop frameworks for evaluating the potential impact of various apps (Hirsh-Pasek et al., 2015). As shown by our research, an app that encourages parents and children to do math together was effective in supporting the math learning of the children of math anxious parents. This app may be more beneficial than math apps that involve the child alone, an important question for future research, particularly because most education apps are designed for children to use on their own and because children are spending increasing amounts of time in front of screens, reducing time for face-to-face interactions (e.g., Common Sense Media, 2017).

While all of these approaches and tools can be useful resources to enhance children's math experiences and to spread positive messages around math, they will have the greatest impact if they are developed in partnership with the parents who will be using them and with multiple sectors of community partners. Creating multisector coalitions that involve families, schools, businesses, health care providers, community organizations, and philanthropists will help ensure that sharing math with young children becomes as much of a focus as sharing books and reading.

ACKNOWLEDGMENTS

We thank the National Science Foundation Spatial Intelligence and Learning Center (Grants SBE-1041707 and SBE-0541957 to Susan C. Levine), the National Science Foundation Science of Learning Collaborative Network Grant #1540741 to Susan C. Levine, and the Heising-Simons Foundation Development and Research in Early Mathematics Education (DREME) Network (support to Susan C. Levine). Research reported in this publication was also supported by the Eunice Kennedy Shriver National Institute of Child Health & Human Development of the National Institutes of Health under Award Number P01HD040605.

REFERENCES

Alexander, L., & Martray, C. (1989). The development of an abbreviated version of the mathematics anxiety rating scale. *Measurement and Evaluation in Counseling and Development*, *22*(3), 143–150. https://doi.org/10.1037/t22970-000.

Alibali, M. W., & DiRusso, A. A. (1999). The function of gesture in learning to count: more than keeping track. *Cognitive Development*, *14*(1), 37–56. https://doi.org/10.1016/S0885-2014(99)80017-3.

Anderson, A. (1997). Families and mathematics: a study of parent-child interactions. *Journal for Research in Mathematics Education*, *28*(4), 484–511. https://doi.org/10.2307/749684.

Ashcraft, M. H. (2002). Math anxiety: personal, educational, and cognitive consequences. *Current Directions in Psychological Science*, *11*(5), 181–185. https://doi.org/10.1111/1467-8721.00196.

Bailey, D. H., Duncan, G. J., Watts, T., Clements, D., & Sarama, J. (2018). Risky business: correlation and causation in longitudinal studies of skill development. *American Psychologist*, *73*, 81–94. https://doi.org/10.1037/amp0000146.

Beilock, S. L., & Carr, T. H. (2005). When high-powered people fail: working memory and "choking under pressure" in math. *Psychological Science*, *16*(2), 101–105. https://doi.org/10.1111/j.0956-7976.2005.00789.

Beilock, S. L., Gunderson, E. A., Ramirez, G., & Levine, S. C. (2010). Female teachers' math anxiety affects girls' math achievement. *Proceedings of the National Academy of Sciences, 107*(5), 1060–1063. https://doi.org/10.1073/pnas.0910967107.

Beilock, S. L., Schaeffer, M. W., & Rozek, C. S. (2017). Understanding and addressing performance anxiety. In *Handbook of Competence and Motivation: Theory and Application*, 155.

Bender, A., & Beller, S. (2012). Nature and culture of finger counting: diversity and representational effects of an embodied cognitive tool. *Cognition, 124*(2), 156–182. https://doi.org/10.1016/j.cognition.2012.05.005.

Berkowitz, T. (2018). The influence of parent math anxiety on children's early math input and knowledge. *Unpublished doctoral dissertation*. Chicago, IL: University of Chicago.

Berkowitz, T., Gibson, D., Monahan, J., & Levine, S. C. (2017). *Math anxiety and parents' use of number words with their children. Paper presented at the Biennial meeting of the Society for Research on Child Development, Austin, TX.*

Berkowitz, T., Schaeffer, M. W., Maloney, E. A., Petersen, L., Gregor, C., Levine, S. C., et al. (2015). Math at home adds up to achievement in school. *Science, 350*(6257), 196–198. https://doi.org/10.1126/science.aac7427.

Blevins-Knabe, B., & Musun-Miller, L. (1998). Number use at home by children and their parents and its relation to mathematical performance. *Early Development and Parenting, 5*(1), 35–45. https://doi.org/10.1002/(SICI)1099-0917(199603)5:1<35::AID-EDP113>3.0.CO;2-0.

Blumenstock, S. R. (2016). *Kindergarten kickstart: Outcomes of a research-based summer pre-K program.*

Briars, D., & Siegler, R. S. (1984). A featural analysis of preschoolers' counting knowledge. *Developmental Psychology, 20*(4), 607. https://doi.org/10.1037/0012-1649.20.4.607.

Butterworth, B. (1999). A head for figures. *Science, 284*(5416), 928–929. https://doi.org/10.1126/science.284.5416.928.

Butterworth, B. (2005). The development of arithmetical abilities. *Journal of Child Psychology and Psychiatry, 46*(1), 3–18. https://doi.org/10.1111/j.1469-7610.2004.00374.x.

Campbell, S., & Collaer, M. (2009). Stereotype threat and gender differences in performance on a novel visuospatial task. *Psychology of Women Quarterly, 33*(4), 437–444. https://doi.org/10.1111/j.1471–6402.2009.01521.x.

Cannon, J., & Ginsburg, H. (2008). "Doing the math": maternal beliefs about early mathematics versus language learning. *Early Education and Development, 19*(2), 238–260. https://doi.org/10.1080/10409280801963913.

Cartmill, E., Pruden, S. M., Levine, S. C., & Goldin-Meadow, S. (2010). In K. Franich, K. M. Iserman, & L. L. Kei (Eds.), *The role of parent gesture in children's spatial language development. Proceedings of the 34th annual Boston University conference on language development* (pp. 70–77). Somerville, MA: Cascadilla Press.

Casad, B. J., Hale, P., & Wachs, F. L. (2015). Parent-child math anxiety and math-gender stereotypes predict adolescents' math education outcomes. *Frontiers in Psychology, 6*, 1–21. https://doi.org/10.3389/fpsyg.2015.01597.

Casey, B. M., Lombardi, C. M., Thomson, D., Nguyen, N. H., Paz, M., Theriault, C. A., et al. (2018). Maternal support of children's early numerical concept learning predicts preschool and first-grade math achievement. *Child Development, 89*(1), 156–173. https://doi.org/10.1111/cdev.12676.

Cheng, Y. L., & Mix, K. S. (2014). Spatial training improves children's mathematics ability. *Journal of Cognition and Development, 15*(1), 2–11. https://doi.org/10.1080/15248372.2012.725186.

Church, R. B., Ayman-Nolley, S., & Mahootian, S. (2004). The role of gesture in bilingual instruction. *International Journal of Bilingual Education and Bilingualism, 7*(4), 303–319. https://doi.org/10.1080/13670050408667815.

Church, R. B., & Goldin-Meadow, S. (1986). The mismatch between gesture and speech as an index of transitional knowledge. *Cognition, 23*(1), 43–71. https://doi.org/10.1016/0010-0277(86)90053-3.

Cimpian, A., Arce, H.-M., Markman, E. M., & Dweck, C. S. (2007). Subtle linguistic cues impact children's motivation. *Psychological Science, 18*(4), 314–316. https://doi.org/10.1111/j.1467-9280.2007.01896.

Civil, M. (1998). In *Bridging in-school mathematics and out-of-school mathematics: a reflection. Paper presented at AERA meeting, San Diego, CA.*

Civil, M., Bratton, J., & Quintos, B. (2005). Parents and mathematics education in a Latino community: redefining parental participation. *Multicultural Education, 13*(2), 60–64.

Claessens, A., & Engel, M. (2013). How important is where you start? Early mathematics knowledge and later school success. *Teachers College Record, 115*, 1–29. Retrieved from: http://www.tcrecord.org/.

Common Sense Media. (2017). *The common sense census: Media use by kids age zero to eight.* https://www.commonsensemedia.org/sites/default/files/uploads/research/csm_zerotoeight_fullreport_release_2.pdf.

Crollen, V., Seron, X., & Noël, M. P. (2011). Is finger-counting necessary for the development of arithmetic abilities? *Frontiers in Psychology, 2*. 242. https://doi.org/10.3389/fpsyg.2011.00242.

Crowder, E. M., & Newman, D. (1993). Telling what they know: the role of gesture and language in children's science explanations. *Pragmatics and Cognition, 1*(2), 341–376. https://doi.org/10.1075/pc.1.2.06cro.

DeFlorio, L., & Beliakoff, A. (2015). Socioeconomic status and preschoolers' mathematical knowledge: the contribution of home activities and parent beliefs. *Early Education and Development, 26*(3), 319–341. https://doi.org/10.1080/10409289.2015.968239.

Di Luca, S., & Pesenti, M. (2008). Masked priming effect with canonical finger numeral configurations. *Experimental Brain Research, 185*(1), 27–39. https://doi.org/10.1007/s00221-007-1132-8.

Domahs, F., Moeller, K., Huber, S., Willmes, K., & Nuerk, H. C. (2010). Embodied numerosity: implicit hand-based representations influence symbolic number processing across cultures. *Cognition, 116*(2), 251–266. https://doi.org/10.1016/j.cognition.2010.05.007.

Duncan, G. J., Dowsett, C. J., Claessens, A., Magnuson, K., Huston, A. C., Klebanov, P., et al. (2007). School readiness and later achievement. *Developmental Psychology, 43*(6), 1428–1446. https://doi.org/10.1037/0012-1649.43.6.1428.

Eason, S. H., Nelson, A. E., Dearing, E., & Levine, S. C. (2017). *How does parents' math anxiety relate to the quality of parent-child talk about number during pretend play? Poster session presented at the 2017.* Brisbane, Australia: International Science of Learning Conference.

Elliott, L., Braham, E. J., & Libertus, M. E. (2017). Understanding sources of individual variability in parents' number talk with young children. *Journal of Experimental Child Psychology, 159*, 1–15. https://doi.org/10.1016/j.jecp.2017.01.011.

Fayol, M., Barrouillet, P., & Marinthe, C. (1998). Predicting arithmetical achievement from neuro-psychological performance: a longitudinal study. *Cognition, 68*(2), B63–B70. https://doi.org/10.1016/S0010-0277(98)00046-8.

Ferrara, K., Hirsh-Pasek, K., Newcombe, N. S., Golinkoff, R. M., & Lam, W. S. (2011). Block talk: spatial language during block play. *Mind, Brain, and Education, 5*(3), 143–151. https://doi.org/10.1111/j.1751-228X.2011.01122.x.

Flevares, L. M., & Perry, M. (2001). How many do you see? The use of nonspoken representations in first- grade mathematics lessons. *Journal of Educational Psychology, 93*(2), 330–345. https://doi.org/10.1037/0022-0663.93.2.330.

Foley, A. E., Herts, J. B., Borgonovi, F., Guerriero, S., Levine, S. C., & Beilock, S. L. (2017). The math anxiety-performance link: a global phenomenon. *Current Directions in Psychological Science*, *26*(1), 52–58. https://doi.org/10.1177/0963721416672463.

Fuson, K. C. (1982). An analysis of thecounting-on solution procedure in addition. In T. P. Carpenter, J. M. Moser, & T. A. Romberg (Eds.), *Addition and subtraction: A cognitive perspective* (pp. 67–81). Hillsdale, NJ: Erlbaum.

Fuson, K. C. (1988). *Children's counting and concepts of number*. New York: Springer-Verlag. https://doi.org/10.1007/978-1-4612-3754-9.

Geary, D. C. (1990). A componential analysis of an early learning deficit in mathematics. *Journal of Experimental Child Psychology*, *49*, 363–383.

Geary, D. C., & Burlingham-Dubree, M. (1989). External validation of the strategy choice model for addition. *Journal of Experimental Child Psychology*, *47*(2), 175–192. https://doi.org/10.1016/0022-0965(89)90028-3.

Geary, D. C., van Marle, K., Chu, F., Rouder, J., Hoard, M. K., & Nugent, L. (2018). Early conceptual understanding of cardinality predicts superior school-entry number system knowledge. *Psychological Science*, *29*, 191–205.

Gelman, R., & Gallistel, C. (1978). *Young children's understanding of numbers*. Cambridge, MA: Harvard University Press.

Gerstmann, J. (1940). Syndrome of finger agnosia, disorientation for right and left, agraphia and acalculia: local diagnostic value. *Archives of Neurology and Psychiatry*, *44*, 398–408.

Gibson, D., Gunderson, E.A., & Levine, S.C. (under review). Causal effects of parent number talk on preschoolers' number knowledge.

Ginsburg, H. P., & Baroody, A. J. (2003). *Test of early mathematics ability* (3rd ed.). Austin, TX: Pro-ed.

Goldin-Meadow, S., Kim, S., & Singer, M. (1999). What the teacher's hands tell the student's mind about math. *Journal of Educational Psychology*, *91*(4), 720–730. https://doi.org/10.1037/0022-0663.91.4.720.

Goldin-Meadow, S., Levine, S., & Jacobs, S. (2014). Gesture's role in learning arithmetic. In *Emerging perspectives on gesture and embodiment in mathematics*. Charlotte, NC: Information Age Publishing.

Gracia-Bafalluy, M., & Noël, M. P. (2008). Does finger training increase young children's numerical performance? *Cortex*, *44*(4), 368–375. https://doi.org/10.1016/j.cortex.2007.08.020.

Graham, T. A. (1999). The role of gesture in children's learning to count. *Journal of Experimental Child Psychology*, *74*(4), 333–355. https://doi.org/10.1006/jecp.1999.2520.

Gunderson, E. A., Gripshover, S. J., Romero, C., Dweck, C. S., Goldin-Meadow, S., & Levine, S. C. (2013). Parent praise to 1- to 3-year-olds predicts children's motivational frameworks 5 years later. *Child Development*, *84*(5), 1526–1541. https://doi.org/10.1111/cdev.12064.

Gunderson, E. A., & Levine, S. C. (2011). Some types of parent number talk count more than others: relations between parents' input and children's cardinal-number knowledge. *Developmental Science*, *14*(5), 1021–1032. https://doi.org/10.1111/j.1467-7687.2011.01050.x.

Gunderson, E. A., Park, D., Maloney, E. A., Beilock, S. L., & Levine, S. C. (2017). Reciprocal relations among motivational frameworks, math anxiety, and math achievement in early elementary school. *Journal of Cognition and Development*, *19*(1), 21–46, https://doi.org/10.1080/15248372.2017.1421538.

Gunderson, E. A., Ramirez, G., Beilock, S. L., & Levine, S. C. (2012). The relation between spatial skill and early number knowledge: the role of the linear number line. *Developmental Psychology*, *48*(5), 1229–1241. https://doi.org/10.1037/a0027433.

Gunderson, E. A., Ramirez, G., Beilock, S. L., & Levine, S. C. (2013). Teachers' spatial anxiety relates to 1st- and 2nd-graders' spatial learning. *Mind, Brain, and Education*, *7*(3), 196–199. https://doi.org/10.1111/mbe.12027.

Gunderson, E. A., Sorhagen, N. S., Gripshover, S. J., Dweck, C. S., Goldin-Meadow, S., & Levine, S. C. (2018). Parent praise to toddlers predicts fourth grade academic achievement via children's incremental mindsets. *Developmental Psychology*, *54*(3), 397–409. https://doi.org/10.1037/dev0000444.

Gunderson, E. A., Spaepen, E., Gibson, D., Goldin-Meadow, S., & Levine, S. C. (2015). Gesture as a window on children's number knowledge. *Cognition*, *144*, 14–28. https://doi.org/10.1016/j.cognition.2015.07.008.

Harackiewicz, J. M., Smith, J. L., & Priniski, S. J. (2016). Interest matters: the importance of promoting interest in education. *Policy Insights from the Behavioral and Brain Sciences*, *3*(2), 220–227. https://doi.org/10.1177/2372732216655542.

Harari, R. R., Vukovic, R. K., & Bailey, S. P. (2013). Mathematics anxiety in young children: an exploratory study. *The Journal of Experimental Education*, *81*(4), 538–555. https://doi.org/10.1080/00220973.2012.727888.

Hawes, Z., Moss, J., Caswell, B., & Poliszczuk, D. (2015). Effects of mental rotation training on children's spatial and mathematics performance: a randomized controlled study. *Trends in Neuroscience and Education*, *4*(3), 60–68. https://doi.org/10.1016/j.tine.2015.05.001.

Hembree, R. (1990). The nature, effects and relief of mathematics anxiety. *Journal for Research in Mathematics Education*, *21*(1), 33–46. https://doi.org/10.2307/749455.

Herts, J. B., Carrazza, C., Braxton, J., Berkowitz, T., Lawrence, C., & Levine, S. C. (2018). In *Discussion tips improve parents' and children's experiences at math museum. To be presented at Midwestern Psychological Association, Chicago, IL.*

Herts, J. B., Rozek, C. S., Schaeffer, M. W., Berkowitz, T., Stallings, W., Beilock, S., et al. (2017). In *Math anxious parents provide lower quality math instruction. Poster presented at the twenty-ninth Association for Psychological Science (APS) annual convention, Boston, MA, USA.*

Hirsh-Pasek, K., Zosh, J. M., Golinkoff, R. M., Gray, J. H., Robb, M. B., & Kaufman, J. (2015). Putting "education" in educational apps: lessons from the science of learning. *Psychological Science in the Public Interest*, *16*(1), 3–34. https://doi.org/10.1177/1529100615569721.

Hopko, D. R., Ashcraft, M. H., Gute, J., Ruggiero, K. J., & Lewis, C. (1998). Mathematics anxiety and working memory: Support for the existence of a deficient inhibition mechanism. *Journal of anxiety disorders*, *12*(4), 343–355. https://doi.org/10.1016/S0887-6185(98)00019-X.

Hyde, J. S., Else-Quest, N. M., Alibali, M. W., Knuth, E., & Romberg, T. (2006). Mathematics in the home: homework practices and mother-child interactions doing mathematics. *The Journal of Mathematical Behavior*, *25*(2), 136–152. https://doi.org/10.1016/j.jmathb.2006.02.003.

Iverson, J. M., & Goldin-Meadow, S. (2005). Gesture paves the way for language development. *Psychological Science*, *16*(5), 367–371. https://doi.org/10.1111/j.0956-7976.2005.01542.x.

Jordan, N. C., Kaplan, D., Ramineni, C., & Locuniak, M. N. (2008). Development of number combination skill in the early school years: when do finger help? *Developmental Science*, *11*(5), 662–668. https://doi.org/10.1111/j.1467-7687.2008.00715.x.

Jordan, N. C., Levine, S. C., & Huttenlocher, J. (1994). Development of calculation abilities in middle- and low-income children after formal instruction in school. *Journal of Applied Developmental Psychology*, *15*(2), 223–240. https://doi.org/10.1016/0193-3973(94)90014-0.

Laski, E., & Siegler, R. S. (2014). Learning from number board games: you learn what you encode. *Developmental Psychology*, *50*(3), 853–864. https://doi.org/10.1037/a0034321.

Lawton, C. A. (1994). Gender differences in wayfinding strategies: relationship to spatial ability and spatial anxiety. *Sex Roles*, *30*(11–12), 765–779. https://doi.org/10.1007/BF01544230.

LeFevre, J., Clarke, T., & Stringer, A. P. (2002). Influences of language and parental involvement on the development of counting skills: comparisons of French- and English-speaking Canadian children. *Early Child Development and Care*, *172*(3), 283–300. https://doi.org/10.1080/03004430212127.

LeFevre, J. A., Skwarchuk, S. L., Smith-Chant, B. L., Fast, L., Kamawar, D., & Bisanz, J. (2009). Home numeracy experiences and children's math performance in the early school years. *Canadian Journal of Behavioural Science*, *41*(2), 55–66. https://doi.org/10.1037/a0014532.

Levine, S. C., Bradley, A., & Mix, K. S. (under review). Messages to advance family math fluency: The importance of family voices.

Levine, S. C., Huttenlocher, J., Taylor, A., & Langrock, A. (1999). Early sex differences in spatial skill. *Developmental Psychology*, *35*(4), 940–949. https://doi.org/10.1037/0012-1649.35.4.940.

Levine, S. C., Jordan, N. C., & Huttenlocher, J. (1992). Development of calculation abilities in young children. *Journal of Experimental Child Psychology*, *53*(1), 72–93. https://doi.org/10.1016/S0022-0965(05)80005-0.

Levine, S. C., Ratliff, K., Cannon, J., & Huttenlocher, J. (2012). Early puzzle play: a predictor of preschoolers' spatial transformation skill. *Developmental Psychology*, *48*(2), 530–542. https://doi.org/10.1037/a0025913.

Levine, S. C., Suriyakham, L., Rowe, M., Huttenlocher, J., & Gunderson, E. (2010). What counts in the development of children's number knowledge? *Developmental Psychology*, *46*(5), 1309–1313. https://doi.org/10.1037/a0019671.

Li, Y., & Geary, D. C. (2013). Developmental gains in visuospatial memory predict gains in mathematics achievement. *PLoS One*, *8*(7), e70160.

Li, Y., & Geary, D. C. (2017). Children's visuospatial memory predicts mathematics achievement through early adolescence. *PLoS One*, *12*(2), e0172046.

Linn, M. C., & Petersen, A. C. (1985). Emergence and characterization of sex differences in spatial ability: a meta-analysis. *Child Development*, *56*(6), 1479–1498. https://doi.org/10.2307/1130467.

Maloney, E. A., Ramirez, G., Gunderson, E. A., Levine, S. C., & Beilock, S. L. (2015). Intergenerational effects of parents' math anxiety on children's math achievement and anxiety. *Psychological Science*, *26*(9), 1480–1488. https://doi.org/10.1177/0956797615592630.

Manolitsis, G., Georgiou, G. K., & Tziraki, N. (2013). Examining the effects of the home literacy and numeracy environment on early reading and math acquisition. *Early Childhood Research Quarterly*, *28*(4), 692–703. https://doi.org/10.10.16/j.ecresq.2013.05.004.

McGlone, M. S., & Aronson, J. (2006). Stereotype threat, identity salience, and spatial reasoning. *Journal of Applied Developmental Psychology*, *27*(5), 486–493. https://doi.org/10.1016/j.appdev.2006.06.003.

McNeil, N. M., Grandau, L., Knuth, E. J., Alibali, M. W., Stephens, A. S., Hattikudur, S., et al. (2006). Middle-school students' understanding of the equal sign: the books they read can't help. *Cognition and Instruction*, *24*(3), 367–385. https://doi.org/10.1207/s1532690xci2403_3.

Mix, K. S., Levine, S. C., Cheng, Y.-L., Young, C., Hambrick, D. Z., Ping, R., et al. (2016). The latent structure of spatial skills and mathematics across development: highly correlated but separate domains. *Journal of Experimental Psychology: General*, *145*(9), 1206–1227. https://doi.org/10.1037/xge0000182.

Mix, K. S., Levine, S. C., Cheng, Y.-L., Young, C. J., Hambrick, D. Z., & Konstantopoulos, S. (2017). The latent structure of spatial skills and mathematics: a replication of the two-factor model. *Journal of Cognition and Development*, *18*(4), 465–492. https://doi.org/10.1080/15248372.2017.1346658.

Mix, K. S., Sandhofer, C. M., Moore, J. A., & Russell, C. (2012). Acquisition of the cardinal word principle: the role of input. *Early Childhood Research Quarterly*, *27*(2), 274–283. https://doi.org/10.1016/j.ecresq.2011.10.003.

Moeller, K., Fischer, U., Link, T., Wasner, M., Huber, S., Cress, U., et al. (2012). Learning and development of embodied numerosity. *Cognitive Processing*, *13*(1), 271–274. https://doi.org/10.1007/s10339-012-0457-9.

Moore, D. S., & Johnson, S. P. (2008). Mental rotation in human infants: a sex difference. *Psychological Science*, *19*, 1063–1066. https://doi.org/10.1111/j.1467-9280.2008.02200.x.

Namy, L. L., Campbell, A. L., & Tomasello, M. (2004). The changing role of iconicity in non-verbal symbol learning: a U-shaped trajectory in the acquisition of arbitrary gestures. *Journal of Cognition and Development*, *5*(1), 37–57. https://doi.org/10.1207/s15327647jcd0501_3.

Neill, S. (1991). *Classroom nonverbal communication*. London: Routledge.

NICHD Early Child Care Research Network. (1999). Child care and mother-child interaction in the first three years of life. *Developmental Psychology*, *35*(6), 1399–1413. https://doi.org/10.1037/0012-1649.35.6.1399.

Nicoladis, E., Pika, S., & Marentette, P. (2010). Are number gestures easier than number words for preschoolers? *Cognitive Development*, *25*(3), 247–261. https://doi.org/10.1016/j.cogdev.2010.04.001.

Niklas, F., & Schneider, W. (2014). Casting the die before the die is cast: the importance of the home numeracy environment for preschool children. *European Journal of Psychology of Education*, *29*(3), 327–345. https://doi.org/10.1007/s10212-013-0201-6.

Noël, M. P. (2005). Finger gnosia: a predictor of numerical abilities in children? *Child Neuropsychology*, *11*(5), 413–430. https://doi.org/10.1080/09297040590951550.

Novack, M. A., Congdon, E. L., Hemani-Lopez, N., & Goldin-Meadow, S. (2014). From action to abstraction: using the hands to learn math. *Psychological Science*, *25*(4), 903–910. https://doi.org/10.1177/0956797613518351.

Perry, M., Church, R. B., & Goldin-Meadow, S. (1988). Transitional knowledge in the acquisition of concepts. *Cognitive Development*, *3*(4), 359–400. https://doi.org/10.1016/0885-2014(88)90021-4.

Pesenti, M., Thioux, M., Seron, X., & Volder, A. D. (2000). Neuroanatomical substrates of Arabic number processing, numerical comparison, and simple addition: a PET study. *Journal of Cognitive Neuroscience*, *12*(3), 461–479. https://doi.org/10.1162/089892900562273.

Petersen, L.A. & Levine, S.C. (in preparation). Sex differences in early parent-child block play.

Petersen, L. A., & McNeil, N. M. (2013). Effects of perceptually rich manipulatives on preschoolers' counting performance: established knowledge counts. *Child Development*, *84*(3), 1020–1033. https://doi.org/10.1111/cdev.12028.

Petersen, L. A., McNeil, N. M., Hall, C. J., Carrazza, C., Devlin, B. L., Tollaksen, A. K., et al. (2014). In P. Bello, M. Guarini, M. McShane, & B. Scassellati (Eds.), *Counting practice with pictures, but not objects, improves children's understanding of cardinality. Proceedings of the thirty-sixth annual conference of the cognitive science society*. Mahwah, NJ: Erlbaum.

Piazza, M., Mechelli, A., Butterworth, B., & Price, C. J. (2002). Are subitizing and counting implemented as separate or functionally overlapping processes? *NeuroImage*, *15*(2), 435–446. https://doi.org/10.1006/nimg.2001.0980.

Pinel, P., Piazza, M., Le Bihan, D., & Dehaene, S. (2004). Distributed and overlapping cerebral representations of number, size, and luminance during comparative judgments. *Neuron*, *41*(6), 983–993. https://doi.org/10.1016/S0896-6273(04)00107-2.

Potter, M. C., & Levy, E. I. (1968). Spatial enumeration without counting. *Child Development, 39*, 265–272. https://doi.org/10.2307/1127377.

Pruden, S. M., & Levine, S. C. (2017). Parents' spatial language mediates a sex difference in preschoolers' spatial-language use. *Psychological Science, 28*(1), 1583–1596. https://doi.org/10.1177/0956797617711968.

Pruden, S. M., Levine, S. C., & Huttenlocher, J. (2012). Children's spatial thinking: does talk about the spatial world matter? *Developmental Science, 14*(6), 1417–1430. https://doi.org/10.1111/j.1467-7687.2011.01088.

Purpura, D. J., Napoli, A. R., Wehrspann, E. A., & Gold, Z. S. (2017). Causal connections between mathematical language and mathematical knowledge: a dialogic reading intervention. *Journal of Research on Educational Effectiveness, 10*(1), 116–137. https://doi.org/10.1080/19345747.2016.1204639.

Quinn, P. C., & Liben, L. S. (2008). A sex difference in mental rotation in young infants. *Psychological Science, 19*(11), 1067–1070. https://doi.org/10.1111/j.1467-9280.2008.02201.x.

Ramani, G. B., Rowe, M. R., Eason, S., & Leech, K. (2015). Math talk during informal learning activities in head start families. *Cognitive Development, 35*, 15–33. https://doi.org/10.1016/j.cogdev.2014.11.002.

Ramani, G. B., & Siegler, R. S. (2008). Promoting broad and stable improvements in low-income children's numerical knowledge through playing number board games. *Child Development, 79*(2), 375–394. https://doi.org/10.1111/j.1467-8624.2007.01131.x.

Ramirez, G., Gunderson, E. A., Levine, S. C., & Beilock, S. L. (2012). Spatial anxiety relates to spatial abilities as a function of working memory in children. *The Quarterly Journal of Experimental Psychology, 65*(3), 474–487. https://doi.org/10.1080/17470218.2011.616214.

Ramirez, G., Gunderson, E. A., Levine, S. C., & Beilock, S. L. (2013). Math anxiety, working memory and math achievement in early elementary school. *Journal of Cognition and Development, 14*(2), 187–202. https://doi.org/10.1080/15248372.2012.664593.

Ramscar, M., Dye, M., Popick, H. M., & O'Donnell-McCarthy, F. (2011). The enigma of number: why children find the meanings of even small number words hard to learn and how we can help them do better. *PLoS One, 6*(7), e22501. https://doi.org/10.1371/journal.pone.0022501.

Reyna, V. F., Nelson, W., Han, P., & Dieckmann, N. F. (2009). How numeracy influences risk comprehension and medical decision making. *Psychological Bulletin, 135*, 943–973.

Richardson, F. C., & Suinn, R. M. (1972). The mathematics anxiety rating scale: psychometric data. *Journal of Counseling Psychology, 19*(6), 551–554. https://doi.org/10.1037/h0033456.

Roth, W.-M., & Welzel, M. (2001). From activity to gestures and scientific language. *Journal of Research in Science Teaching, 38*(1), 103–136. https://doi.org/10.1002/1098-2736(200101)38:1<103::AID-TEA6>3.0.CO;2-G.

Rozek, C. S., Svoboda, R. C., Harackiewicz, J. M., Hulleman, C. S., & Hyde, J. S. (2017). Utility-value intervention with parents increases students' STEM preparation and career pursuit. *Proceedings of the National Academy of Sciences of the United States of America, 114*, 909–914. https://doi.org/10.1073/pnas.1607386114.

Saxe, G. B. (1977). A developmental analysis of notational counting. *Child Development, 48*, 1512–1520. https://doi.org/10.2307/1128514.

Saxe, G. B., Guberman, S. R., & Gearhart, M. (1987). Social processes in early number development. *Monographs of the Society of Research in Child Development, 52(2)*, I–162. Serial N. 216. https://doi.org/10.2307/1166071.

Saxe, G. B., & Kaplan, R. (1981). Gesture in early counting: a developmental analysis. *Perceptual and Motor Skills, 53*(3), 851–854. https://doi.org/10.2466/pms.1981.53.3.851.

Schaeffer, M. W., Rozek, C. S., Berkowitz, T., & Levine, S. C., & Beilock, S. L. (in press). Disassociating the relation between parents' math anxiety and children's math achievement: Long-term effects of a math app intervention. *Journal of Experimental Psychology: General*.

Schaeffer, M. W., Rozek, C. S., Maloney, E. A., Berkowitz, T., Levine, S. C., & Beilock, S. L. (under review). Math anxious elementary school teachers undermine students' math learning.

Siegler, R., & Shrager, S. (1984). Strategy choice in addition and subtraction: how do children know what to do. In *Origins of cognitive skills* (pp. 229–294). Hillsdale, NJ: Lawrence Erlbaum Associates.

Siegler, R. S., & Jenkins, E. (1989). *How children discover new strategies*. Hillsdale, NJ: Lawrence Erlbaum Associates.

Siegler, R. S., & Ramani, G. B. (2009). Playing linear number board games—but not circular ones—improves low income preschoolers' numerical understanding. *Journal of Experimental Psychology*, *101*(3), 545–560. https://doi.org/10.1037/a0014239.

Simpkins, S. D., Fredricks, J. A., & Eccles, J. S. (2012). Charting the Eccles' expectancy-value model from mothers' beliefs in childhood to youths' activities in adolescence. *Developmental Psychology*, *48*(4), 1019–1032. https://doi.org/10.1037/a0027468.

Skwarchuk, S. L. (2009). How do parents support children's preschool numeracy experiences at home? *Early Childhood Education Journal*, *37*(3), 189–197. https://doi.org/10.1007/s10643-009-0340-1.

Skwarchuk, S. L., Sowinski, C., & LeFevre, J. A. (2014). Formal and informal home learning activities in relation to children's early numeracy and literacy skills: the development of a home numeracy model. *Journal of Experimental Child Psychology*, *121*, 63–84. https://doi.org/10.1016/j.jecp.2013.11.006.

Starkey, P., & Klein, A. (2000). Fostering parental support for children's mathematical development: an intervention with head Start families. *Early Education and Development*, *11*(5), 659–680. https://doi.org/10.1207/s1556693eed1105 7.

Suriyakham, L. W. (2007). *Input effects on the development of the cardinality principle: Does gesture count?* [Doctoral dissertation]. University of Chicago.

Susperreguy, M. I., & Davis-Kean, P. E. (2016). Maternal math talk in the home and math skills in preschool children. *Early Education and Development*, *27*(6), 841–857. https://doi.org/10.1080/10409289.2016.1148480.

Tabors, P. O., Beals, D. E., & Weizman, Z. O. (2001). "You know what oxygen is?" Learning new words at home. In D. K. Dickinson & P. O. Tabors (Eds.), *Beginning literacy with language: Young children learning at home and school* (pp. 93–110). Baltimore: Paul H. Brookes.

U.S. Department of Education (2016). *STEM 2026: A vision for innovation in STEM education*. Washington, DC: Office of Innovation and Improvement.

Uttal, D. H., Meadow, N. G., Tipton, E., Hand, L. L., Alden, A. R., Warren, C., et al. (2012). The malleability of spatial skills: a meta-analysis of training studies. *Psychological Bulletin*, *139*(2), 352–402. https://doi.org/10.1037/a0028446.

Valenzeno, L., Alibali, M. W., & Klatzky, R. (2003). Teachers' gestures facilitate students' learning: a lesson in symmetry. *Contemporary Educational Psychology*, *28*(2), 187–204. https://doi.org/10.1016/S0361-476X(02)00007-3.

Vandermaas-Peeler, M., Boomgarden, E., Finn, L., & Pittard, C. (2012). Parental support of numeracy during a cooking activity with four-year-olds. *International Journal of Early Years Education*, *20*(1), 78–93. https://doi.org/10.1080/09669760.2012.663237.

Vandermaas-Peeler, M., Ferretti, L., & Loving, S. (2012). Playing the lady bug game: parent guidance of young children's numeracy activities. *Early Child Development and Care*, *182*(10), 1289–1307. https://doi.org/10.1080/03004430.2011.609617.

Verdine, B. N., Golinkoff, R. M., Hirsh-Pasek, K., Newcombe, N. S., Filipowicz, A. T., & Chang, A. (2014). Deconstructing building blocks: preschoolers' spatial assembly performance relates to early mathematical skills. *Child Development*, *85*(3), 1062–1076. https://doi.org/10.1111/cdev.12165.

Wai, J., Lubinski, D., & Benbow, C. P. (2009). Spatial ability for STEM domains: aligning over 50 years of cumulative psychological knowledge solidifies its importance. *Journal of Educational Psychology*, *101*(4), 817. https://doi.org/10.1037/a0016127.

Wiese, H. (2003). *Numbers, language, and the human mind.* Cambridge: Cambridge University Press. https://doi.org/10.1017/CBO9780511486562.

Wynn, K. (1990). Children's understanding of counting. *Cognition*, *36*(2), 155–193. https://doi.org/10.1016/0010-0277(90)90003-3.

Young, C., Cartmill, E., Levine, S. C., & Goldin-Meadow, S. (2014). In *Gesture and speech input are interlocking pieces: the development of children's jigsaw puzzle assembly ability. Proceedings of the annual meeting of the cognitive science society, 36.* Retrieved from: https://escholarship.org/uc/item/3h2198h1.

ZenoMath. (2017). *Summary report of 6 discussion groups developed by Zeno and Purple Group.* 15 November 2017.

Zukow-Goldring, P., Romo, L., & Duncan, K. R. (1994). Gestures speak louder than words: achieving consensus in Latino classrooms. In A. Alvarez & P. del Rio (Eds.), *vol. 4. Education as cultural construction: Exploration in socio-cultural studies* (pp. 227–239). Madrid, Spain: Fundacio Infancia y Aprendizage.

Chapter 6

From Cognition to Curriculum to Scale☆

Julie Sarama and Douglas H. Clements
Morgridge College of Education, Marsico Institute, University of Denver, Denver, CO, United States

INTRODUCTION

Most developers and publishers claim that their curricula are based on research, but few explicate their claims. In this chapter, we briefly assess the state of affairs regarding "research-based curricula" and present a model to mitigate weaknesses in the field that is based on coordinated interdisciplinary research ranging from cognitive science to scale-up. We describe an example in early mathematics.

Finding a curriculum that does not claim a research basis is difficult, but these claims are often vacuous, citing theories or empirical results vaguely (Clements, 2007, 2008; Clements & Sarama, 2013; Kinzie, Whittaker, McGuire, Lee, & Virginia, 2015). For example, they often cite research evidence relevant to the beginning or end of the curriculum development process. That is, at the beginning, "research-based" often indicates asserting that the curriculum was built upon broad theoretical frameworks or, with little specificity, "research on students' thinking." Such a research-to-practice model alone is inadequate, because it includes a one-way translation of research results to principles to instructional designs and therefore is often

☆This research was supported by the Institute of Education Sciences, U.S. Department of Education through Grants R305A120813, R305K05157, and R305A110188. The opinions expressed are those of the authors and do not represent views of the U.S. Department of Education. Although the research is concerned with theoretical issues, not particular curricula, a small component of the intervention used in this research have been published by the authors and their collaborators on the project, who thus could have a vested interest in the results. Researchers from an independent institution oversaw the research design, data collection, and analysis and confirmed findings and procedures. The authors wish to express appreciation to the school districts, teachers, and students who participated in this research.

flawed in its presumptions, insensitive to changing goals in the content area, and unable to contribute to a revision of the theory and knowledge on which it is built (Clements, 2007).

At the other end, research validated may mean that effectiveness of the finished curriculum was evaluated. Not only does this leave out critical stages of a scientific research and development process (Battista & Clements, 2000; Clements & Battista, 2000; Doabler et al., 2014), but the research designs are often weak (Munter, Cobb, & Shekell, 2016). In the area of early mathematics, for example, of 78 elementary school programs evaluated, less than 10% had valid evidence of effectiveness and four of those had only "potentially positive" effects on achievement (Doabler et al., 2014). Why this state of affairs continues is explained by the confluence of many factors, such as instrumentalist views of mathematics and knowledge acquisition as simple transmission, a skepticism or rejection of mathematics curricula in the earliest years, lack of standards for curriculum development, a bias against design sciences, such as curriculum development in particular, in academe, and limited involvement of and communication between, relevant parties (for elaboration, see Clements & Sarama, 2013).

This is not to say that there have been no viable attempts to build valid research-based curricula. There are many (for lists of examples, see Clements, 2008; Day-Hess & Clements, 2017). However, they remain relatively small in number and frequently do not explicate the methods and findings of the development process.

To address these weaknesses, close the gap between research and practice, and increase the impact of research on the field (Cai et al., 2017), we need scientific approaches to the conceptualization, design, creation, implementation, and scale-up of curricula that are not just "based on" or "validated by" research but that were constructed, refined, and evaluated with a comprehensive program of research and development (Clements, 2007, 2008; Clements & Sarama, 2013).

THE CURRICULUM RESEARCH FRAMEWORK (CRF)

Based on a review of research and expert practice (Clements, 2008), we constructed and tested a framework for the construct of research-based curricula. The goal was to promote a valid scientific curriculum development program that addresses two basic questions—about effects and conditions—in three domains: practice, policy, and theory. For example, a curriculum development program should address not only the practical question of whether the curriculum is effective in helping children achieve specific learning goals, but also the conditions under which it is effective. Theoretically, the research program should also address why it is effective and why certain sets of conditions decrease or increase the curriculum's effectiveness.

We developed the Curriculum Research Framework (CRF, Clements, 2007), which identifies three categories and ten phases of research and development, along with methods appropriate for each. A core feature of the CRF is that it is grounded in coordinated interdisciplinary research ranging from cognitive science to early childhood and mathematics education to implementation science to scale-up (the final scale-up phrase is complex and has its own elaborated model, Sarama & Clements, 2013).

Each phase must yield positive results to proceed to the next. This process can reveal weaknesses that have to be addressed and reevaluated (or the project halted, saving resources before large-scale evaluations are conducted, most likely yielding little to no benefits). This approach has higher validity than others for the same reason: Construct validity tests are more frequent and more trustworthy. For example, if research on students' thinking and learning in the goal domain is not carefully reviewed or conducted, it is considerably less likely that later phases of development (curricula, professional development, implementation, etc.) will be successful.

The CRF and Early Mathematics

We first implemented the CRF in the field of early mathematics, given its importance (Clements & Sarama, 2014; Sarama & Clements, 2009) and the low use of mathematics curricula in the earliest years of schooling in the United States. For example, US teachers tend to use emerging curricula, whereas those in China use mathematics-specific curricula (Li, Chi, DeBey, & Baroody, 2015).>

The CRF Enacted

As stated, the CRF includes ten phases for asserting that a curriculum is based on research, which can be ordered by the chronology of typical curriculum development, although they are cyclic or recursive (Clements, 2007, 2008; Clements & Sarama, 2013). In the remainder of this section, we briefly describe each phase and then illustrate how we enacted that phase in the *Building Blocks* research and development project, a NSF-funded early childhood mathematics research and curriculum development project that was the first to be based on the CRF.

Category I: A Priori Foundations

The first category is that of *a priori foundations*. Here, the nature of the phase is a focused version of the research-to-practice model. That is, the extant research is reviewed and implications for the nascent curriculum development effort drawn. The questions asked regard what is already known that can be applied to an anticipated curriculum, concerning psychology, education, systemic change, and so forth in general (phase 1), the specific subject matter content, including the role it would play in students' knowledge development (phase 2), and pedagogy, including the effectiveness of certain types of activities (phase 3).

A general guideline across these evaluation phases is that equity issues (Confrey & Lachance, 2000) be considered. For example, considerable thought should be given to the students who are envisioned as users and who participate in field tests; a convenience sample is often inappropriate. Systemic sociocultural issues should be considered as well (Tate, 1997). For *Building Blocks*, we used research on and conducted all field tests with two populations: Children from low-resource communities and children with special needs.

Phase 1. General A Priori Foundation

Developers review broad philosophies, theories, and empirical results on learning and teaching. Based on theory and research on early childhood learning and teaching (Clements & Sarama, 2007a), we determined that Building Blocks' basic approach would be finding the mathematics in, and developing mathematics from, children's activity. That is, we wanted to "mathematize" everyday activities, such as puzzles, songs, moving, and building. For example, teachers might help children mathematize moving their bodies in many ways. Children might count their steps as they walk. They might also move in patterns: step, step, hop; step, step, hop.... They might do both, counting as they walk, "one, two, three, four, five six, ..." These examples show that mathematizing means representing and elaborating everyday activities mathematically. Children create models of everyday situations with mathematical objects, such as numbers and shapes; mathematical actions, such as counting or transforming shapes; and their structural relationships—and use those understandings to solve problems. They learn to increasingly see the world through mathematical lenses.

Phase 2. Subject Matter A Priori Foundation

Developers review research and consult with experts to identify topics that make a substantive contribution to children's mathematical development, are generative in children's development of future mathematical understanding, and are interesting to children. We determined the topics that fit those criteria by considering what mathematics is culturally valued (e.g., standards from domain-specific organizations and states) and empirical research on what constituted the core ideas and skill areas of mathematics for young children (Clements, Sarama, & DiBiase, 2004). We then organized for the development of learning trajectories in the domains of number (subitizing, counting, sequencing, arithmetic), geometry (matching, naming, building, and combining shapes), patterning, and measurement.

Phase 3. Pedagogical A Priori Foundation

Developers review empirical findings on making activities educationally effective—motivating and efficacious—to create general guidelines for the generation of activities. As an example, research using computer software with young children (Clements & Swaminathan, 1995; Sarama & Clements, in press) showed that preschoolers can use computers effectively and that software can be made more effective by employing animation, children's voices, and clear feedback. Although such software is only a small component of the Building Blocks curriculum, it makes a significant contribution, because research was used in its development, giving the developers information on how to make the software targeted and effective.

Another issue that should be considered is for whom the curriculum is intended (e.g., sophisticated reform-oriented teacher, reform-oriented reform teacher, traditional teacher). Is it intended to be "ahead of where the teacher is" or fit the teachers' current practice (Martin A. Simon, personal communication, May 28, 2002)? We planned that *Building Blocks* would be considerably "ahead" of the teachers because most had little preparation in mathematics education. However, to reduce the unfamiliarity for them, we also connected all aspects of the curriculum to typical early childhood educational practice whenever this was consistent with (e.g., traditional scheduling) and especially when it strengthened (e.g., an emphasis on child development and processes, not just products) the mathematics.

Category II: Learning Model and Learning Trajectory

Within the second category is the most extensive and intensive development phase, in which developers' structure activities in accordance with theoretically and empirically based models of children's thinking. This phase involves the creation of research-based learning trajectories—One for every major topic. For this paper, we focus on just one of the many topics from the Building Blocks curriculum, subitizing, or the quick recognition of a number of items in a set without counting, from the Latin "to arrive suddenly."

Phase 4. Structure According to Specific Learning Model and Learning Trajectory

The question is how the curriculum can be constructed to be consistent with, and build upon, students' thinking and learning, which are posited to have characteristics and developmental courses that are not arbitrary, and therefore not equally amenable to various instructional approaches or curricular routes (this is based on our arching theory of hierarchic interactionalism, to which space constraints allow only short references, but see Sarama & Clements, 2009). What distinguishes phase 4 from phase 3, which concerns pedagogical *a prior* foundations, is not only the focus on the child's learning, rather than teaching strategies alone, but also on the iterative nature of its application.

Fig. 1 diagrams the influence of research that creates a chain of development around this core component of learning trajectories (Confrey, Gianopulos, McGowan, Shah, & Belcher, 2017). We first discuss the three rounded rectangles that represent three fields of research that support the development of each initial (or "Hypothetical") Learning Trajectory.

Mathematical progressions contribute to the identification of goals. We posit that worthwhile goals are based on the big ideas of mathematics: those that are mathematically central and coherent, consistent with children's thinking, and generative of future learning (i.e., they are part of a coherent mathematical progression, Clements & Conference Working Group, 2004; van Marle et al., 2018). The fundamental importance of cardinal understanding of whole numbers needs no justification and children's first cardinal meanings for number words may be labels for small sets of subitized objects, even if they determined the labels by counting (Fuson, 1992b; Slusser & Sarnecka, 2011; Steffe, Thompson, & Richards, 1982). Subitizing introduces basic ideas of cardinality—"how many," ideas of "more" and "less," ideas of parts and wholes and their relationships, beginning arithmetic, and, in general, ideas of quantity. Developed well, these are related, forming webs of connected ideas that are the building blocks of mathematics through elementary, middle, and high school and beyond. Finally, the Common Core State Standards explicitly describes subitizing: "Students choose, combine, and apply effective strategies for answering quantitative questions, including quickly recognizing the cardinalities of small sets of objects" (National Governor's Association Center for Best Practices & Council of Chief State School Officers, 2010, p. 9).

Psychological research. Psychologically oriented research, from cognitive science, developmental psychology, education, and so forth, was particularly critical and extensive for the topic of subitizing. Indeed, we note that this research has shown that innate competencies that underlie subitizing are the first quantification mechanisms. For example, children as young as 6 months of age and probably younger are sensitive to number. They habituate to 1 vs. 2 or 3 objects (Antell & Keating, 1983; P. Starkey, Spelke, & Gelman, 1990). For example, shown repeated sets of 3, they eventually "get used to" the number, even as color, size, and arrangements change, and become more attentive only when a set with a different number such as 2 is shown. This indicates that infants are sensitive to small numerosities of a set of items before they are taught number words, counting, or finger patterns. Such research contributes, from a psychological perspective, to the importance of the *goal* of subitizing.

However, the main contribution of psychologically oriented research for all topics is to the creation of the developmental progression of the learning trajectories. Given the importance of this to the creation of a learning trajectory, we review the research in some detail to illustrate that creation

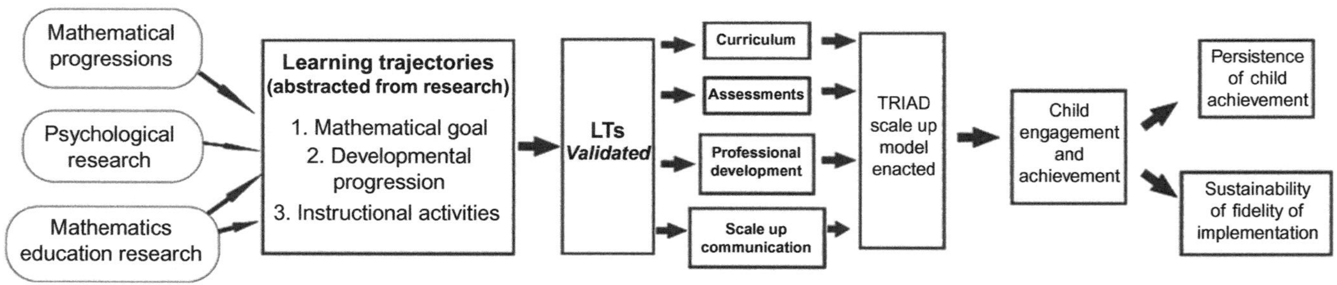

FIG. 1 The logic model for the Phase 4 of the Curriculum Research Framework.

(for comprehensive reviews, see Butterworth, 2010; Clements, Sarama, & Mac Donald, 2017; Hannula, Lepola, & Lehtinen, 2010; Sarama & Clements, 2009). We begin with research on the nature of subitizing.

Subitizing was initially defined in psychology (Kaufman, Lord, Reese, & Volkmann, 1949). Subitizing activity was differentiated from estimation as a unique form of visual number discrimination characterized by speed, exactness, and degree of confidence. For example, individuals identified sets of five or fewer objects quickly (<40 ms/item in a perceptual field) in their recall times, accurately, and with confidence, and had higher accuracy rates than sets of more than five (Kaufman et al., 1949). Klahr found that subitizing did not rely on an encoding process, but in fact was an encoding process, explaining such different recall times when individuals subitized items between one and five. Agreeing that subitizing was a more "basic" skill than counting (Klahr & Wallace, 1976; Schaeffer, Eggleston, & Scott, 1974), Klahr (1973a) hypothesized that after items were encoded through subitizing activity, individuals stored matched patterned stimuli to numerical thinking structures in their long-term memory. Supporting this hypothesis, Fitzhugh (1978) found that some children could subitize sets of one or two but were not able to count them. None of these very young children, however, were able to count any sets that they could not subitize. Fitzhugh concluded that subitizing is a necessary precursor to counting. This research also began to define subitizing as supported by preattentional mechanisms (Klahr, 1973b; Trick & Pylyshyn, 1994) and a form of numerical encoding system (Klahr, 1973a).

However, others questioned whether subitizing is really a numerical process or rather a general sense of quantity. That is, some suggested that infants in "number" experiments may be responding to continuous quantities such as contour length, area, mass, or density rather than discrete number (Feigenson, Carey, & Spelke, 2002; Tan & Bryant, 2000). For example, infants dishabituated to changes in contour length when the number of objects was held constant, but they did not dishabituate to changes in number when contour length was held constant (Clearfield & Mix, 1999). There are, then, various empirical findings and theoretical models. Fig. 2 illustrates several. The "actions-on-objects" of each theory is a depiction of the mental actions, or processes, and what mental representation those processes act upon.

The scan paths view is that subitizing is the recognition of spatial patterns built through motion (Chi & Klahr, 1975; Glasersfeld, 1982; Klahr & Wallace, 1976). Numerical subitizing requires a subsequent reflective abstraction, which occurs when the child abstracts the mental from the sensory-motor contexts and is capable of reflecting on these actions.

The "accumulator" perspective views subitizing as a numerical process enabled by a mental functional equivalent of a number line that operates on both simultaneous and sequential items (cf. Huntley-Fenner, 2001). A pacemaker emits equivalent pulses at a constant rate. When a unitized item (e.g., a block taken as "one") is encountered, a pulse is allowed to pass through a

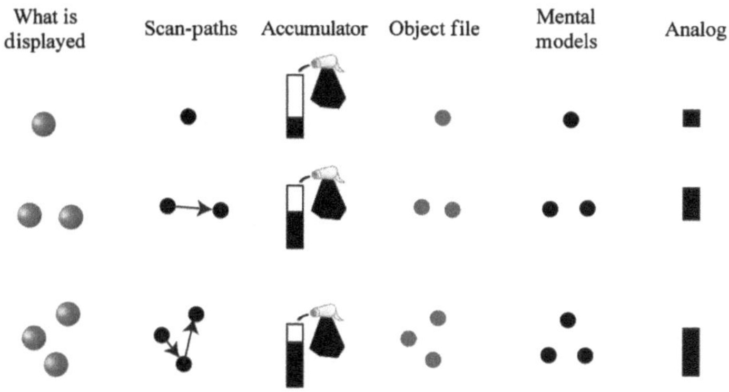

FIG. 2 Illustrations of theoretical models for subitizing. *(Adapted from Clement & Sarama, 2014).*

gate, entering an accumulator—metaphorically like a squirt of water entering a tall glass. The gradations on the accumulator estimate the number in the collection of units, similar to height indicating the numerosity of squirts in the glass (Meck & Church, 1983). This model does not require that the accumulator has an exact representation of number (see also Feigenson, Dehaene, & Spelke, 2004), consistent with research indicating that children younger than 3 years tend not to represent any numbers except 1 and 2 precisely (Antell & Keating, 1983; Baroody, Lai, & Mix, 2005; Feigenson, Carey, & Hauser, 2002; Mix, Huttenlocher, & Levine, 2002).

The next theory in Fig. 2 holds that humans create "object files" that store data on each object's properties and then can use these object files to respond to various situations. Thus some situations can be addressed by using the objects' individuation or separateness as objects, and others can be addressed by using the analog properties of these objects, such as contour length (Feigenson, Carey, & Spelke, 2002). For example, children might use parallel-processed individuation for very small collections, but continuous extent when storage for individuation is exceeded.

The mental models view postulates that children represent numbers nonverbally and approximately, then nonverbally but exactly, and eventually via verbal, counting-based processes (Huttenlocher, Jordan, & Levine, 1994; Mix et al., 2002). Children cannot initially differentiate between discrete and continuous quantities, but represent both approximately using one or more perceptual cues such as contour length (Mix et al., 2002). Children gradually develop the ability to individuate objects, providing the ability to build notions of discrete number. About the age of 2 years, they develop representational, or symbolic, competence, allowing them to create mental models of collections, which they can retain, manipulate (move), add to or subtract from, and so forth (although the model does not adequately describe how cardinality is ultimately cognized and how comparisons are made). The symbolization differentiates this view from the related "object files" theory. Early nonverbal capabilities then provide a basis for the development of verbally based numerical and arithmetic knowledge (young children are more successful on nonverbal than verbal versions of number and arithmetic tasks, Huttenlocher et al., 1994; Jordan, Hanich, & Uberti, 2003; Jordan, Huttenlocher, & Levine, 1992; Jordan, Huttenlocher, & Levine, 1994; Levine, Jordan, & Huttenlocher, 1992).

The abstract model postulates an evolutionarily based module—a distinct mental component that is dedicated to a particular process or task and is unavailable for general processing. A number perception module would process number-based information directly, but only generates approximations of the quantity for quantities above the subitizing range (Dehaene, 1997). This approximate number system is thought to guide the development of children's initial understanding of cardinality (van Marle et al., 2018) and eventually contributes to the development of whole number counting. Researchers have used findings from both humans and nonhuman animals to support this position (Gallistel & Gelman, 2005).

For our learning model, we created a synthesis of these positions. What infants quantify are collections of rigid objects; that is, they do not quantify sequences of sounds and events or materials that are nonrigid and noncohesive such as water (Huntley-Fenner, Carey, & Solimando, 2002). Such quantifications, including number, begin as an undifferentiated, innate notion of "amount." Object individuation, which occurs early in preattentive processing (and is a general, not numerical, process—we are not arguing that all subitizing is preattentive, cf. Moore & Ashcraft, 2015), helps lay the groundwork for differentiating discrete from continuous quantity. Multiple systems are employed, including an object file system that stores information about the objects, some or all of which is used depending on the situation, and an estimator (accumulator) mechanism that stores analog quantitative information only (Feigenson, Carey, & Spelke, 2002; Gordon, 2004; Johnson-Pynn, Ready, & Beran, 2005). This estimator may also include a set of number filters (a cognitive scheme that detects the numerosity of a small group), each tuned to an approximate number of objects (e.g., 2), although they overlap (Nieder, Freedman, & Miller, 2002). The child encountering small sets opens object files for each in parallel. If the situation elicits quantitative comparisons, continuous extent is usually retrieved and used. For example, by about a half-year of age, infants may represent very small numbers (1 or 2) as individuated objects (close to the "mental models" view). However, large numbers in which continuous extent varies or is otherwise not reliable (McCrink & Wynn, 2004) may be processed by the analog estimator as a collection of binary impulses (as are event sequences later in development, see the "analog" column of Fig. 2), but not by exact enumeration (Shuman & Spelke, 2005) by a brain region that processes quantity (size and number, undifferentiated, Pinel, Piazza, Le Bihan, & Dehaene, 2004). Without language support, these are inaccurate processes for numbers above two (Gordon, 2004).

To compare quantities, correspondences are processed. Initially, these are inexact estimates comparing the results of two estimators, depending on the ratio between the sets (Johnson-Pynn et al., 2005). Once the child can represent objects mentally, they can also make exact correspondences between these nonverbal representations, and eventually develop a quantitative notion of that comparison (e.g., 'not just that ●●● is more than ●●', but also that it contains one more ●).

Even these correspondences, however, do not necessarily imply a cardinal representation of the collection (a representation of the collection as a specific numerosity). That is, we still must distinguish between noncardinal representations of a collection and explicit cardinal representations. Indeed, a neuroimaging study found that brain regions that represent numerical magnitude also represent spatial magnitude, such as the relations between sizes of objects, and thus may not be numerical in function (Pinel et al., 2004). Only for the latter does the individual apply an integration operation (Steffe & Cobb, 1988) to create a composite with some numerical index. This integration operation uses present cognitive schemes to project and reorganize actions so they are considered mathematical objects. Some claim that the accumulator yields a cardinal output; however, it may be quantitative, and even numerical in some situations, and still—because it indexes a collection using an abstract, cross-modality system for numerical magnitude (cf. Lourenco & Longo, 2011; Shuman & Spelke, 2005)—it may lack an explicit cardinality. This system would not necessarily differentiate between ordinal and cardinal interpretations. Comparisons, such as correspondence mapping, might still be performed, but only at an implicit level (cf. Sandhofer & Smith, 1999). (It is possible to index a numerical label without attributing explicit cardinality. For example, lower animal species seem to have some perceptual number abilities, but only birds and primates also have shown the ability to connect a perceived quantity with a written mark or auditory label, Davis & Perusse, 1988.) In this view, only with experience representing and naming collections is an explicit cardinal representation created.

In summary, early quantitative abilities exist, but they may not initially constitute systems that can be said to have an explicit number concept. Instead, they may be premathematical, foundational abilities (cf. Clements et al., 2004) that develop and integrate slowly, in a piecemeal fashion (Baroody, Benson, & Lai, 2003). The explicit, cultural, numeral-based sense of number develops in interaction with, but does not replace (indeed, may always be based on, Gallistel & Gelman, 2005) the analog sense of number.

We next turn toward the cognitive *development* of subitizing. A main point is that it does develop. In contrast to what might be expected from a nativist perspective, many innate systems such as subitizing develop based on experience. *Perceptual subitizing* (Clements, 1999; see also theoretical justification in Karmiloff-Smith, 1992) is closest to the original definition of subitizing: Recognizing a number without consciously using other mental or mathematical processes and then naming it. Thus perceptual subitizing employs a preattentional, encoding quantitative process but adds an intentional numerical process; that is, infant sensitivity to number is not (yet) perceptual subitizing. The term "perceptual" applies only to the quantification mechanism as phenomenologically experienced by the person; the intentional numerical labeling, of course, makes the cognitive act conceptual.

A main aspect of the developmental progression for subitizing is, of course, the size of the numbers. As we have seen, children first perceptually subitize sets of only one or two accurately. Indeed, in laboratory research, children can accurately differentiate one from "more than one" at about 33 months of age (Wynn, 1992). Between 35 and 37 months, they differentiate between one and two, but not larger numbers. A few months later, at 38–40 months, they identify three as well. After about 42 months, they can subitize all numbers that they can count, four and higher, at about the same time. However, research in natural, child-initiated settings shows that the development of these abilities can occur much earlier, with children working on one and two around their second birthdays or earlier (Mix, Sandhofer, & Baroody, 2005). Further, some children may begin with "two" rather than "one" (Spelke, 2017). These studies suggest that language and social interactions interact with internal factors in development, as well as showing that number knowledge develops in levels, over time (see also Gordon, 2004). Most studies suggest that children begin recognizing "one," then "one" and "two," then "three" and then "four," whereupon they learn to count and know other numbers (Le Corre, Van de Walle, Brannon, & Carey, 2006).

Complementary research from the field of mathematics education has documented that most kindergartners appear to have good competence recognizing two and three, with most recognizing four and some recognizing higher numbers (note that different tasks were used, some of which did not limit time, which we call "recognition of number" rather than subitizing, so wide ranges are expected). A recent study of low-income children beginning pre-K, using a short-exposure subitizing task, report 2%–14% accuracy for 3, 0%–5% for 4, and virtually no competence with 5, 8, or 10 (Sarama & Clements, 2011). Thus children appear to be most confident with very small numbers, but those from less advantaged environments may not achieve the same skills levels as their more advantaged peers. Some special populations find subitizing particularly difficult. Only a minority (31%) of children with moderate mental handicaps (chronological ages 6–14 years) and a slight majority (59%) of children with mild mental handicaps (ages 6–13) successfully subitize sets of three and four (Baroody, 1986; see also Butterworth, 2010). Some children with learning disabilities could not subitize even at 10 years of age (Chu, van Marle, & Geary, 2013; Geary, Hoard, & Bailey, 2012; Koontz & Berch, 1996). Early deficits in spatial pattern recognition may be the source of difficulty (Ashkenazi, Mark-Zigdon, & Henik, 2013). Subitizing in preschool is a better predictor of later mathematics success for children with ASD (autism spectrum disorder) than for typically developing children (Titeca, Roeyers, Josephy, Ceulemans, & Desoete, 2014).

A second type of subitizing, *conceptual subitizing* (Clements, 1999), plays an advanced organizing role, as seeing "10" on a pair of dice by recognizing the two collections (via perceptual subitizing) and composing them as units of units (Steffe & Cobb, 1988). There is empirical evidence for this distinction (see also the later term groupitizing in G. S. Starkey & McCandliss, 2014; Trick & Pylyshyn, 1994). Some research suggests that only the smallest numbers, perhaps up to three, are actually perceptually recognized; thus sets of 1–3 may be perceptually recognized, sets of 3 to about 6 may be composed (i.e., subitized sets are combined) with the person not necessarily aware of the process. Conceptual subitizing as we use the term refers to recognition in which the person uses partitioning strategies explicitly. Composing and decomposing are combining and separating operations that help children develop generalized part-whole relations, one of the most important accomplishments in arithmetic (National Research Council, 2001).

MacDonald and Wilkins (2016) elaborated on types of subitizing that may serve as a transition from perceptual to conceptual subitizing, in which children justified a total number (such as five) "because I saw three and two," but without and explicit use of the subgroups nor a recognition of the part-whole relationships involved. They also identified two types of conceptual subitizing. In rigid conceptual subitizing, children may subitize "two, two, and one" each time they are shown a wide variety of "five." They then develop flexible conceptual subitizing, as they build different ways of composing items (e.g., two and three; four and one).

The spatial arrangement of sets also influences how difficult they are to subitize (Mandler & Shebo, 1982). People from the primary grades to college usually find rectangular arrangements easiest, followed by linear, circular, and scrambled arrangements (Beckwith & Restle, 1966; Wang, Resnick, & Boozer, 1971). The only change across these ages is rectangular arrangements were much faster for the oldest students, who could multiply. Certain arrangements are easier for specific numbers. Arrangements yielding a better "fit" for a given number are easier (Brownell, 1928). Children make fewer errors for ten dots than for eight with the "domino five" arrangement, but fewer errors for eight dots for the "domino four" arrangement.

For young children, however, neither of these arrangements is easier for any number of dots. Indeed, children 2–4 years old show no differences between any arrangements of four or fewer items (Potter & Levy, 1968). For larger numbers, the linear arrangements are easier than rectangular arrangements. It may be that many preschool children do not use decomposing (conceptual subitizing). They can learn to conceptually subitize, though older research suggested that first graders' limit for subitizing scrambled arrangements is about four or five (Dawson, 1953). If the arrangement does not lend itself to grouping, people of any age have more difficulty with larger sets (Brownell, 1928). They also take more time with larger sets (Beckwith & Restle, 1966).

Another categorization involves the different types of things people can subitize. Spatial patterns such as those on dice are just one type. Other patterned modalities include finger and rhythmic patterns.

Fig. 3 presents the subitizing trajectory; here we focus on the first three columns that document the developmental progression, including the usual age of acquisition (first column; importantly-probably without effective interventions), and descriptions and illustrations (second column) of each of the levels of thinking, and the mental actions on objects we hypothesize account for each level of thinking.

Educationally oriented research contributes to the third component of learning trajectories, provision of instructional experiences. Regardless of the precise mental processes in the earliest years, subitizing appears to be phenomenologically distinct from counting and other means of quantification and deserves differentiated educational consideration. Supporting this assertion, there is little or no relationship between children's performance on counting and subitizing tasks (Pepper & Hunting, 1998) and educational interventions have been shown to be effective (Clements & Sarama, 2008, 2013; Hannula, 2005), the topic to which we now turn.

What kinds of educational experiences are most effective? We first consider informal settings and strategies. Frequent naming of the numbers in groups appears essential. That is, the human facility with language links relations between different representations and thus making early premathematical cognition numeric (Gordon, 2004; Wiese, 2003). Even early grammatical structures for plurals ("dogs") and quantifiers ("some," "all," and "a") provide a framework that allows quantitative ideas to develop (Carey, 2004). More directly, experience naming collections helps children create explicit cardinal representations. This is a prolonged process. Children may initially make word-word mappings between requests for counting or numbers (e.g., "how many?") to number words only after they have learned several number words (Sandhofer & Smith, 1999). Then they label some (small number) cardinal situations with the corresponding number word, that is, map the number word to the numerosity property of the collection. They begin this phase even before 2 years of age, but for some time, this applies mainly to the word "two," a bit less to "one," and with considerable less frequency, "three" and "four" (Fuson, 1992a; Wagner & Walters, 1982). Thus meaningful learning of number words may cause the transition to exact numerical representations (Baroody et al., 2005). This may provide the basis for understanding cardinality and other counting principles, as well as arithmetic ideas (Geary et al., 2012, 2018).

Such experiences promote not only subitizing competence, but motivation and habit of applying it. Such a tendency to spontaneously focus on numerosity is a distinct, mathematically significant process (Hannula, 2005;

The *goal* connects learning trajectories to the important *big ideas* in mathematics. The relevant big idea is that numbers can be used to tell how many, describe order, and measure; they involve numerous relations and can be represented in various ways. The specific goals include understanding that subitizing can be used to tell how many (perceptual subitizing), that a quantity can consist of parts and can be broken apart (decomposed) into the parts, and that those parts can be combined (composed) to form the whole (conceptual subitizing). Specific objectives are for children to achieve those understandings and fluency in perceptual and conceptual subitizing.

Age* (yrs)	Developmental progression	Cognitive actions on objects	Sample instructional tasks
0–1	**Preexplicit number.** Within the first year, the child is sensitive to small numbers, but does not have explicit, intentional knowledge of number.	An initial bootstrap: Implicit sensitivity to quantity with perceptual input. An object file system stores information about the objects (some or all of which is used depending on the situation). An estimator (accumulator-type) mechanism stores analog quantitative information. If the situation elicits quantitative comparisons, continuous extent is retrieved and used except in extreme circumstances (e.g., 1 and maybe 2 are processed as individuated objects; numbers in which continuous extent varies or is otherwise not reliable are processed by the analog estimator as a collection of binary impulses. To compare, bootstrap processes make a correspondence between estimators.	*Noticing collections*. Provide a rich sensory environment, use words such as *more,* and use actions of adding objects.
1–2	**Small-collection namer**. Names groups of 1 to 2, sometimes 3. Shown a pair of shoes, says, "Two shoes."	Mental schemes (number filters) act on perceptions of collections of 1-3. Eventually, mental, nonverbal representations are developed of each object in such collections. The become	*Board games–small numbers*. Play board games with a special die (number cube) or spinner that shows only 1, 2, and 3 dots (then add 0 to it).

		exact and associated with the verbal labels ("two" then "one" then "three," etc.). To compare, two such representations can be placed in mental correspondence.	
3	**Maker of small collections** . Nonverbally makes a small collection (no more than 4, usually 1–3) with the same number as another collection via mental model (i.e., not necessarily by matching). Might also be verbal. When shown a collection of 3, makes another collection of 3.	Mental representations can be maintained and direct physical actions so that one perceived object corresponds to each represented object.	*Get the number*. Ask children to get the right number of crackers or some other item for a small number of children.
4	**Perceptual subitizer to 4.** Instantly recognizes collections up to 4 briefly shown, and verbally names the number of items. When shown objects briefly, says, "Four."	Schemes act on perceptual input (including collections, but also sounds, etc.) to identify sets of 0 to 4 (schemes may use lower-level schemes for 1-3 and combine them to recognize 4), each of which is associated with the verbal number name.	*Snapshots*. Play "Snapshots" with collections of one to four objects, arranged in line or other simple arrangement, asking children to respond verbally with the number name. Start with the smaller quantities and easier arrangements, moving to those of moderate difficulty only when children are fully competent and confident. 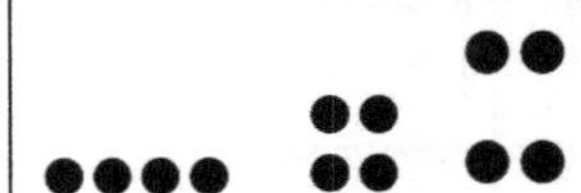 easy medium transitional
5	**Perceptual subitizer to 5.** Instantly recognizes briefly shown collections up to 5 items	Schemes act on perceptual input (including collections, but also sounds, etc.) to identify sets of 0 to 5 (schemes may use	*Snapshots*: Play "Snapshots" with dot cards, starting with easy arrangements,

	and verbally names the number of items. Recognizes and uses spatial and numeric structures beyond the situations in which they were already experienced. Shown 5 objects briefly, says, "Five."	lower-level schemes for 1-3 and combine them to recognize 4; note that visual patterns used in the following levels have beginning to be constructed), each of which is associated with the verbal number name.	moving to more difficult arrangements when children are able to do so.
5	**Conceptual subitizer to 5**. Verbally labels all arrangements to about 5, when shown only briefly. Asked "Why?" says, "Because I saw 3 and 2, and so I said 5."	An executive process determines whether an existing scheme can quantify the perceptual input; if so, that is used. If not, gestalt visual principals are used to partition the collection to identify two or more sets that existing schemes can quantify; the results of these are combined with pattern matching to known compositions.	*Snapshots*. Use different arrangements of the various modifications of "Snapshots" to develop conceptual subitizing and ideas about addition and subtraction. The goal is to encourage children to see the addends and the sum. 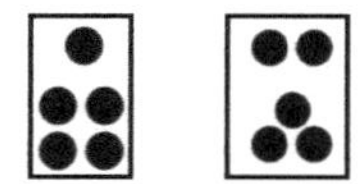
5	**Conceptual subitizer to 10**. Verbally labels most briefly shown arrangements to 6, then up to 10, using groups. "In my mind, I made two groups of 3 and one more, so 7."	As in the previous level, with the addition of other compositions.	*Snapshots*. Play "Snapshots" with larger quantities to develop ideas about addition and subtraction. 
6	**Conceptual subitizer to 20**. Verbally labels structured arrangements up to 20, shown only briefly, using groups. "I saw three 5s, so 5, 10, 15."	As in the previous level, with the addition of other compositions and explicit knowledge of teens as ten and some number more.	*Ten frame addition snapshots*: Briefly show 2 ten frames to help children visualize addition combinations. 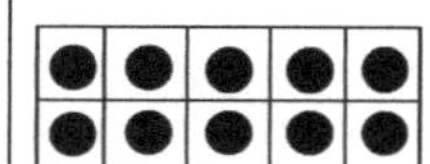

			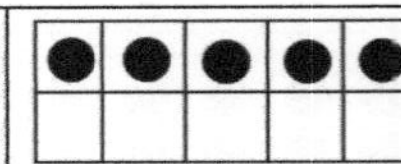
7	**Conceptual subitizer with place value and skip counting.** Verbally labels structured arrangements, shown only briefly, using groups, skip counting, and place value. "I saw groups of 10s and 2s, so 10, 20, 30, 40, 42, 44,. . . 44!"	As in the previous level, with the addition of other compositions and explicit place value knowledge.	*Ten frame addition snapshots*. Briefly show several ten frames. 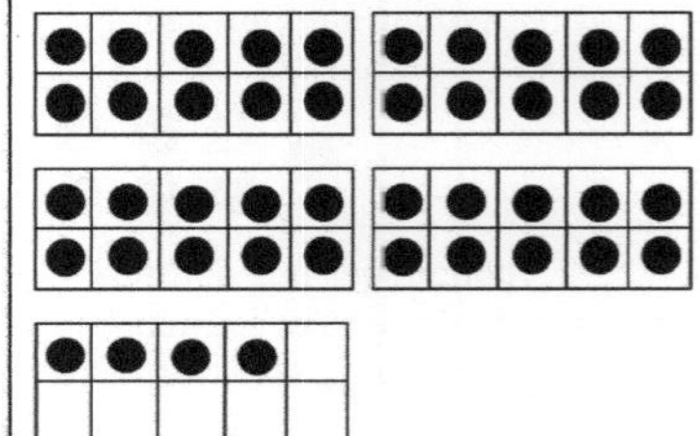
8	**Conceptual subitizer with place value and multiplication.** Verbally labels structured arrangements shown only briefly, using groups, multiplication, and place value. "I saw groups of 10s and 3s, so I thought, 3 tens is 30 and 4 threes is 12, so 42 in all."	As in the previous level, with the addition of other compositions and explicit knowledge of multiplication and place value.	*Snapshots with dots*. Play "Snapshots" with structured groups that support the use of increasingly sophisticated mental strategies and operations. 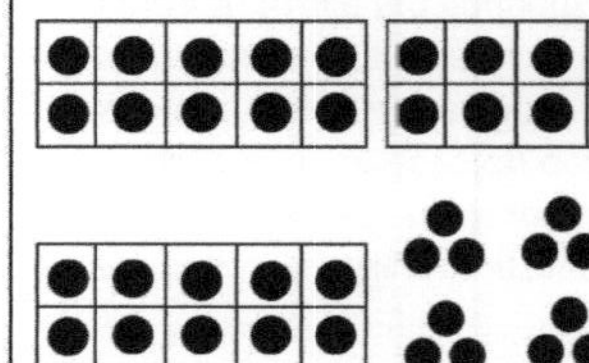

*The ages are estimates based on extant research and are provided only as a general guide—age levels are strongly dependent on experience.

FIG. 3 A learning trajectory for recognition of number and subitizing.

Hannula et al., 2010). Children's failure to focus on numerosity is not due to their lack of competences related to task demands (Lehtinen & Hannula, 2006). They simply have not developed the habit of focusing on numerosity, which is why it is important that all caregivers use number words throughout the day (LeFevre, Fast, et al., 2009; LeFevre, Skwarchuk, et al., 2009).

Turning to more formal school contexts, our theory of hierarchic interactionalism posits that "Instruction based on learning consistent with natural developmental progressions is more effective, efficient, and generative for the child than learning that does not follow these paths" (Sarama & Clements, 2009, p. 24). Therefore our learning trajectory moves from small numbers to greater numbers, but also from tasks that encourage mastery of perceptual subitizing before conceptual subitizing. Within each broad level, tasks following the research on types of objects and spatial arrangements of those objects (see the last column in Fig. 3).

To elaborate, because the goal is for the child to perceive the group as a quantity, we discourage the use of perceptually distinct objects (some teachers say, "I will use a yellow, a blue, and a red counter to help them remember"). We instead promote clear object-background contrast with identical objects (e.g., block circles on white background).

For arrangements, we follow the research described previously: With preschoolers, we begin with linear arrangements up to four or five, then move to rectangular arrangements, as well as dice and domino patterns.

Transitions to larger numbers are carefully planned. For instance, Logan and Zbrodoff (2003) found that the space between these groups of "twos" and "threes" afforded individuals more effective subitizing of four or more items (cf. Gebuis & Reynvoet, 2011), providing the transitional arrangements discovered by MacDonald and Wilkins (2016).

To continue to promote conceptual subitizing, we present arrangements that separate small sets, such as two and two separated symmetrically to see the set as "four" but also "two and two." Symmetrical orientations appear to "free" children's working memory resources. Individuals' subitizing activity also has been found to be affected by the space between the items (Gebuis & Reynvoet, 2011); increasing this space may support young children (MacDonald, 2015; MacDonald & Shumway, 2016; MacDonald & Wilkins, 2016, 2017). We present many arrangements of each number that suggest different partitions of that number and move to extensive use of "five and ten frames."

Computer technology can also contribute, especially software that moves children forward (or backward) along the learning trajectory automatically based on children's performance (Clements & Sarama, 2007/2016). Tasks can be designed to fit the trajectory precisely. Errors are immediately detected and feedback given (Fig. 3; see other examples a learning trajectories.org). In these ways, work with computers can provide a unique and substantial contribution to children's learning. Teaching and learning also should consider a wide variety of social and contextual aspects (Aguirre et al., 2017; Confrey & Lachance, 2000; Secada, 1992; Tate, 1997).

Formative evaluation and validation of the learning trajectories (see Fig. 1) involves multiple methodologies, including consistency with extant research as just presented, statistical tests such as Rasch modeling (Clements, Sarama, & Liu, 2008; Sarama & Clements, 2011), and teaching experiments (Clements & Sarama, 1999; Sarama & Clements, 2004). All these were used in validating the learning trajectory in Fig. 3. Of course, formative evaluation of each learning trajectory continues through the following four phases.

Category III: Evaluation

The third category, evaluation, includes phases in which empirical evidence is collected to evaluate the learning trajectories and the synthesis of them into the complete curriculum. The objective is to evaluate the appeal, usability, and effectiveness of instantiations of the curriculum. Phase 5 focuses on questions of marketability. Phases 6–8 involve formative research, asking whether the curriculum is usable by, and effective with, teachers and students in expanding social contexts (with teachers familiar, and then new, to the materials), and, especially, how the curriculum can be improved.

Phase 5. Market Research

Market research is usually considered commercially oriented research about the customer, what the customer wants, and what they will buy. Typically, prototype materials are presented to "focus groups." Publishers' names and the results are hidden. Instead, we reveal all interviews and surveys we give to teachers, including their reaction to the subitizing learning trajectory—especially the activities.

The following three phases are also forms of formative evaluation. In contrast to market research, these phases often involve repeated test-and-revise cycles.

Phase 6. Formative Research: Small Group

Pilot testing with individuals or small groups of students is to be conducted on curricular components (e.g., a particular activity or software environment) or on sequences such as multiple subitizing activities (e.g., sequences of contiguous levels from Fig. 3). Early interpretive work evaluates components using a mix of model (or hypothesis) testing and model generation strategies, including clinical interviews, teaching experiences, and microethnographic approaches. The objective is to understand the meaning that students give to the curriculum objects and activities.

Evaluation focuses on consonance between the actions of the students and the learning model. If there are discrepancies, either the mental model, or the way in which this model is instantiated in the curriculum, should be altered. In all cases, are students' actions-on-objects enactments of their cognitive operations in the way the model posits are the focus. For example, do children engage in the activity Snapshots at the level of *Conceptual Subitizer to 5* verbally label the whole and then the parts in ways the arrangements suggested? We found that largely so, although we found it useful if teachers encouraged children to name "different ways you could see it." Are children who experience activities at level n more prepared to engage with tasks at level $n + 1$? This we found to be consistent with observations of children's behaviors; further, although some children showed signs of developing two levels during the same period of time, few "skipped" more than two. Thus developers use the cognitive model and learning trajectories as guides, and the software and activities as catalysts, to create more refined models of particular students or groups of students. Simultaneously, the developer describes what elements of the teaching and learning environment were observed as having contributed to, or hindered, student learning. As an example for subitizing, the *Snapshots* activity was written to have teachers secretly hide a number of objects under a cloth. We found two problems: (a) Teachers found it took too long to prepare each set, and (b) more important, we observed many teachers placing the objects one at a time—and children would count. This harmed their subitizing. Both problems were solved by switching to having patterns premade (with black circle stampers) on large numbers of paper plates.

Often this is the most iterative research-design phase; sometimes evaluation and redesign may cycle in quick succession, within a week to prepare modifications for another classroom, and sometimes as much as every 24 hours. Activities may be completed reconstituted, with edited or newly created ones tried the next day.

With so many research and development processes happening, and so many possibilities, extensive documentation is required. Documentation must allow researchers to relate findings to specific components and characteristics of the curriculum. Video or audio recordings (for later microgenetic analysis) and field notes are collected. This documentation also helps evaluate those components of the design that were based on intuition, esthetics, and subconscious beliefs.

Phase 7. Formative Research: Single Classroom

Although teachers are ideally involved in all phases, a special emphasis here is the process of curricular enactment. For example, a goal of the curriculum may be to help teachers interpret students' thinking about the activities and the content they are designed to teach; support teachers' learning of that content, especially which is probably new to teachers; and provide guidance regarding the external representation of content that the materials use. So, classroom-based teaching experiments help track and evaluate student learning, with the goal of making sense of the curricular activities as they are experienced by individual students. At the same time, the class is observed for information concerning the usability and effectiveness of the curriculum. Ethnographic participant observation is used heavily because we wish to research the teacher and students as they interact. Thus the focus is on how the materials are used and how the teacher guides students through the activities. This phase often involves teachers working closely with the developers. That is, the class may be taught either by a team including one of the developers and the teacher, or by a teacher familiar with and intensively involved in curricula development. The goal is to examine learning in the context of the curriculum with teachers who can enact it with a high fidelity of implementation, as opposed to ascertaining how the curriculum works in classrooms in general, which is one focus of the next phase.

As one simple example for subitizing, we observed that although children were recognizing visual arrangements adequately, they seemed unable to transfer this ability to other activities (e.g., in a patterning activity in which children repeatedly "stomp, stomp, stomp, jump.... stomp, stomp, stomp, jump...," not all children recognized the number of stomps). Therefore we mixed in temporal and kinesthetic subitizing tasks as soon as children gained competence with the visual presentations of a given number. This is significant because creating and using these patterns through conceptual subitizing helps children develop abstract number and arithmetic strategies. For example, children use temporal patterns when counting on. "I knew there were three more so I just said, nine ... ten, eleven, twelve" (rhythmically gesturing three times, one "beat" with each count). They use finger patterns to figure out addition problems. For example, for 3 + 2, a child might put up a finger pattern they know as three, then put up two more (rhythmically—up, up) and then recognize the resulting finger pattern as "five." Children who cannot subitize are handicapped in learning such arithmetic processes (Butterworth, 2010; Hannula et al., 2010).

Phase 8. Formative Research: Multiple Classrooms

In several classrooms, the entire class is observed for information about the effectiveness and usability of the curriculum, but more emphasis is placed on the usability by such teachers. Innovative materials often provide less support for teachers than the traditional materials with which they are familiar (Burkhardt, Fraser, & Ridgway, 1990), so this phase is especially important for curricula that are different than what teachers are used to. The goal of this phase is to ascertain if the supporting materials are flexible enough to support multiple situations, various modes of instruction, and different modes and styles of management. A new use of the Snapshot activity was created during this phase as we observed creative teachers placing large subitizing cards near the doorway to as children were lining up.

The next two phases involve summative research, with the goal of evaluating the effectiveness (e.g., in affecting teaching practices and ultimately student learning) of the curriculum, now in its complete form, as it is implemented in realistic contexts. These two phases (9 and 10) differ from each other most markedly on the characteristic of scale.

Phase 9. Summative Research: Small Scale

This phase evaluates what can actually be achieved with typical teachers under realistic circumstances (Burkhardt et al., 1990) and may overlap in practice with phase 8. In multiple classrooms (2 to about 10), pre- and posttest (standardized instruments), experimental or quasi-experimental designs using measures of learning are often used, in conjunction with, and to complement, methodologies previously described. Relevant here is that the Building Blocks group not only outperformed the control group with large effects (0.85 SD for the number test), but that the largest of all the effects was on subitizing (1.56 SD, Clements & Sarama, 2007b).

Phase 10. Summative Research: Large Scale

With any curriculum, but especially one that differs from the familiar for teachers, evaluations must be conducted on a large scale (after considering issues of ethics and practical consequences, see Lester Jr. & Wiliam, 2002). Such research should use an embedded mixed methods design with as broad set of instruments to assess the impact of the implementation on participating children, teachers, program administrators, and parents, as well as document the fidelity of the implementation of the curriculum across diverse contexts. The objective should be to measure and analyze the critical variables, including contextual variables (e.g., settings, such as urban/suburban/rural; type of program; class size; teacher characteristics; student/family characteristics) and implementation variables (e.g., engagement in professional development opportunities; fidelity of implementation; leadership, such as principal leadership, as well as support and availability of resources, funds, and time; peer relations at the school; "convergent perspectives" of the developers, school administrators, and teachers in a cohort; and incentives used). A randomized experiment might be used to provide an assessment of the average impact of being exposed to the curriculum with embedded qualitative analyses.

We have created a model (TRIAD, for Technology-enhanced, research-based instruction, assessment, and professional development) and have evaluated it extensively with positive results, although effects decline longitudinally (Clements, Sarama, Spitler, Lange, & Wolfe, 2011; Sarama & Clements, 2013).

The TRIAD model holds that professional development should be ongoing, intentional, reflective, focused on mathematics content knowledge and children's thinking and on learning trajectories, grounded in particular curriculum materials, situated in the classroom and the school. To realize this, we conducted both repeated (e.g., >10) full-day sessions of training in regular meetings and frequent coaching. Training included all three components of each learning trajectory, the goal, the developmental progression, and the instructional activities and strategies (as in Fig. 3). To understand the goal, teachers study the mathematical content; examples for subitizing include the concept of cardinality and partitioning numbers. A key instructional use of learning trajectories is in formative assessment along the developmental progression. We worked with teachers to study the developmental progression for subitizing, analyze multiple video segments illustrating each level and discuss the mental "actions on objects" that constitute the defining cognitive components of each level; order tasks corresponding to those levels; and practice diagnosis in teams, with a couple of teachers exemplifying behaviors of children at different levels, and one teacher identifying the level of each (we used an online application; an update to it can be seen at learningtrajectories.org). Further, teachers need training in understanding, administering, and especially using data from new assessment strategies (Foorman, Santi, & Berger, 2007). TRIAD training focuses mainly on the curriculum-embedded assessment of Building Blocks' Small Group Record Sheets.

Formative assessment requires more than identifying children's levels of thinking. Teachers must select and modify instructional activities and strategies that are appropriate and effective for each level. To learn about instructional tasks and strategies, teachers practice the curriculum's activities, but also analyze them to establish and justify their connection to a particular level of the developmental progression (as in Fig. 3).

Across all forms of professional development, the focus is on children's thinking and learning. Conversations about children learning serves as way to address implementation issues. Although early mathematics is often an uncomfortable topic for early childhood educators, the newness of the learning trajectories for all participants helps establish a sense of shared learning and community. Each session in the last third of professional development includes scheduled time to discuss "learning stories" (Perry, Dockett, & Harley, 2007). Teachers show their record keeping on small group record sheets, and sometimes videos, and discuss their use of learning trajectories in teaching children, including challenges, questions, and successes. These discussions promote peer learning and collaboration and also motivate peers to solve implementation difficulties.

CONCLUSIONS AND FUTURE DIRECTIONS

We described our Curriculum Research Framework and illustrated its instantiation with the Building Blocks curriculum, focusing on the goal of subitizing. As diagrammed in Fig. 1, we believe the success it had stemmed from the complementary research bases, and especially the establishment of a set of cognitively grounded learning trajectories that contributed to all aspects of the projects. Future research should critically evaluate the veracity of these beliefs.

Future research and development might also evaluate the CRF's implementation with other grade levels and other topics (see Doabler et al., 2014;

Kinzie et al., 2015). Just as important, our research designs could not identify which components of the CRF and TRIAD models and their instantiations are critical. Such research would be theoretically and practically useful.

The specific contribution of the learning trajectories per se needs to be disentangled and identified. In a present project, we are addressing this issue. In an IES-funded project entitled, "Evaluating the Efficacy of Learning Trajectories in Early Mathematics", we are testing the efficacy of learning trajectories in a series of eight randomized clinical trials testing different aspects of LTs. These experiments will determine whether LTs are more efficacious than other approaches in supporting young children's learning.

On a practical side, with funding from the Heising-Simons Foundation and the Bill and Melinda Gates Foundation, we are developing a technology-based tool for teachers and teacher trainers that extends a resource we created for the TRIAD evaluation. The Learning and Teaching with Learning Trajectories tool, or LT^2, is a new, free resource for early mathematics (see www.LearningTrajectories.org). LT^2 provides learning-trajectories-based math resources for teachers, caregivers, and parents. LT^2 runs on all technological platforms, addresses new ages—birth to age 8 years—and including new alignments with standards and assessments, as well as new software for children from—their everyday activities, including art, stories, puzzles, and games.

REFERENCES

Aguirre, J., Herbel-Eisenmann, B. A., Celedón-Pattichis, S., Civil, M., Wilkerson, T., Stephan, M., et al. (2017). Equity within mathematics education research as a political act: moving from choice to intentional collective professional responsibility. *Journal for Research in Mathematics Education, 48*(2), 124–147.

Antell, S. E., & Keating, D. P. (1983). Perception of numerical invariance in neonates. *Child Development, 54*, 695–701.

Ashkenazi, S., Mark-Zigdon, N., & Henik, A. (2013). Do subitizing deficits in developmental dyscalculia involve pattern recognition weakness? *Developmental Science, 16*(1), 35–46. https://doi.org/10.1111/j.1467-7687.2012.01190.x.

Baroody, A. J. (1986). Counting ability of moderately and mildly handicapped children. *Education and Training of the Mentally Retarded, 21*, 289–300. December.

Baroody, A. J., Benson, A. P., & Lai, M.-L. (2003). *Early number and arithmetic sense: A summary of three studies.* Tampa, FL.

Baroody, A. J., Lai, M.-L., & Mix, K. S. (2005, December). In *Changing views of young children's numerical and arithmetic competencies. Paper presented at the National Association for the Education of Young Children, Washington, DC.*

Battista, M. T., & Clements, D. H. (2000). Mathematics curriculum development as a scientific endeavor. In A. E. Kelly & R. A. Lesh (Eds.), *Handbook of research design in mathematics and science education* (pp. 737–760). Mahwah, NJ: Erlbaum.

Beckwith, M., & Restle, F. (1966). Process of enumeration. *Journal of Educational Research, 73*, 437–443.

Brownell, W. A. (1928). *The development of children's number ideas in the primary grades.* Chicago: Department of Education, University of Chicago.

Burkhardt, H., Fraser, R., & Ridgway, J. (1990). The dynamics of curriculum change. In I. Wirszup & R. Streit (Eds.), *Vol. 2. Developments in school mathematics around the world* (pp. 3–30). Reston, VA: National Council of Teachers of Mathematics.

Butterworth, B. (2010). Foundational numerical capacities and the origins of dyscalculia. *Trends in Cognitive Sciences, 14*, 534–541.

Cai, J., Morris, A., Hohensee, C., Hwang, S., Robison, V., & Hiebert, J. (2017). Making classroom implementation an integral part of research. *Journal for Research in Mathematics Education, 48*(4), 342–347.

Carey, S. (2004). Bootstrapping and the origin of concepts. *Daedulus, 133*(1), 59–68. https://doi.org/10.1162/001152604772746701.

Chi, M. T. H., & Klahr, D. (1975). Span and rate of apprehension in children and adults. *Journal of Experimental Child Psychology, 19*, 434–439.

Chu, F. W., van Marle, K., & Geary, D. C. (2013). Quantitative deficits of preschool children at risk for mathematical learning disability. *Frontiers in Psychology, 4*, 195. https://doi.org/10.3389/fpsyg.2013.00195.

Clearfield, M. W., & Mix, K. S. (1999). *Infants use contour length—not number—to discriminate small visual sets*. Paper presented at the Society for Research in Child Development, Albuquerque, NM.

Clements, D. H. (1999). Subitizing: what is it? Why teach it? *Teaching Children Mathematics, 5*, 400–405.

Clements, D. H. (2007). Curriculum research: toward a framework for 'research-based curricula. *Journal for Research in Mathematics Education, 38*(1), 35–70.

Clements, D. H. (2008). Linking research and curriculum development. In L. D. English (Ed.), *Handbook of international research in mathematics education* (2nd ed., pp. 589–625). New York, NY: Taylor & Francis.

Clements, D. H., & Battista, M. T. (2000). Designing effective software. In A. E. Kelly & R. A. Lesh (Eds.), *Handbook of research design in mathematics and science education* (pp. 761–776). Mahwah, NJ: Erlbaum.

Clements, D. H., & Conference Working Group. (2004). Part one: major themes and recommendations. In D. H. Clements, J. Sarama, & A.-M. DiBiase (Eds.), *Engaging young children in mathematics: Standards for early childhood mathematics education* (pp. 1–72). Mahwah, NJ: Erlbaum.

Clements, D. H., & Sarama, J. (1999). *Preliminary report of building blocks—Foundations for Mathematical Thinking, Pre-Kindergarten to grade 2: Research-based materials development*. [NSF Grant No. ESI-9730804]. Buffalo, NY: State University of New York at Buffalo.

Clements, D. H., & Sarama, J. (2007/2016). *Building blocks software [computer software]*. Columbus, OH: McGraw-Hill Education.

Clements, D. H., & Sarama, J. (2007a). Early childhood mathematics learning. In F. K. Lester Jr., (Ed.), *Vol. 1. Second handbook of research on mathematics teaching and learning*, (pp. 461–555). New York, NY: Information Age Publishing.

Clements, D. H., & Sarama, J. (2007b). Effects of a preschool mathematics curriculum: summative research on the *Building Blocks* project. *Journal for Research in Mathematics Education, 38*(2), 136–163.

Clements, D. H., & Sarama, J. (2008). Experimental evaluation of the effects of a research-based preschool mathematics curriculum. *American Educational Research Journal, 45*(2), 443–494. https://doi.org/10.3102/0002831207312908.

Clements, D. H., & Sarama, J. (2013). Rethinking early mathematics: what *is* research-based curriculum for young children? In L. D. English & J. T. Mulligan (Eds.), *Reconceptualizing early mathematics learning* (pp. 121–147). Dordrecht, Germany: Springer.

Clements, D. H., & Sarama, J. (2014). *Learning and teaching early math: The learning trajectories approach* (2nd ed.). New York, NY: Routledge.

Clements, D. H., Sarama, J., & DiBiase, A.-M. (2004). *Engaging young children in mathematics: Standards for early childhood mathematics education*. Mahwah, NJ: Erlbaum.

Clements, D. H., Sarama, J., & Liu, X. (2008). Development of a measure of early mathematics achievement using the Rasch model: the research-based early maths assessment. *Educational Psychology*, *28*(4), 457–482. https://doi.org/10.1080/01443410701777272.

Clements, D. H., Sarama, J., & Mac Donald, B. L. (2017). Subitizing: the neglected quantifier. In N. Anderson & M. W. Alibali (Eds.), *Constructing number: Merging perspectives from psychology and mathematics education*. Gateway East, Singapore: Springer.

Clements, D. H., Sarama, J., Spitler, M. E., Lange, A. A., & Wolfe, C. B. (2011). Mathematics learned by young children in an intervention based on learning trajectories: a large-scale cluster randomized trial. *Journal for Research in Mathematics Education*, *42*(2), 127–166.

Clements, D. H., & Swaminathan, S. (1995). Technology and school change: new lamps for old? *Childhood Education*, *71*, 275–281.

Confrey, J., Gianopulos, G., McGowan, W., Shah, M., & Belcher, M. (2017). Scaffolding learner-centered curricular coherence using learning maps and diagnostic assessments designed around mathematics learning trajectories. *ZDM*, 1–18. https://doi.org/10.1007/s11858-017-0869-1.

Confrey, J., & Lachance, A. (2000). Transformative teaching experiments through conjecture-driven research design. In A. E. Kelly & R. A. Lesh (Eds.), *Handbook of research design in mathematics and science education* (pp. 231–265). Mahwah, NJ: Erlbaum.

Davis, R. B., & Perusse, R. (1988). Numerical competence in animals: definitional issues, current evidence, and a new research agenda. *Behavioral and Brain Sciences*, *11*, 561–579.

Dawson, D. T. (1953). Number grouping as a function of complexity. *The Elementary School Journal*, *54*, 35–42.

Day-Hess, C. A., & Clements, D. H. (2017). The DREME network: research and interventions in early childhood mathematics. *Advances in Child Development and Behavior*, *53*, 1–42. https://doi.org/10.1016/bs.acdb.2017.03.002.

Dehaene, S. (1997). *The number sense: How the mind creates mathematics*. New York, NY: Oxford University Press.

Doabler, C. T., Clarke, B., Fien, H., Baker, S. K., Kosty, D. B., & Cary, M. S. (2014). The science behind curriculum development and evaluation: taking a design science approach in the production of a tier 2 mathematics curriculum. *Learning Disability Quarterly*, *38*(2), 97–111. https://doi.org/10.1177/0731948713520555.

Feigenson, L., Carey, S., & Hauser, M. (2002). The representations underlying infants' choice of more: object files versus analog magnitudes. *Psychological Science*, *13*, 150–156.

Feigenson, L., Carey, S., & Spelke, E. S. (2002). Infants' discrimination of number vs. continuous extent. *Cognitive Psychology*, *44*, 33–66.

Feigenson, L., Dehaene, S., & Spelke, E. S. (2004). Core systems of number. *Trends in Cognitive Sciences*, *8*, 307–314.

Fitzhugh, J. I. (1978). The role of subitizing and counting in the development of the young children's conception of small numbers. *Dissertation Abstracts International*, *40*, 4521B–4522B.

Foorman, B. R., Santi, K. L., & Berger, L. (2007). Scaling assessment-driven instruction using the internet and handheld computers. In B. Schneider & S.-K. McDonald (Eds.), *2. Scale up in practice* (pp. 69–89). Lanham, MD: Rowman & Littlefield.

Fuson, K. C. (1992a). Research on learning and teaching addition and subtraction of whole numbers. In G. Leinhardt, R. Putman, & R. A. Hattrup (Eds.), *Handbook of research on mathematics teaching and learning* (pp. 53–187). Mahwah, NJ: Erlbaum.

Fuson, K. C. (1992b). Research on whole number addition and subtraction. In D. A. Grouws (Ed.), *Handbook of research on mathematics teaching and learning* (pp. 243–275). New York, NY: Macmillan.

Gallistel, C. R., & Gelman, R. (2005). Mathematical cognition. In K. Holyoak & R. Morrison (Eds.), *Cambridge handbook of thinking and reasoning* (pp. 559–588). Cambridge, UK: Cambridge University Press.

Geary, D. C., Hoard, M. K., & Bailey, D. H. (2012). Fact retrieval deficits in low achieving children and children with mathematical learning disability. *Journal of Learning Disabilities, 45*(4), 291–307. https://doi.org/10.1177/0022219410392046.

Geary, D. C., Marle, K., Chu, F., Rouder, J., Hoard, M. K., & Nugent, L. (2018). Early conceptual understanding of cardinality predicts superior school-entry number system knowledge. *Psychological Science, 29*(2), 191–205.

Gebuis, T., & Reynvoet, B. (2011). Generating nonsymbolic number stimuli. *Behavior Research Methods, 43*(4), 981–986.

Glasersfeld, E. v. (1982). Subitizing: the role of figural patterns in the development of numerical concepts. *Archives de Psychologie, 50*, 191–218.

Gordon, P. (2004). Numerical cognition without words: evidence from Amazonia. *Science, 306*, 496–499.

Hannula, M. M. (2005). *Spontaneous focusing on numerosity in the development of early mathematical skills*. Turku, Finland: University of Turku.

Hannula, M. M., Lepola, J., & Lehtinen, E. (2010). Spontaneous focusing on numerosity as a domain-specific predictor of arithmetical skills. *Journal of Experimental Child Psychology, 107*, 394–406.

Huntley-Fenner, G. (2001). Children's understanding of numbers is similar to adults' and rats': numerical estimation by 5-7-year-olds. *Cognition, 78*, 27–40.

Huntley-Fenner, G., Carey, S., & Solimando, A. (2002). Objects are individuals but stuff doesn't count: perceived rigidity and cohesiveness influence infants' representations of small groups of discrete entities. *Cognition, 85*, 203–221.

Huttenlocher, J., Jordan, N. C., & Levine, S. C. (1994). A mental model for early arithmetic. *Journal of Experimental Psychology: General, 123*, 284–296.

Johnson-Pynn, J. S., Ready, C., & Beran, M. (2005). In *Estimation mediates preschoolers: numerical reasoning: evidence against precise calculation abilities. Paper presented at the Biennial Meeting of the Society for Research in Child Development, Atlanta, GA*. April.

Jordan, N. C., Hanich, L. B., & Uberti, H. Z. (2003). Mathematical thinking and learning difficulties. In A. J. Baroody & A. Dowker (Eds.), *The development of arithmetic concepts and skills: Constructing adaptive expertise* (pp. 359–383). Mahwah, NJ: Erlbaum.

Jordan, N. C., Huttenlocher, J., & Levine, S. C. (1992). Differential calculation abilities in young children from middle- and low-income families. *Developmental Psychology, 28*, 644–653.

Jordan, N. C., Huttenlocher, J., & Levine, S. C. (1994). Assessing early arithmetic abilities: effects of verbal and nonverbal response types on the calculation performance of middle- and low-income children. *Learning and Individual Differences, 6*, 413–432.

Karmiloff-Smith, A. (1992). *Beyond modularity: A developmental perspective on cognitive science*. Cambridge, MA: MIT Press.

Kaufman, E. L., Lord, M. W., Reese, T. W., & Volkmann, J. (1949). The discrimination of visual number. *American Journal of Psychology, 62*, 498–525.

Kinzie, M. B., Whittaker, J. V., McGuire, P., Lee, Y., & Virginia, U. o. (2015). Research on curricular development for pre-kindergarten mathematics and science. *Teachers College Record, 117*(070304), 1–40.

Klahr, D. (1973a). A production system for counting, subitizing, and adding. In W. G. Chase (Ed.), *Visual information processing* (pp. 527–544). New York, NY: Academic Press.

Klahr, D. (1973b). Quantification processes. In W. G. Chase (Ed.), *Visual information processing* (pp. 3–31). New York, NY: Academic Press.

Klahr, D., & Wallace, J. G. (1976). *Cognitive development: An information-processing view.* Mahwah, NJ: Erlbaum.

Koontz, K. L., & Berch, D. B. (1996). Identifying simple numerical stimuli: processing inefficiencies exhibited by arithmetic learning disabled children. *Mathematical Cognition*, *2*, 1–23.

Le Corre, M., Van de Walle, G. A., Brannon, E. M., & Carey, S. (2006). Re-visiting the competence/performance debate in the acquisition of counting as a representation of the positive integers. *Cognitive Psychology*, *52*(2), 130–169.

LeFevre, J.-A., Fast, L., Sowinski, C., Osana, H., Skwarchuk, S.-L., & Quian, N. M. (2009). *Who's counting?: Numeracy and literacy and literary practices of early learning and child care.* Ottawa, Ontario, Canada: Canadian Council on Learning.

LeFevre, J.-A., Skwarchuk, S.-L., Smith-Chant, B. L., Fast, L., Kamawar, D., & Bisanz, J. (2009). Home numeracy experiences and children's math performance in the early school years. *Canadian Journal of Behavioural Science*, *41*, 55–66.

Lehtinen, E., & Hannula, M. M. (2006). Attentional processes, abstraction and transfer in early mathematical development. In L. Verschaffel, F. Dochy, M. Boekaerts, & S. Vosniadou (Eds.), *Vol. 49. Instructional psychology: Past, present and future trends. Fifteen essays in honour of Erik De Corte* (pp. 39–55). Amsterdam: Elsevier.

Lester Jr., F. K., Jr., & Wiliam, D. (2002). On the purpose of mathematics education research: making productive contributions to policy and practice. In L. D. English (Ed.), *Handbook of international research in mathematics education* (pp. 489–506). Mahwah, NJ: Erlbaum.

Levine, S. C., Jordan, N. C., & Huttenlocher, J. (1992). Development of calculation abilities in young children. *Journal of Experimental Child Psychology*, *53*, 72–103.

Li, X., Chi, L., DeBey, M., & Baroody, A. J. (2015). A study of early childhood mathematics teaching in the United States and China. *Early Education and Development*, *26*(3), 450–478. https://doi.org/10.1080/10409289.2015.994464.

Logan, G. D., & Zbrodoff, N. J. (2003). Subitizing and similarity: toward a pattern-matching theory of enumeration. *Psychonomic Bulletin & Review*, *10*(3), 676–682.

Lourenco, S. F., & Longo, M. R. (2011). Origins and development of generalized magnitude representation. In S. Dehaene & E. M. Brannon (Eds.), *Space, time and number in the brain: Searching for the foundations of mathematical thought* (pp. 225–244). San Diego, CA: Elsevier Academic Press. https://doi.org/10.1016/b978-0-12-385948-8.00015-3.

MacDonald, B. L. (2015). Ben's perception of space and subitizing activity: a constructivist teaching experiment. *Mathematics Education Research Journal*, *27*(4), 563–584. https://doi.org/10.1007/s13394-015-0152-0.

MacDonald, B. L., & Shumway, J. F. (2016). Subitizing games: assessing preschool children's number understanding. *Teaching Children Mathematics*, *22*(6), 340–348.

MacDonald, B. L., & Wilkins, J. L. M. (2016). Seven types of subitizing activity characterizing young children's mental activity. In S. Marx (Ed.), *Qualitative research in STEM* (pp. 256–286). New York, NY: Routledge.

MacDonald, B. L., & Wilkins, J. L. M. (2017). Seven types of subitizing activity characterizing young children's mental activity. In S. Marx (Ed.), *Qualitative research in STEM* (pp. 256–286). New York, NY: Routledge.

Mandler, G., & Shebo, B. J. (1982). Subitizing: an analysis of its component processes. *Journal of Experimental Psychology: General*, *111*, 1–22.

McCrink, K., & Wynn, K. (2004). Large number addition and subtraction by 9-month-old infants. *Psychological Science*, *15*, 776–781.

Meck, W. H., & Church, R. M. (1983). A mode control model of counting and timing processes. *Journal of Experimental Psychology: Animal Behavior Processes*, *9*, 320–334.

Mix, K. S., Huttenlocher, J., & Levine, S. C. (2002). *Quantitative development in infancy and early childhood*. New York, NY: Oxford University Press.

Mix, K. S., Sandhofer, C. M., & Baroody, A. J. (2005). Number words and number concepts: the interplay of verbal and nonverbal processes in early quantitative development. In R. Kail (Ed.), *Vol. 33. Advances in child development and behavior* (pp. 305–345). New York, NY: Academic Press.

Moore, A. M., & Ashcraft, M. H. (2015). Children's mathematical performance: five cognitive tasks across five grades. *Journal of Experimental Child Psychology*, *135*, 1–24. https://doi.org/10.1016/j.jecp.2015.02.003.

Munter, C., Cobb, P., & Shekell, C. (2016). The role of program theory in evaluation research: a consideration of the what works clearinghouse standards in the case of mathematics education. *American Journal of Evaluation*, *37*(1), 7–26. https://doi.org/10.1177/1098214015571122.

National Governor's Association Center for Best Practices & Council of Chief State School Officers. (2010). *Common core state standards*. Washington, DC: Council of Chief State School Officers and the National Governors Association Center for Best Practices.

National Research Council. (2001). *Adding it up: Helping children learn mathematics*. Washington, DC: National Academy Press.

Nieder, A., Freedman, D. J., & Miller, E. K. (2002). Representation of the quantity of visual items in the primate prefrontal cortex. *Science*, *297*, 1708–1711.

Pepper, K. L., & Hunting, R. P. (1998). Preschoolers' counting and sharing. *Journal for Research in Mathematics Education*, *29*, 164–183.

Perry, B., Dockett, S., & Harley, E. (2007). Learning stories and children's powerful mathematics. *Early Childhood Research & Practice*, *9*(2), 117–134.

Pinel, P., Piazza, D., Le Bihan, D., & Dehaene, S. (2004). Distributed and overlapping cerebral representations of number, size, and luminance during comparative judgments. *Neuron*, *41*, 983–993.

Potter, M., & Levy, E. (1968). Spatial enumeration without counting. *Child Development*, *39*, 265–272.

Sandhofer, C. M., & Smith, L. B. (1999). Learning color words involves learning a system of mappings. *Developmental Psychology*, *35*, 668–679.

Sarama, J., & Clements, D. H. (2004). Building Blocks for early childhood mathematics. *Early Childhood Research Quarterly*, *19*, 181–189.

Sarama, J., & Clements, D. H. (2009). *Early childhood mathematics education research: Learning trajectories for young children*. New York, NY: Routledge.

Sarama, J., & Clements, D. H. (2011). Mathematics knowledge of low-income entering preschoolers. *Far East Journal of Mathematical Education*, *6*(1), 41–63.

Sarama, J., & Clements, D. H. (2013). Lessons learned in the implementation of the TRIAD scale-up model: teaching early mathematics with trajectories and technologies. In T. G. Halle, A. J. Metz, & I. Martinez-Beck (Eds.), *Applying implementation science in early childhood programs and systems* (pp. 173–191). Baltimore, MD: Brookes.

Sarama, J., & Clements, D. H. (in press). Promoting a good start: technology in early childhood mathematics. In E. Arias, J. Cristia, & S. Cueto (Eds.), Promising models to improve primary mathematics learning in Latin America and the Caribbean using technology. Washington, DC: Inter-American Development Bank.

Schaeffer, B., Eggleston, V. H., & Scott, J. L. (1974). Number development in young children. *Cognitive Psychology*, *6*, 357–379.

Secada, W. G. (1992). Race, ethnicity, social class, language, and achievement in mathematics. In D. A. Grouws (Ed.), *Handbook of research on mathematics teaching and learning* (pp. 623–660). New York, NY: Macmillan.

Shuman, M., & Spelke, E. S. (2005, April). In *The development of numerical magnitude representation. Paper presented at the Biennial Meeting of the Society for Research in Child Development, Atlanta, GA*.

Slusser, E. B., & Sarnecka, B. W. (2011). Find the picture of eight turtles: a link between children's counting and their knowledge of number word semantics. *Journal of Experimental Child Psychology*, *110*(1), 38–51.

Spelke, E. S. (2017). Core knowledge, language, and number. *Language Learning and Development*, *13*(2), 147–170.

Starkey, G. S., & McCandliss, B. D. (2014). The emergence of "groupitizing" in children's numerical cognition. *Journal of Experimental Child Psychology*, *126*, 120–137. https://doi.org/10.1016/j.jecp.2014.03.006.

Starkey, P., Spelke, E. S., & Gelman, R. (1990). Numerical abstraction by human infants. *Cognition*, *36*, 97–128.

Steffe, L. P., & Cobb, P. (1988). *Construction of arithmetical meanings and strategies*. New York, NY: Springer-Verlag.

Steffe, L. P., Thompson, P. W., & Richards, J. (1982). Children's counting in arithmetical problem solving. In T. P. Carpenter, J. M. Moser, & T. A. Romberg (Eds.), *Addition and subtraction: A cognitive perspective*. Mahwah, NJ: Erlbaum.

Tan, L. S. C., & Bryant, P. E. (2000). The cues that infants use to distinguish discontinuous quantities: evidence using a shift-rate recovery paradigm. *Child Development*, *71*, 1162–1178.

Tate, W. F. (1997). Race-ethnicity, SES, gender, and language proficiency trends in mathematics achievement: an update. *Journal for Research in Mathematics Education*, *28*, 652–679.

Titeca, D., Roeyers, H., Josephy, H., Ceulemans, A., & Desoete, A. (2014). Preschool predictors of mathematics in first grade children with autism spectrum disorder. *Research in Developmental Disabilities*, *35*(11), 2714–2727. https://doi.org/10.1016/j.ridd.2014.07.012.

Trick, L. M., & Pylyshyn, Z. W. (1994). Why are small and large numbers enumerated differently? A limited-capacity preattentive stage in vision. *Psychological Review*, *101*, 80–102.

van Marle, K., Chu, F. W., Mou, Y., Seok, J. H., Rouder, J. N., & Geary, D. C. (2018). Attaching meaning to the number words: contributions of the object tracking and approximate number systems. *Developmental Science*, *21*(1), e12495. https://doi.org/10.1111/desc.12495.

Wagner, S. W., & Walters, J. (1982). A longitudinal analysis of early number concepts: from numbers to number. In G. E. Forman (Ed.), *Action and thought* (pp. 137–161). New York, NY: Academic Press.

Wang, M., Resnick, L. B., & Boozer, R. F. (1971). The sequence of development of some early mathematics behaviors. *Child Development*, *42*, 1767–1778.

Wiese, H. (2003). Iconic and non-iconic stages in number development: the role of language. *Trends in Cognitive Sciences*, *7*, 385–390.

Wynn, K. (1992). Children's acquisition of the number words and the counting system. *Cognitive Psychology*, *24*, 220–251.

Chapter 7

Development of Mathematical Language in Preschool and Its Role in Learning Numeracy Skills

David J. Purpura, Amy R. Napoli and Yemimah King
Department of Human Development and Family Studies, Purdue University, West Lafayette, IN, United States

INTRODUCTION

Young children's acquisition of mathematics skills is tightly linked to the development of other academic and cognitive skills during the early years. In particular, a growing body of evidence highlights the important role that language skills—especially mathematical language (e.g., words and concepts such as *many, most, few, fewest*)—play in young children's acquisition of early mathematical knowledge. In this chapter we (1) discuss the connections between general language and mathematics development, (2) define mathematical language and distinguish it from mathematics talk, (3) present the correlational and experimental evidence supporting the role mathematical language plays in early numeracy acquisition and existing mechanisms for intervention, and finally (4) discuss remaining unanswered questions about the mechanisms underlying this relation and future directions regarding this line of research.

CONNECTIONS BETWEEN EARLY MATHEMATICS DEVELOPMENT AND GENERAL LANGUAGE

Mathematics proficiency is an academic and economic gatekeeper that provides a foundation for achievement and career skills (Geary, 1994; Jordan, Hanich, & Uberti, 2003; National Mathematics Advisory Panel [NMAP], 2008). Children with early mathematics deficits typically develop their skills at a slower rate than their more advanced peers (Aunola, Leskinen, Lerkkanen, & Nurmi, 2004) and experience lifelong academic and career difficulties (Adelman, 1999;

Mathematical Cognition and Learning, Vol. 5. https://doi.org/10.1016/B978-0-12-815952-1.00007-4

Duncan et al., 2007; Evan, Gray, & Olchefske, 2006; McGregor, 1994). Although early mathematics skills form the foundation for the acquisition of later mathematics skills, these early skills do not develop in isolation from other academic and cognitive domains—particularly language and literacy skills (LeFevre et al., 2010; Purpura, Hume, Sims, & Lonigan, 2011). For example, LeFevre et al. (2010) indicated that of the three main pathways by which early mathematics skills develop (quantitative, visual spatial, and linguistic), the strongest and most stable pathway was the linguistic pathway which was broadly defined to include both language and phonological skills. Furthermore, children who have difficulty in one domain have a high likelihood of experiencing difficulties in the other area (Barberisi, Katusic, Colligan, Weaver, & Jacobsen, 2005).

Difficulties in Both Domains

Among the approximately 5%–8% of children who are diagnosed with a mathematics learning disability, a large proportion also meet criteria for a reading disability (Geary, 2004). Notably, children with both mathematics and reading difficulties experience more significant difficulties in mathematics than children who only have either mathematics difficulties or reading difficulties alone (Hanich, Jordan, Kaplan, & Dick, 2001; Jordan & Hanich, 2000; Jordan & Levine, 2009; Mann Koepke & Miller, 2013; Willcutt et al., 2013). Donlan (2007) suggests that there may be an interaction between linguistic and visuospatial neural systems during development, resulting in children with specific language impairments also having difficulties in mathematics. Moreover, verbal skills are necessary for children to succeed in calculation tasks that require knowledge of number words (Donlan, 2007; Jordan, Huttenlocher, & Levine, 1994; Lewis, Hitch, & Walker, 1994). For example, language comprehension is a necessary component of word problems, and children who struggle with both mathematics and reading tend to have particular difficulties with story problems (Fuchs et al., this volume; Jordan & Levine, 2009; Powell, Fuchs, Fuchs, Cirino, & Fletcher, 2009). Ultimately, children who tend to have co-occurring mathematics/reading difficulties are at an increased risk for experiencing persistent difficulties in mathematics (Jordan & Levine, 2009; Mann Koepke & Miller, 2013), which suggests that deficits in language and literacy skills may either lead to difficulties in mathematics or exacerbate existing difficulties, above and beyond any domain-general deficits (e.g., working memory) that may slow learning in both areas.

Early Connections Between Mathematics and Literacy Skills

Though a large body of evidence indicates that there are strong relations between mathematics and literacy skills in elementary school (Hooper, Roberts, Sideris, Burchinal, & Zeisel, 2010; Romano, Babchishin, Pagani, & Kohen, 2010) it is also evident that this relation begins early in life (Davidse, De Jong, & Bus, 2014). Notably, Spelke (2003) suggests that early language skills help children refine their concepts of basic number. Specifically, by building and applying the constructs of language such as grammatical rules to the underlying approximate number system, children are able to intuit specific quantities, their relations, and their positioning relative to other quantities (Spelke, 2017). Even at the prereading level, emergent literacy skills such as phonological awareness (Krajewski, Schneider, & Niedling, 2008), language (Hooper et al., 2010; LeFevre et al., 2010), and print knowledge (Geary & van Marle, 2016; Neumann, Hood, Ford, & Neumann, 2013; Piasta, Purpura, & Wagner, 2010) are individually predictive of mathematics skills. Yet, though each of these three components of emergent literacy skills has been found to be independently related to mathematics performance, little work focused on identifying if the relations were unique and specific to each component, or if some components were acting as a proxy measure for the other components of literacy development.

To begin to address this issue, Purpura et al. (2011) assessed 69 preschool students on a broad measure of mathematics skills and on measures of phonological awareness, language, and print knowledge. These children were then assessed 1 year later on three mathematics measures (two broad measures that included formal and informal skills and one that just included formal skills). The results indicated that language and print knowledge each accounted for between 3% and 9% of the *unique variance* in predicting the two broader measures of mathematics, and phonological awareness (the core deficit in reading difficulties) was not a significant predictor of mathematics skills. None of the emergent literacy domains were significant predictors of formal-symbolic mathematics skills. These findings indicate that emergent literacy skills—specifically language and print knowledge—have a specific relation with early informal mathematics skills but not with formal skills.

Although Purpura et al. (2011) and others (Austin, Blevins-Knabe, Ota, Rowe, & Lindauer, 2011; LeFevre et al., 2010) have shown that both language and print knowledge in general are related to informal numeracy and numeral knowledge and that expressive vocabulary is a significant predictor of almost all individual aspects of early numeracy (Purpura & Ganley, 2014), the distinct manner in which these two literacy components are related to specific aspects of early mathematical performance had not previously been evaluated. Austin and colleagues indicated that, though early language and print knowledge were related to early mathematics performance, it was likely in a developmental manner where children first obtain an understanding of receptive language and letter awareness before obtaining an understanding of early math concepts. To further evaluate this theory, Purpura and Napoli (2015) conducted a study to determine the distinct relation of vocabulary skills and print knowledge to components of early mathematics skills. To do this, 180 preschool children were assessed on a battery of targeted early mathematics and emergent literacy skills. It was found that the relation between language and numeral knowledge is fully mediated by informal numeracy skills (e.g., those mathematics skills that do not require the use of Arabic numerals) and the relation between informal numeracy skills and numeral knowledge skills is partially mediated by print knowledge. These findings indicate that language is likely to be more strongly connected to mathematical development early on and print-based skills may subsequently be more strongly connected. In all, this series of studies suggest that the association between language skills and mathematics development is particularly strong in the early years.

Language Interventions and Mathematics Skills

Although research suggests that language skills are related to a broad array of early mathematics skills, interventions that focus only on language development do not appear to improve mathematics skills (Jordan, Glutting, Dyson, Hassinger-Das, & Irwin, 2012). In a randomized control study, Jordan et al. (2012) tested the effectiveness of an 8-week-long intervention to build number sense knowledge in kindergartners at risk for mathematical deficits. Children in this study were randomly assigned to a number sense intervention group, a language intervention group, or a business-as-usual control group. The number sense intervention condition focused on number recognition, number sequencing, verbal subitizing, counting on fingers, number list activities, written number activities, number sets, problem solving and operations, and linear number board game. The language intervention condition focused on teaching 43 vocabulary words (e.g., general vocabulary and quantitative vocabulary) from eight different storybooks. The results showed that children in the number sense intervention condition had significantly higher scores on mathematics assessments by the end of the intervention than children in the language intervention and control groups. Further, the language intervention group did not significantly differ from the control group on posttest assessments of mathematics skills. This study suggests that interventions that focus on general vocabulary are not sufficient for improving children's mathematics skills. Rather, it may be more effective to teach math-related vocabulary along with number sense concepts. It may also be more effective to focus only on mathematics-specific vocabulary (i.e., mathematical language) rather than including general vocabulary that is unrelated to mathematics because the language related to mathematics is highly content specific (Harmon, Hedrick, & Wood, 2005).

WHAT IS CONTENT-SPECIFIC MATHEMATICAL LANGUAGE?

In contrast to mathematics knowledge (or numeracy; the ability to work with exact numbers such as counting, comparing, adding), mathematical language is a child's understanding of the key words and concepts used in early math. Specifically, mathematical language consists of terms that are used to describe quantity or spatial relations (e.g., *more*, *fewer*, *near*, *below*). Mathematical language is different from "number talk," which usually relates to mathematical input in domains such as counting, cardinality, and calculation (Boonen, Kolkman, & Kroesbergen, 2011; Klibanoff, Levine, Vasilyeva, & Hedges, 2006). Unlike number talk, mathematical language does not include references to specific numbers or direct teaching of mathematics skills, and is related to children's knowledge of specific terms rather than their production of mathematics-focused language.

Two specific aspects of mathematical language have been identified as potentially important for early mathematical learning: quantitative language (Barner, Chow, & Yang, 2009) and spatial language (Ramani, Zippert, Schweitzer, & Pan, 2014). Quantitative language includes terms such as "more than," "less than," "many," and "fewer." Understanding these terms allows children to make and describe comparisons between groups or numbers. For example, knowing that the term "more" can mean an increase in quantity ("give me more") or can be used in comparative statements ("five is more than two"). This knowledge may allow children to refine their understanding of quantity more precisely as approximate number words seem to develop prior to children's acquisition of more exact cardinal number knowledge (e.g., children may understand that "five" refers to a quantity but not know the exact quantity; Gunderson, Spaepen, & Levine, 2015). Spatial language consists of words that describe the dimension of objects, locations and directions, and relations between objects, including words such as *near* and *above* (Cannon, Levine, & Huttenlocher, 2007). Understanding these spatial terms may allow children to talk about relations between physical objects and between numbers on a number line as well as develop spatial skills that have been found to be important for mathematics development (Mix & Cheng, 2012). For example, spatial words such as "before" and "after" are often connected to the number sequence (e.g., "five comes *after* four"). Children's production of spatial language is positively related to their performance on spatial problem-solving tasks (Pruden, Levine, & Huttenlocher, 2011), as well as joint block play (Ramani et al., 2014). A list of the words typically considered to comprise these two types of mathematical language are presented in Table 1.

CORRELATIONAL AND EXPERIMENTAL EVIDENCE ON THE RELATIONS BETWEEN MATHEMATICAL LANGUAGE AND MATHEMATICS PERFORMANCE

Mathematical language is one of the strongest predictors of mathematics development during the preschool (Purpura & Logan, 2015) and early elementary school years (Toll & Van Luit, 2014a). Notably, in Purpura and Logan's study, the numeracy skills, mathematical language, executive functioning, and rapid automatized naming of 114 preschoolers were assessed in the fall and spring of the school year. Using both mixed-effect regressions, it was found that mathematical language was the strongest predictor of

TABLE 1 **Quantitative and Spatial Language Words Used in Prior Studies**

Quantitative Terms	Spatial Terms
A couple	Above
A lot	After
Add	Around
Big, bigger, biggest	Away
Different	Back
Enough	Behind
Few, fewer, fewest	Before
Less than	Below
Least	Between
Many	Bottom
Minus	Close, closer, closest
More, Most	Down
Same	End
Several	Far, farther, farthest
Similar	First
Small, smaller, smallest	Forward
Some	Front
Subtract	High, higher, highest
Take away	In
	Into
	Inside
	Last
	Long, longer, longest
	Middle
	Near
	On
	Out, outside
	Over
	Through
	Top
	Toward
	Under

numeracy development (i.e., set comparison, numeral comparison, one-to-one correspondence, number order, identifying numerals, ordinality, and number combinations) during preschool. When utilizing quantile regression which allows for examining predictors of the dependent variable across the spectrum of ability, mathematical language was found to be the best predictor at the 50th and 75th percentiles of ability. Furthermore, mathematical language has also been found to be more proximal to mathematics performance than are general language skills (Purpura & Reid, 2016; Toll & Van Luit, 2014b). Even though general language often predicts mathematics performance (LeFevre et al., 2010; Purpura et al., 2011), when mathematical language is also included in the model, general language is no longer a specific predictor—this is likely because general language acts as a proxy for more precise domain-specific language, including mathematical language.

Young children who do not understand specific aspects of mathematical language such as comparative words (e.g., more) struggle to acquire early mathematics skills such as cardinal number knowledge (Barner et al., 2009), which can lead to later difficulties in mathematics development. In fact, in a recent study, Purpura, Day, Napoli, and Hart (2017) found that among a wide range of academic and cognitive skills assessed at the beginning of preschool (e.g., mathematical language, numeracy skills, vocabulary, executive functioning, approximate number system, print knowledge, general vocabulary, and phonological awareness), mathematical language was the strongest and most consistent classifier of the children who would perform the lowest on a numeracy measure at the end of preschool—it was even a better classifier than beginning of preschool numeracy performance. These findings suggest that children who enter preschool with lower mathematical language skills are the most likely to struggle to acquire numeracy skills during the preschool years. This was likely because children with lower mathematical language skills may not have had access to the mathematical instruction occurring in classrooms because they did not understand the language used to discuss these concepts, whereas other children who entered preschool low on numeracy skills but not low on mathematical language knowledge may have been able to benefit more from instruction.

Children's interactions with adults are central to the acquisition of early numeracy skills and mathematical language. The more mathematical talk and mathematical language that children hear at school and at home, the more likely it is that they have higher mathematical skills. For instance, studies have shown that interactions that include mathematical language, both with caregivers and teachers, predict children's mathematics outcomes (Chard et al., 2008; Clements & Sarama, 2011; Gunderson & Levine, 2011; Levine, Suriyakham, Rowe, Huttenlocher, & Gunderson, 2010; Ramani, Rowe, Eason, & Leech, 2015). It has been postulated that mathematical language-rich interactions are important in developing children's early mathematics skills (Clements, Baroody, & Sarama, 2013) and children from families with low SES are particularly at risk for delayed acquisition of mathematical language because they experience significantly less language-rich environments than their middle and higher SES peers (Hart & Risely, 1995). This language deficit is especially apparent in their exposure to mathematical language (Levine et al., 2010) and manifests into large differences ($d=0.80$) in knowledge of mathematical language concepts between children from families with low SES and middle/high SES by the time they reach preschool (Purpura & Reid, 2016). Thus children from families of lower SES are, from early ages, less likely to have acquired mathematical language skills.

Interventions to Improve Mathematical Language

A strong emphasis on mathematical language is believed to be a critical component of the success of early mathematics instruction (NCTM, 2006), and researchers have emphasized the importance of including a mathematical language component in general mathematics instruction (Chard et al., 2008; Clements & Sarama, 2011). However, to date only a few intervention studies have focused on directly improving children's mathematical language skills (Hassinger-Das, Jordan, & Dyson, 2015; Jennings, Jennings, Richey, & Dixon-Krauss, 1992; Powell & Driver, 2015), and these studies have provided strong evidence that children's mathematical language can be improved through intervention. For example, Hassinger-Das et al. (2015) randomly assigned kindergarten children to a number sense plus mathematical language intervention, a number sense only intervention, or a business-as-usual control group. They found that children in the mathematical language group outperformed the other groups on immediate posttest measures of mathematical language, but not general mathematics measures. Yet, at a delayed posttest, the mathematical language group showed significant improvement on mathematical skills in comparison to the other conditions. These findings suggest that although some benefits of mathematical language instruction may not have been immediately found, children may have been provided with access to the content of instruction by gaining a better understanding of the language components of mathematics which allowed them to further develop their mathematics skills. Importantly, and similar to research conducted by Jennings et al. (1992) and Powell and Driver (2015), the mathematical language intervention targeted both mathematical knowledge and numeracy skills. Though these studies provide evidence of the malleability of mathematical language skills, they cannot be used to disentangle the effects of mathematical language specifically, without the additive component of mathematical knowledge. Critically, despite prior correlational evidence that mathematical language is predictive of mathematical development, these correlational and intervention studies cannot be used to make causal assertions regarding this relation.

To address the issue of whether or not mathematical language is causally related to numeracy development, Purpura, Napoli, Wehrspann, and Gold (2017) conducted an intervention study that provided mathematical language instruction with no additional numeracy instruction. Forty-seven preschool children from Head Start centers were randomly assigned to a storybook reading intervention or a business-as-usual control group. Children were pre- and posttested on measures of numeracy, mathematical language, and general vocabulary. The mathematical language storybook intervention was administered for approximately 15 min per day, 2–3 days per week, for 7 weeks (*M* sessions = 14.5). For the first 6 weeks, interventionists read one book per week to small groups of 2–3 children at a time. The foci of books alternated weekly between quantitative and spatial language, though both quantitative and spatial terms were highlighted when possible. During the seventh week, children were allowed to choose one of the six books for each session to review content and to ensure prolonged engagement. An eighth week was used to make up sessions with students who missed earlier sessions. A key aspect of this intervention was that it was designed to include only mathematical language terms (e.g., more, less, near, far) and not numeracy skill content (e.g., number names, counting, addition) to assess the causal relation between mathematical language and numeracy skills. Over the course of the intervention, children were exposed to 56 different mathematical language terms either as part of the story or as questions asked by the interventionists.

Dialogic reading (Arnold & Whitehurst, 1994; Lonigan, Anthony, Bloomfield, Dyer, & Samwel, 1999) was used as the primary instructional framework of this intervention in order to promote children's own use of mathematical language and more fully engage them with the content. The utility of dialogic reading for improving children's language skills is likely derived from the emphasis on eliciting verbal responses and descriptions from children (Lever & Sénéchal, 2011). It was hypothesized that this framework would be effective for improving mathematical language as mathematical curricula rich in interaction are effective for improving both mathematics and language skills (Sarama, Lange, Clements, & Wolfe, 2012; see also Hassinger-Das et al., 2015). After the intervention was completed, the results of the study indicated that the intervention had significant effects on both mathematical language skills (Hedge's $g = 0.42$) and numeracy skills (Hedge's $g = 0.32$), but not general vocabulary. These findings indicate that promoting children's mathematical language skills also positively affects their general numeracy skills (e.g., a measure that broadly assesses children's skills in one-to-one correspondence, numeral identification, set comparison, numeral comparison, number order, ordinal number knowledge, story problems, and formal addition). It is likely that exposure to mathematical language terms in an interactive context where children repeatedly heard the words and were also encouraged to use the words themselves allowed children to form an understanding of the terms. As such, it appears that mathematical language may causally underlie numeracy development.

Despite the positive findings from the Purpura, Napoli, et al. (2017) intervention study, it is unclear which aspect of mathematical language—quantitative or spatial—was driving the positive transfer to numeracy skills. Our work only examined effects of the intervention on the broad measure of mathematical language. However, this question can begin to be answered by examining the intervention's impact on the quantitative and spatial mathematical language terms separately. When the mathematical language measure was separated into quantitative (six items) and spatial domains (10 items), these new exploratory analyses indicated that the effects on mathematical language were primarily found for quantitative mathematical language (Hedge's $g = 0.80$; Fig. 1) and not spatial language (Hedge's $g = 0.20$; Fig. 2). These findings suggest that instruction on

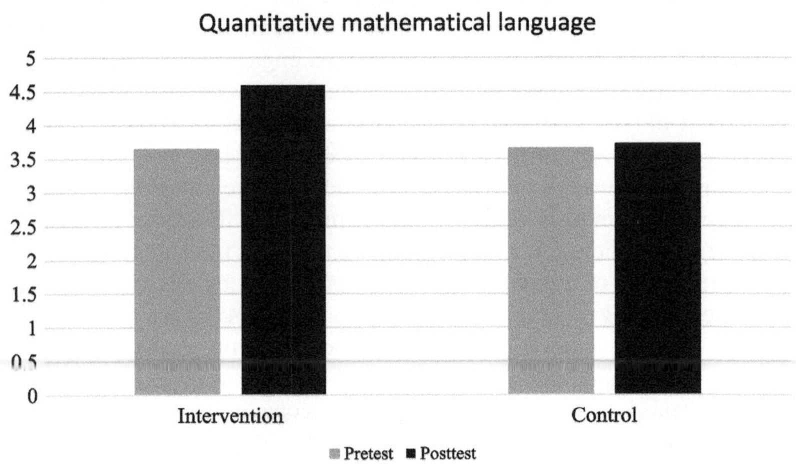

FIG. 1 Pre- and posttest mathematical language scores by condition for the *quantitative* mathematical language items.

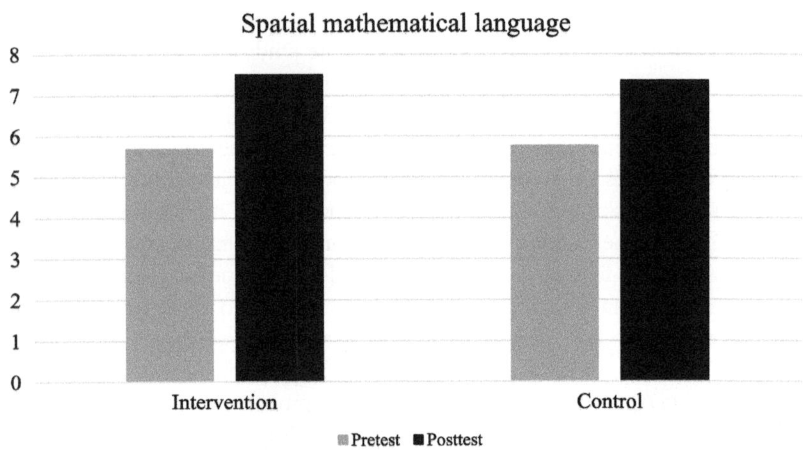

FIG. 2 Pre- and posttest mathematical language scores by condition for the *spatial* mathematical language items.

quantitative mathematical language, but not spatial mathematical language, underlies the early development of numeracy skills.

Furthermore, by examining the pretest and posttest scores of each group in Figs. 1 and 2, it can be seen that for spatial mathematical language, both groups improve by about two points from pre- to posttest. In contrast, for quantitative mathematical language, only the intervention group improved, suggesting that children may not have acquired these quantitative terms over the course of normal schooling. Ultimately, it is likely that quantitative language affects numeracy development because it provides approximate benchmarks of quantitative relations (Gunderson et al., 2015) by which children can then develop more exact knowledge of specific numbers and relations about numbers (Shusterman, Slusser, Halberda, & Odic, 2016). However, given that both quantitative and spatial language were taught in an integrative manner, the more specific claim of whether it is instruction in quantitative language or spatial language that underlies this relation cannot yet be answered fully. A number of critical questions regarding impacts of mathematical language instruction and mechanisms by which it improves children's mathematics skills need to be evaluated through future research.

UNANSWERED QUESTIONS

There is a growing body of evidence that suggests mathematical language may be a key component in young children's acquisition of early mathematics skills. Not only is it one of the strongest predictors of numeracy development during preschool and elementary school (Purpura & Logan, 2015; Toll & Van Luit, 2014b) and the strongest classifier of later risk status for mathematics difficulties during preschool (Purpura, Day, et al., 2017), but it also appears to causally contribute to numeracy development (Purpura, Napoli, et al., 2017). However, four critical gaps need to be addressed to better understand how mathematical language is related to mathematical skills over time and provide appropriate instruction in this area, including: (1) Investigating the way mathematical language instruction interacts with mathematics content instruction, (2) examining the relation between different aspects of mathematical language (e.g., *more, fewer* versus *before, after*) to different aspects of mathematics (e.g., numeracy versus geometry), (3) determining how to best assess mathematical language skills across the years, and (4) developing methods to provide effective mathematical language instruction. Following, we detail these key issues and future directions.

Mathematical Language and Numeracy Instruction

It seems evident that quantitative mathematical language underlies early numeracy development; however, the nature of the relation and how these two constructs develop needs to be better understood. Notably, most common references to providing rich mathematical language environments encourage "mathematics instruction to be rich in mathematical language" (Clements et al., 2013; NCTM, 2006). Understanding mathematical language terms facilitates broader conceptual knowledge of numeracy (Dunston & Tyminski, 2013) which ultimately should lead to greater gains in numeracy skills (e.g., enumerating quantities, understanding the relations between quantities or numerals, and using informal and formal arithmetic operations). Likely, this is because understanding the quantitative terms provides an approximate understanding of numbers, which then enables children to more readily learn numeracy concepts with direct engagement of numeracy instruction. Suggestions that embedding mathematical language into mathematical knowledge instruction may be more beneficial than mathematical knowledge instruction alone have been partially supported by empirical evidence with elementary-age children (Fuchs et al., this volume; Hassinger-Das et al., 2015; Powell & Driver, 2015). Both of these studies with elementary students compared the combination of mathematical language and numeracy instruction, numeracy instruction alone, and a control (note that though the two studies had different mathematical content focus, the type of comparison groups was the same). Hassinger-Das et al. found that the combined intervention led to significantly improved numeracy skills compared to the numeracy intervention alone only at the delayed posttest, whereas, Powell and Driver found no significant differences between the conditions on addition fluency skills.

The findings from existing studies are limited in applicability to younger children for three reasons. First, the specific numeracy content is different across grades, and early informal numeracy skills (e.g., counting, comparison) appear to be more language based than formal skills (e.g., addition, subtraction; Purpura & Napoli, 2015) that were targeted in these studies. Second, the findings from the Powell and Driver (2015) study are based on a measure of addition fluency. If mathematical language broadens conceptual knowledge of skills, the gains may not be manifest in a timed measure at immediate posttest. Third, neither Power and Driver nor Hassinger-Das et al. (2015) contrasted the interventions with a mathematical language only condition which has been shown to be sufficient for the development of early numeracy skills (Purpura, Napoli, et al., 2017). Importantly, is quantitative mathematical language necessary for the development of numeracy skills? Is it simply sufficient? Or, is it complementary where instruction of quantitative language enhances subsequent numeracy instruction? Ultimately, delineating the way in which quantitative language and numeracy instruction jointly or independently affect numeracy development is a necessary step in developing methods for maximizing the effectiveness of early numeracy instruction.

Spatial Language

Quantitative language appears to drive the relation between mathematical language and early numeracy skills, but the role of spatial language in the development of mathematical skills is an important area for future research.

Understanding spatial terms may allow children to talk about relations between physical objects and how to manipulate shapes and objects, as well as develop spatial skills that have been found to be important for mathematical development (Mix & Cheng, 2012). For example, many spatial words such as "near" and "above" are often used in instruction related to early geometry skills. Though spatial language is unlikely to be directly related to numeracy skills, it may be related to more proximal mathematics skills such as geometry (Casey et al., 2015) and spatial skills (Borriello & Liben, 2017; Miller, Vlach, & Simmering, 2017).

Mathematical Language Measures

As researchers continue to investigate the role that mathematical language plays in mathematics development, there is a need to develop valid and reliable ways of assessing the understanding of mathematical language (Raghubar & Barnes, 2017). At the present time, no published standardized measure for mathematical language exists. Some work has been done to develop a measure of mathematical language for children in preschool (Purpura & Logan, 2015) and first grade (Powell & Nelson, 2017). However, significant work needs to be conducted to link these assessment frameworks as well as expand them beyond existing grades. Without the development of a uniform measure across these ages, it will be difficult to evaluate if mathematical language plays a continuous role in early mathematics development, or if this role is more foundational in nature.

Developing Methods for Mathematical Language Instruction

Developing methods to enhance young children's math skills before early difficulties become long-standing problems, particularly for children from families of lower SES, is important for enabling academic success. Despite the substantial evidence supporting math language as a critical foundation for early math skills development and the general consensus in major policy documents that math language should be incorporated into preschool math instruction (Clements et al., 2013; National Council of Teachers of Mathematics [NCTM], 2006; Purpura, Napoli, et al., 2017), there are limited available and empirically supported resources for preschool teachers to effectively implement math language instruction in the classroom. Though the concept of math language instruction may outwardly appear relatively simple, merely encouraging teachers to use more math language is not sufficient for improving children's skills because there is significant variability in the quality of how these language concepts are presented and using a large range of math language may be negatively related to math skill acquisition because it may detract from more targeted language and concepts (Boonen et al., 2011; Lansdell, 1999).

Even within effective empirically supported early mathematics curricula such as *Building Blocks* (Clements & Sarama, 2013) there is little specific structured emphasis on mathematical language concepts built into the lessons.

Though teachers are instructed to "...develop children's language and vocabulary by describing what individual children are doing..." in this curriculum, there is little direction for exactly how to do this language-enhanced instruction. There are a few lessons that target comparative skills, but they are mainly limited to concepts of "more." Parents and teachers often do not have the tools or training to provide such targeted instruction (Ginsburg, Lee, & Boyd, 2008). As a consequence, there is a critical need to develop and evaluate easy-to-implement methods for teachers to provide math language instruction to their preschool students in an engaging manner.

CONCLUSIONS AND FUTURE DIRECTIONS

Mathematical language during the preschool years is a critical component of early mathematics development. Children who acquire more advanced quantitative language skills also develop their numeracy skills in tandem. Further research is needed to examine how mathematical language instruction can be combined with mathematical knowledge instruction to enhance children's early learning opportunities, as well as to identify which components, or combination of components, are most important for building early mathematical knowledge. Subsequently, building on that knowledge to develop effective instructional methods in this area is needed to enable teachers to maximize children's learning of these concepts during preschool. Ultimately, the support of a causal relation between mathematical language and general mathematics knowledge begins to fill a critical gap in the research literature and complements and enhances prior correlational evidence, but more work is needed to better understand the underlying mechanisms driving these connections and how best to support young children's learning of these terms and concepts.

REFERENCES

Adelman, C. (1999). *Answers in the toolbox: Academic intensity, attendance patterns and bachelor's degree attainment.* Washington, DC: U.S. Department of Education.

Arnold, D. H., & Whitehurst, G. J. (1994). Accelerating language development through picture book reading: a summary of dialogic reading and its effect. In D. K. Dickinson (Ed.), *Bridges to literacy: Children, families, and schools* (pp. 103–128). Malden, MA: Blackwell Publishing.

Aunola, K., Leskinen, E., Lerkkanen, M. K., & Nurmi, J. E. (2004). Developmental dynamics of math performance from preschool to grade 2. *Journal of Educational Psychology, 96*, 699–713. https://doi.org/10.1037/0022-0663.96.4.699.

Austin, A. M. B., Blevins-Knabe, B., Ota, C., Rowe, T., & Lindauer, S. L. K. (2011). Mediators of preschoolers' early mathematics concepts. *Early Child Development and Care, 181*, 1181–1198. https://doi.org/10.1080/03004430.2010.520711.

Barberisi, M. J., Katusic, S. K., Colligan, R. C., Weaver, A. L., & Jacobsen, S. J. (2005). Math learning disorder: incidence in a population-based birth cohort, 1976–1982, Rochester, Minn. *Ambulatory Pediatrics, 5*, 281–289.

Barner, D., Chow, K., & Yang, S.-J. (2009). Finding one's meaning: a test of the relation between quantifiers and integers in language development. *Cognitive Psychology, 58*, 195–219. https://doi.org/10.1016/j.cogpsych.2008.07.001.

Boonen, A. J. H., Kolkman, M. E., & Kroesbergen, E. H. (2011). The relation between teachers' mathematics talk and the acquisition of number sense within kindergarten classrooms. *Journal of School Psychology, 49*, 281–299. https://doi.org/10.1016/j.jsp.2011.03.002.

Borriello, G. A., & Liben, L. S. (2017). Encouraging maternal guidance of preschoolers' spatial thinking during block play. *Child Development*. Advance online publication. https://doi.org/10.1111/cdev.12779.

Cannon, J., Levine, S., & Huttenlocher, J. (2007). *A system for analyzing children and caregivers' language about space in structured and unstructured contexts*. Spatial Intelligence and Learning Center (SILC) technical report. Retrieved from http://www.silccenter.org/media/silc_pdfs/resources/testsandinstruments/Spatial%20Language%20Coding%20Manual%201-10-12.pdf.

Casey, B. M., Pezaris, E., Fineman, B., Pollock, A., Demers, L., & Dearing, E. (2015). A longitudinal analysis of early spatial skills compared to arithmetic and verbal skills as predictors of fifth-grade girls' mathematics reasoning. *Learning and Individual Differences, 40*, 90–100. https://doi.org/10.1016/j.lindif.2015.03.028.

Chard, D. J., Baker, S. K., Clarke, B., Jungjohann, K., Davis, K., & Smolkowski, K. (2008). Preventing early mathematics difficulties: the feasibility of a rigorous kindergarten mathematics curriculum. *Learning Disability Quarterly, 31*, 11–20. https://doi.org/10.2307/30035522.

Clements, D. H., Baroody, A. J., & Sarama, J. (2013). *Background research on early mathematics: Background research for the National Governors Association (NGA) center project on early mathematics*. Washington, DC: National Governors Association. Retrieved from http://www.nga.org/cms/home/nga-center-for-best-practices/meeting–webcast-materials/page-edu-meetings-webcasts/col2-content/main-content-list/strengthening-early-mathematics.html.

Clements, D. H., & Sarama, J. (2011). Early childhood mathematics intervention. *Science, 333*, 968–970. https://doi.org/10.1126/science.1204537.

Clements, D. H., & Sarama, J. (2013). *Building blocks*. (Vols. 1 and 2). Columbus, OH: McGraw-Hill Education.

Davidse, N. J., De Jong, M. T., & Bus, A. G. (2014). Explaining common variance shared by early numeracy and literacy. *Reading and Writing, 27*, 631–648. https://doi.org/10.1007/s11145-013-9465-0.

Donlan, C. (2007). Mathematical development in children with specific language impairments. In D. B. Berch & M. M. Mazzocco (Eds.), *Why is mathematics so hard for some children? The nature and origins of mathematical learning difficulties and disabilities* (pp. 151–172). Baltimore, MD: Brookes Publishing Company.

Duncan, G. J., Dowsett, C. J., Claessens, A., Magnuson, K., Huston, A. C., Klebanov, P., et al. (2007). School readiness and later achievement. *Developmental Psychology, 43*, 1428–1446. https://doi.org/10.1037/0012-1649.43.6.1428.

Dunston, P., & Tyminski, A. (2013). What's the big deal about vocabulary? *Mathematics Teaching in the Middle School, 19*, 38–45.

Evan, A., Gray, T., & Olchefske, J. (2006). *The gateway to student success in mathematics and science*. Washington, DC: American Institutes for Research.

Geary, D. C. (1994). *Children's mathematical development: Research and practical applications*. Washington, DC: American Psychological Association.

Geary, D. C. (2004). Mathematics and learning disabilities. *Journal of Learning Disabilities, 37*, 4–15.

Geary, D. C., & van Marle, K. (2016). Young children's core symbolic and non-symbolic quantitative knowledge in the prediction of later mathematics achievement. *Developmental Psychology, 52*, 2130–2144. https://doi.org/10.1037/dev0000214.

Ginsburg, H. P., Lee, J. S., & Boyd, J. S. (2008). Mathematics education for young children: what it is and how to promote it. *Social Policy Report, 22*, 3–23.

Gunderson, E. A., & Levine, S. C. (2011). Some types of parent number talk count more than others: relations between parents' input and children's cardinal-number knowledge. *Developmental Science, 14*, 1021–1032. https://doi.org/10.1111/j.1467-7687.2011.01050.x.

Gunderson, E. A., Spaepen, E., & Levine, S. C. (2015). Approximate number word knowledge before the cardinal principle. *Journal of Experimental Child Psychology, 130*, 35–55. https://doi.org/10.1016/j.jecp.2014.09.008.

Hanich, L. B., Jordan, N. C., Kaplan, D., & Dick, J. (2001). Performance across different areas of mathematical cognition in children with learning difficulties. *Journal of Educational Psychology, 93*, 615–626. https://doi.org/10.1037/0022-0663.93.3.615.

Harmon, J. M., Hedrick, W. B., & Wood, K. D. (2005). Research on vocabulary instruction in the content areas: implications for struggling readers. *Reading and Writing Quarterly, 21*, 261–280. https://doi.org/10.1080/10573560590949377.

Hart, B., & Risley, T. R. (1995). *Meaningful differences in the everyday experience of young American children.* Baltimore, MD: Paul H. Brookes Publishing Co.

Hassinger-Das, B., Jordan, N. C., & Dyson, N. (2015). Reading stories to learn math: mathematics vocabulary instruction for children with early numeracy difficulties. *The Elementary School Journal, 116*, 242–264. https://doi.org/10.1086/683986.

Hooper, S. R., Roberts, J., Sideris, J., Burchinal, M., & Zeisel, S. (2010). Longitudinal predictors of reading and mathematics trajectories through middle school for African American versus Caucasian students across two samples. *Developmental Psychology, 46*, 1018–1029. https://doi.org/10.1037/a0018877.

Jennings, C. M., Jennings, J. E., Richey, J., & Dixon-Krauss, L. (1992). Increasing interest and achievement in mathematics through children's literature. *Early Childhood Research Quarterly, 7*, 263–276. https://doi.org/10.1016/0885-2006(92)90008-M.

Jordan, N. C., Glutting, J., Dyson, N., Hassinger-Das, B., & Irwin, C. (2012). Building kindergartners' number sense: a randomized controlled study. *Journal of Educational Psychology, 104*, 647–660. https://doi.org/10.1037/a0029018.

Jordan, N. C., & Hanich, L. B. (2000). Mathematical thinking in second-grade children with different forms of LD. *Journal of Learning Disabilities, 33*, 567–578. https://doi.org/10.1177/002221940003300605.

Jordan, N. C., Hanich, L. B., & Uberti, H. Z. (2003). Mathematical thinking and learning difficulties. In A. J. Baroody & A. Dowker (Eds.), *The development of arithmetic concepts and skills: Constructing adaptive expertise.* Mahwah, NJ: Lawrence Earlbaum Associates Publishers.

Jordan, N. C., Huttenlocher, J., & Levine, S. C. (1994). Assessing early arithmetic abilities: effects of verbal and nonverbal response types on the calculation performance of middle-and-low income children. *Learning and Individual Differences, 6*, 413–432. https://doi.org/10.1016/1041-6080(94)90003-5.

Jordan, N. C., & Levine, S. C. (2009). Socioeconomic variation, number competence, and mathematics learning difficulties in young children. *Developmental Disabilities Research Reviews, 15*, 60–68. https://doi.org/10.1002/ddrr.46.

Klibanoff, R. S., Levine, S. C., Vasilyeva, M., & Hedges, L. V. (2006). Preschool children's mathematical knowledge: the effect of teacher "mathematics talk" *Developmental Psychology, 42*, 59–69. https://doi.org/10.1037/0012-1649.42.1.59.

Krajewski, K., Schneider, W., & Niedling, G. (2008). On the importance of working memory, intelligence, phonological awareness, and early quantity-number-competencies for the successful transition from kindergarten to elementary school. *Psychologie in Erziehung und Unterricht, 55*, 100–113.

Lansdell, J. M. (1999). Introducing young children to mathematical concepts: problems with 'new' terminology. *Educational Studies, 25*, 327–333. https://doi.org/10.1080/03055699997837.

LeFevre, J.-A., Fast, L., Skwarchuk, S.-L., Smith-Chant, B. L., Bisanz, J., Kamawar, D., et al. (2010). Pathways to mathematics: longitudinal predictors of performance. *Child Development, 81*, 1753–1767. https://doi.org/10.1111/j.1467-8624.2010.01508.x.

Lever, R., & Sénéchal, M. (2011). Discussing stories: on how a dialogic reading intervention improves kindergartners' oral narrative construction. *Journal of Experimental Child Psychology, 108*, 1–24. https://doi.org/10.1016/j.jecp.2010.07.002.

Levine, S. C., Suriyakham, L. W., Rowe, M. L., Huttenlocher, J., & Gunderson, E. A. (2010). What counts in the development of young children's number knowledge? *Developmental Psychology, 46*, 1309–1319. https://doi.org/10.1037/a0019671.

Lewis, C., Hitch, G. J., & Walker, P. (1994). The prevalence of specific arithmetic difficulties and specific reading difficulties in 9- to 10-year-old boys and girls. *Journal of Child Psychology and Psychiatry, 35*, 283–292. https://doi.org/10.1111/j.1469-7610.1994.tb01162.x.

Lonigan, C. J., Anthony, J. L., Bloomfield, B. G., Dyer, S. M., & Samwel, C. S. (1999). Effects of two shared-reading interventions on emergent literacy skills of at-risk preschoolers. *Journal of Early Intervention, 22*, 306–322. https://doi.org/10.1177/105381519902200406.

Mann Koepke, K., & Miller, B. (2013). At the intersection of mathematics and reading disabilities: introduction to the special issue. *Journal of Learning Disabilities, 46*, 483–489. https://doi.org/10.1177/0022219413498200.

McGregor, E. (1994). Economic development and public education: strategies and standards. *Educational Policy, 8*, 252–271. https://doi.org/10.1177/0895904894008003001.

Miller, H. E., Vlach, H. A., & Simmering, V. R. (2017). Producing spatial words is not enough: understanding the relation between language and spatial cognition. *Child Development., 88*, 1966–1982. Advance online publication. https://doi.org/10.1111/cdev.12664.

Mix, K. S., & Cheng, Y. L. (2012). The relation between space and math: developmental and educational implications. *Advances in Child Development and Behavior, 42*, 197–243. https://doi.org/10.1016/B978-0-12-394388-0.00006-X.

National Council of Teachers of Mathematics (2006). *Curriculum focal points for prekindergarten through grade 8 mathematics*. Reston, VA: National Council of Teachers of Mathematics. Retrieved from https://www2.bc.edu/solomon-friedberg/mt190/nctm-focal-points.pdf.

National Mathematics Advisory Panel (2008). *Foundations for success: The final report of the National Mathematics Advisory Panel*. US Department of Education.

Neumann, M. M., Hood, M., Ford, R. M., & Neumann, D. L. (2013). Letter and numeral identification: their relationship with early literacy and numeracy skills. *European Early Childhood Research Journal, 21*, 489–501. https://doi.org/10.1080/1350293X.2013.845438.

Piasta, S. B., Purpura, D. J., & Wagner, R. (2010). Fostering alphabet knowledge development: a comparison of two instructional approaches. *Reading and Writing, 23*, 607–626. https://doi.org/10.1007/s11145-009-9174-x.

Powell, S. R., & Driver, M. K. (2015). The influence of mathematics vocabulary instruction embedded within addition tutoring for first-grade students with mathematics difficulty. *Learning Disability Quarterly, 38*, 221–233. https://doi.org/10.1177/0731948714564574.

Powell, S. R., Fuchs, L. S., Fuchs, D., Cirino, P. T., & Fletcher, J. M. (2009). Do word-problem features differentially affect problem difficulty as a function of students' mathematics difficulty with and without reading difficulty? *Journal of Learning Disabilities*, *42*, 99–110. https://doi.org/10.1177/0022219408326211.

Powell, S. R., & Nelson, G. (2017). An investigation of the mathematics-vocabulary knowledge of first-grade students. *The Elementary School Journal*, *117*, 664–686. https://doi.org/10.1086/691604.

Pruden, S. M., Levine, S. C., & Huttenlocher, J. (2011). Children's spatial thinking: does talk about the spatial world matter? *Developmental Science*, *14*, 1417–1430. https://doi.org/10.1111/j.1467-7687.2011.01088.x.

Purpura, D. J., Day, E. A., Napoli, A. R., & Hart, S. A. (2017). Identifying domain-general and domain-specific predictors of low mathematics performance: a classification and regression tree analysis. *Journal of Numerical Cognition*, *3*, 365–399. https://doi.org/10.5964/jnc.v.3i2.53.

Purpura, D. J., & Ganley, C. M. (2014). Working memory and language: skill-specific or domain-general relations to mathematics? *Journal of Experimental Child Psychology*, *122*, 104–121. https://doi.org/10.1016/j.jecp.2013.12.009.

Purpura, D. J., Hume, L. E., Sims, D. M., & Lonigan, C. J. (2011). Early literacy and early numeracy: the value of including early literacy skills in the prediction of numeracy development. *Journal of Experimental Child Psychology*, *110*, 647–658. https://doi.org/10.1016/j.jecp.2011.07.004.

Purpura, D. J., & Logan, J. A. R. (2015). The nonlinear relations of the approximate number system and mathematical language to early mathematics development. *Developmental Psychology*, *51*, 1717–1724. https://doi.org/10.1037/dev0000055.

Purpura, D. J., & Napoli, A. R. (2015). Early numeracy and literacy: untangling the relation between specific components. *Mathematical Thinking and Learning*, *17*, 197–218. https://doi.org/10.1080/10986065.2015.1016817.

Purpura, D. J., Napoli, A. R., Wehrspann, E. A., & Gold, Z. S. (2017). Causal connections between mathematical language and mathematical knowledge: a dialogic reading intervention. *Journal of Research on Educational Effectiveness*, *10*, 116–137. https://doi.org/10.1080/19345747.2016.1204639.

Purpura, D. J., & Reid, E. E. (2016). Mathematics and language: individual and group differences in mathematical language skills in young children. *Early Childhood Research Quarterly*, *36*, 259–268. https://doi.org/10.1016/j.ecresq.2015.12.020.

Raghubar, K. P., & Barnes, M. A. (2017). Early numeracy skills in preschool-aged children: a review of neurocognitive findings and implications for assessment and intervention. *The Clinical Neuropsychologist*, *31*, 329–351. https://doi.org/10.1080/13854046.2016.1259387.

Ramani, G. B., Rowe, M. L., Eason, S. H., & Leech, K. A. (2015). Math talk during informal learning activities in head start families. *Cognitive Development*, *35*, 15–33. https://doi.org/10.1016/j.cogdev.2014.11.002.

Ramani, G. B., Zippert, E., Schweitzer, S., & Pan, S. (2014). Preschool children's joint block building during a guided play activity. *Journal of Applied Developmental Psychology*, *35*, 326–336. https://doi.org/10.1016/j.appdev.2014.05.005.

Romano, E., Babchishin, L., Pagani, L. S., & Kohen, D. (2010). School readiness and later achievement: replication and extension using a nationwide Canadian survey. *Developmental Psychology*, *46*, 995–1007. https://doi.org/10.1037/a0018880.

Sarama, J., Lange, A. A., Clements, D. H., & Wolfe, C. B. (2012). The impacts of an early mathematics curriculum on oral language and literacy. *Early Childhood Research Quarterly*, *27*, 489–502. https://doi.org/10.1016/j.ecresq.2011.12.002.

Shusterman, A., Slusser, E., Halberda, J., & Odic, D. (2016). Acquisition of cardinal principle coincides with improvement in approximate number system acuity in preschoolers. *PLoS One*. *11*(4), e0153072. https://doi.org/10.1371/journal.pone.0153072.

Spelke, E. S. (2003). What makes us smart? Core knowledge and natural language. In D. Gentner & S. Goldin-Meadow (Eds.), *Language in mind* (pp. 277–311). Cambridge, MA: MIT Press.

Spelke, E. S. (2017). Core knowledge, language, and number. *Language Learning and Development*, *13*, 147–170. https://doi.org/10.1080/15475441.2016.1263572.

Toll, S. W. M., & Van Luit, J. E. H. (2014a). Explaining numeracy development in weak performing kindergartners. *Journal of Experimental Child Psychology*, *124*, 97–111. https://doi.org/10.1016/j.jecp.2014.02.001.

Toll, S. W. M., & Van Luit, J. E. H. (2014b). The developmental relationship between language and low early numeracy skills throughout kindergarten. *Exceptional Children*, *81*, 64–78. https://doi.org/10.1177/0014402914532233.

Willcutt, E. G., Petrill, S. A., Wu, S., Boada, R., DeFries, J. C., Olson, R. K., et al. (2013). Comorbidity between reading disability and mathematics disability: concurrent psychopathology, functional impairment, and neuropsychological functioning. *Journal of Learning Disabilities*, *46*, 500–516.

Chapter 8

Early Numeracy Skills Learning and Learning Difficulties—Evidence-based Assessment and Interventions

Pirjo Aunio
Special Needs Education, Faculty of Educational Sciences, University of Helsinki, Helsinki, Finland

INTRODUCTION

Children's mathematical competencies provide a foundation for later occupational opportunities and for coping with the complexities of life as an adult (Hakkarainen, Holopainen, & Savolainen, 2013; Korhonen, Linnanmäki, & Aunio, 2014). Early numeracy skills are the footing for children's mathematics learning at school, as these predict later achievement in mathematics (Geary et al., 2017; Jordan, Glutting, & Ramineni, 2010). Low performance in early numeracy skills often leads to learning difficulties in mathematics.

There are several different terms used to describe learning difficulties in mathematics, such as low performance in mathematics, learning difficulties in mathematics, mathematical learning disability, dyscalculia, mathematics disorder, and many more. These various terms reflect differences in the mathematics achievement cutoff scores used to define groups and different origins of the problems, ranging from neurological dysfunctions to inappropriate opportunities to learn and practice mathematical skills (e.g., low socioeconomic status of the child's family; Ansari, 2015; Mazzocco, 2009). Geary (2013) suggests that children who score at or below the 10th percentile on standardized mathematics achievement tests for at least two consecutive academic years are categorized as having an MLD (Mathematical learning disability). He further suggests that all children scoring between the 11th and 25th percentiles, inclusive, across two consecutive years are classed as LA (Low Achievers). This is a useful approach for education, but it depends on well-developed assessment scales that are accessible to educators. Teachers

Mathematical Cognition and Learning, Vol. 5. https://doi.org/10.1016/B978-0-12-815952-1.00008-6

not only have to identify children with mathematics difficulties, but they have to adjust their instruction to support the learning of these children, preferably using evidence-based intervention programs if they are available.

Finnish educational authorities have invested in efforts over the last 10 years to boost teachers' levels of knowledge of individual learning differences in early reading and mathematical skills. The emphasis has been on the early identification of learning difficulties and early intervention. At the same time, there has been a nation-wide initiative toward the Responsiveness to Intervention model and general (Tier 1), intensified (Tier 2) and special educational support (Tier 3) in the national education system (National Core Curriculum for Basic Education, 2014/2016). Along with these changes, the National Ministry of Education and Culture has financially supported researchers to produce evidence-based knowledge for educators to use in the classroom and to produce assessment tools and intervention programs for children with mathematical and reading difficulties.

Our research groups in Niilo Mäki Instituutti (University of Jyväskylä) and University of Helsinki have designed two Web services for educators, namely, LukiMat (www.lukimat.fi) and ThinkMath (https://thinkmathglobal.com). LukiMat has three parts: reading, mathematics, and assessment of learning. All parts focus on basic skills learning for 5- to 10-year-olds. Together these services provide educators with evidence-based knowledge about learning and learning difficulties in reading and mathematics, in addition to assessment scales and computerized intervention programs (e.g., Graphogame, the Number Race). ThinkMath provides educators with knowledge about how to efficiently teach children with learning difficulties and hands-on intervention materials to be used with 5- to 8-year-olds, who have problems with learning early mathematical skills. In this chapter, I will describe our research on the evidence-based knowledge, assessment, and intervention tools for teachers.

EARLY NUMERACY SKILLS ARE IMPORTANT FOR FUTURE

Early numeracy provides the foundation for later math learning (Aunio & Räsänen, 2016; Merkley & Ansari, 2016). Critical early numeracy skills include number line estimation and differences in numerical magnitudes (LeFevre et al., 2010; Merkley & Ansari, 2016; Muldoon, Towse, Simms, Perra, & Menzies, 2013), recognition and naming of number symbols (Göbel, Watson, Lervåg, & Hulme, 2014; Pinto, Bigozzi, Tarchi, Vezzani, & Accorti Gamannossi, 2016), numerical relational and counting skills (Aunio & Niemivirta, 2010; Purpura & Reid, 2016), basic addition and subtraction skills and early arithmetical word problem-solving skills (Jordan, Kaplan, Oláh, & Locuniak, 2006). In their recent study, Geary et al. (2017) suggest that all of these skills are first dependent on children's understanding of the cardinal value of number words and numerals by the time they are 4 years old.

Low performance in early numeracy skills is a risk factor for later mathematical learning difficulties (Jordan et al., 2006; Morgan, Farkas, & Wu, 2009) and can be observed during children's average classroom activities. In the classroom, these children have not yet memorized the count sequence, make errors when counting collections of objects, have difficulties comparing the magnitudes of numbers, and have difficulties with basic addition and subtraction (Aunio & Niemivirta, 2010; Desoete, Stock, Schepense, Baeyens, & Roeyers, 2009; Jordan et al., 2006). These differences exist before formal schooling (Aunio, Hautamäki, Sajaniemi, & Van Luit, 2009) and without intervention (Missall, Mercer, Martinez, & Casebeer, 2012).

It is important that researchers communicate with educators in an accessible way. We developed the core factor model of mathematics skills for 5- to 8-year-olds with this in mind (Aunio & Räsänen, 2016). It aims to be a working model for the educators by presenting them with an overview of the most important skills (i.e., core skills) that develop in early childhood and in the early grades. The model also highlights individual differences in early mathematical skills development. This model was based on a systematic literature review of longitudinal studies of mathematical development in this age range. We also analyzed the assessment batteries designed for identifying children with potential learning difficulties in mathematics. We were able to categorize skills into four main groups of numerical factors that are most crucial to the development of mathematical skills: (1) symbolic and nonsymbolic number sense, (2) understanding mathematical relations, (3) counting skills, and (4) basic skills in arithmetic (Table 1; Aunio & Räsänen, 2016). Based on this model we have produced assessment and intervention materials and published them in LukiMat and ThinkMath services.

IDENTIFYING CHILDREN AT RISK FOR MATHEMATICAL LEARNING DIFFICULTIES

One of the practical challenges in education is to identify at-risk children as early as possible. To do so requires the use of valid and reliable assessment scales. There have been discussions about what is the most efficient way to identify children at risk for learning difficulties in mathematics. One possibility is to use Curriculum-based Measurement (CBM) framework as it combines two well-known educational assessment traditions. The first is the conventional achievement testing paradigm, with documentation to support reliability and validity and with normative frameworks to compare a child's performance with national norms. The second is behavioral measurement's time series displays that permit scores for the same individual to be compared at different time points (see Fuchs, 2017).

For instance, Purpura, Reid, Eiland, and Baroody (2015) argued for brief CBM measurements, such as Child Math Assessment—Abbreviated (e.g., Starkey, Klein, & Wakeley, 2004), Brief Research-based Early Mathematics Assessment (Weiland et al., 2012), and Preschool Early Numeracy Skills Screener—Brief version (Purpura et al., 2015). The benefit of a short assessment is ease of use, but other measures are needed to follow at-risk children's learning

TABLE 1 Core Numerical Skills for Learning Mathematics in Children aged 5–8 Years.

Core Numerical Factor	Core Numerical Skills Included in Factors
(1) Symbolic and nonsymbolic number sense	Understanding magnitude embedded in the number word sequence
	Subitising
(2) Understanding mathematical relational skills	One-to-one correspondence related to counting skills
	Basic arithmetic principles (incl. Associativity, commutativity, inversion)
	Base-10 arithmetic strategies (e.g., place value understanding)
(3) Counting skills	Knowledge of number symbols
	Number (word) sequence skills
	Enumeration skills
(4) Basic skills in arithmetic	Addition skills
	Subtraction skills

and to inform instruction and intervention. The monitoring tools can be done several times during the school year and they can help establish whether and to what extent children have acquired key skills (Lei, Wu, DiPerna, & Morgan, 2009). Lei et al. (2009) found that early mathematics CBM tasks are reliable and have potential as screening and progress monitoring tools. However, Laracy, Hojnoski, and Dever (2016) reported positive but low predictive validity with CBM measure (IGDIs-EN) and recommend that none of these measures should be used alone to determine risk of later poor performance. Their results also suggest caution in selecting cutoff scores for instructional or intervention purposes, which is also critical in studies focusing on mathematical learning difficulties (Geary, 2013).

Fuchs (2004) describes two ways of making CBM. One is a curriculum sampling approach, which is tightly linked to curriculum, and requires different assessments for each school year. The second one is the Robust indicator method, which relies more on indicators representing the core competence in mathematics and having good predictive power relative to later mathematics learning. Our approach follows the Robust indicator method of designing CBM. We have developed screening tools for teachers to help identify children who are at risk for learning difficulties. The skills measured originate in the Core Factor model (Aunio & Räsänen, 2016) and are predictive of later mathematics learning. We have concentrated on developing measurements that include several tasks for each of the four factors. This means that our approach is not a brief measurement (i.e., which could be done in 10 min with children), nor is it a pure progress monitoring tool, which can be repeated several times during a school term (Purpura et al., 2015). Instead, we designed scales that include several skills, work as screeners and include norms to support the decision making.

We have devised two sets of screening instruments, one already available for Finnish educators and the other is being designed for Norwegian and South African educators. The first one, LukiMat scales, focuses on core skill factors: symbolic and nonsymbolic number sense, understanding mathematical relations, counting skills, and basic skills in arithmetic. They are designed to be used with kindergarten, first- and second-grade children. Teachers conduct the 30–45-min assessments which can be done individually or in groups. There are three scales that are administered three times a year: in the beginning of the school year (August–September), in the middle of the school year (November–December), and in the end of the school year (April–May). Hellstrand, Korhonen, Räsänen, Linnanmäki, and Aunio (in revision) studied the psychometric features of the LukiMat scales with 1139 children (552 boys) from kindergarten ($n=361$), first grade ($n=321$), and second grade ($n=457$) and found that early numeracy skills were divided into four reliable ($\alpha=0.91$–0.95) factors (symbolic and nonsymbolic number sense, understanding mathematical relations, counting skills, and basic skills in arithmetic) that were all important in all three age groups (Table 2). The findings suggest that a focus on four core numerical skills will be a good working model for educators and researchers in structuring their educational assessments and for supporting instruction and research on children at risk for learning difficulties. The assessment batteries with teacher's manuals, children's materials, and norms are published in the LukiMat web service. We have provided norms for three measurements points in kindergarten, first and second grade. Educators have reported that the assessment is easy to use with groups of children and the results help them design their teaching and educational support (Tier 1 and Tier 2).

The second series of assessment batteries, the ThinkMath-math scales, were primarily developed to measure the effects of early numeracy interventions in Finland (Mononen & Aunio, 2014, 2016). We have also begun developing these for use in South Africa (Aunio et al., in revision; Aunio, Mononen, Ragpot, & Törmänen, 2016) and Norway (Lopez-Pedersen, Mononen, Korhonen, and Aunio, in preparation). The ThinkMath scale includes tasks that assess numerical relational skills, counting skills, and simple arithmetical word problem-solving skills. The scales are administered by teachers to groups of students and take about 30–45 min to complete.

TABLE 2 Core Numerical Skills and Tasks in LukiMat Assessment Batteries

Skills Measured	Content of Tasks
Symbolic and nonsymbolic number sense	Comparing magnitudes, approximate counting
Understanding mathematical relational skills	Seriation, comparison, classification, one-to-one correspondence
	Basic arithmetic principles: additive composition, commutativity, associativity, inversion
	Understanding mathematical symbols
	Understanding place-value and base-ten system
Counting skills	Counting up or back, skip, count from given number
	Number identification, recognition, and writing
	Counting numerosity of a set, counting part of a whole
Basic skills in arithmetic	Addition and subtraction
	Verbal story problems; symbols
Applying understanding of place-value and base-ten system in counting (several skills measured)	Counting sums of money

In South Africa, we used the kindergarten scale in first grade, as the level of performance was estimated to be low. The correlation between ThinkMath total score and Listening Comprehension scale was statistically significant ($r=0.52$, $P<.01$) indicating that language and early numeracy performance were strongly related (see also Fuchs et al., this volume). We further investigated language effects by dividing children based on whether English was their primary language ($n=79$) or English as a second language (ESL; $n=188$). We found clear differences between the groups, indicating that it is not wise to use the same norms for all children. Fig. 1 (and Table 3) illustrates the differences between English as a home language and ESL children in five age groups. The preliminary results from our Norwegian study revealed the same three factors as were found in South Africa, namely, numerical relational skills, counting skills, and simple arithmetical word

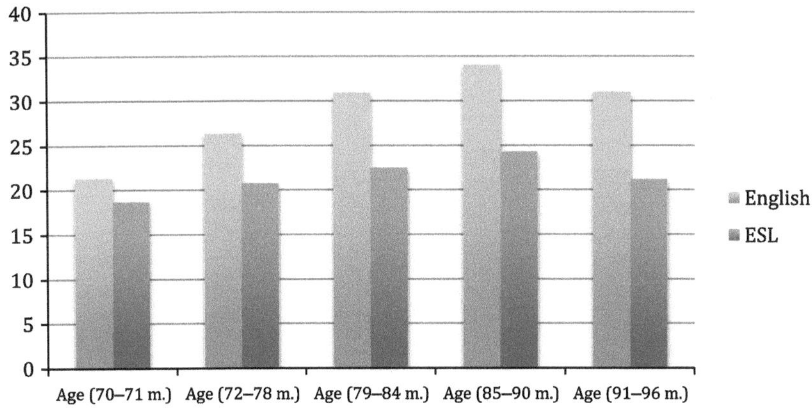

FIG. 1 South Africa ThinkMath total score in age groups and in ESL and non-ESL group.

problems. Again, there was a significant correlation with listening comprehension total score and ThinkMath total score ($r = 0.30$, $P < .01$).

To sum up it is clear that three clusters of skills are important for children's mathematical development from kindergarten to second grade: numerical relational skills (comparison concepts with quantities and comparison of numbers), counting skills (number sequences forwards and backwards with missing number, and number word–quantity–number symbol relations), and simple arithmetical word problems (verbal addition and subtraction problems). With good screening scales that assess each of these three skills, it is possible for educators to identify children at risk for learning difficulties in mathematics. However, we definitely need more longitudinal studies to demonstrate how core skills develop and relate to later mathematical skills learning. There are also other future challenges, for instance, how to develop evidence-based short screening and deeper monitoring scales so that all educators will have access to them. In addition, it is important to further investigate the relation between language and mathematical performance (Fuchs et al., this volume; Pupura et al., this volume; Purpura, Hume, Sims, & Lonigan, 2011).

EARLY NUMERACY INTERVENTIONS FOR LOW-PERFORMING CHILDREN

Meta-analyses of group-based interventions for children with mathematical learning difficulties show that explicit instructional procedures are more effective than other instructional approaches (Baker, Gersten, & Lee, 2002; Dennis et al., 2016; Gersten et al., 2009; Kroesbergen & Van Luit, 2003; Mononen, Aunio, Koponen, & Aro, 2014; Swanson, Hoskyn, & Lee, 1999). Explicit interventions include, for instance, breaking the instruction logical sequences, clearly presenting subject matter, guided practice, independent

TABLE 3 South Africa ThinkMath Total Score in Age Groups and in ESL and Non-ESL Group

			ThinMath Total Score					
Group	*Age*		*English (Home Language)*			*ESL*		
			n	M	*SD*	n	M	*SD*
1	70–71 m.	<6 years	3	21.33	11.68	6	18.67	9.63
2	72–78 m.	6 y. 0 m.–6y. 6 m.	17	26.35	9.43	61	20.72	8.41
3	79–84 m.	6 y. 7 m.–7 y. 0 m.	32	30.91	9.81	76	22.46	8.07
4	85–90 m.	7 y. 1 y.–7 y. 6 m.	19	34.05	9.08	35	24.31	10.15
5	91–96 m.	7 y. 7 m.–8 y. 0 m.	7	31.00	9.45	6	21.17	10.30

Note: ELS is English as a second language.

practice, evaluating student learning on a regular basis, and providing feedback. Explicit instruction often includes using concrete-representational-abstract (CRA) sequence which has been found to be beneficial for learning (Miller, Butler, & Lee, 1998; Mononen et al., 2014; Xin & Jitendra, 1999). Peer-assisted instruction has been found to be effective with younger students (Baker et al., 2002; Kunsch, Jitendra, & Sood, 2007). The effects of CAI in interventions for children with learning difficulties in mathematics are controversial; some have found support (Kroesbergen & Van Luit, 2003; Miller et al., 1998; Mononen et al., 2014) and others are more critical about its effectiveness (Dennis et al., 2016; Räsänen, Salminen, Wilson, Aunio, & Dehaene, 2009).

Previous research shows that research-based intervention programs that are provided with care and whose effectiveness has been investigated produce better classroom learning than non-research-based interventions (Jacob & Parkinson, 2015; Slavin & Lake, 2008). However, it can be quite difficult to compare the efficacy of various intervention programs. The challenge originates from the fact that intervention programs and studies can vary in many ways. There can be differences in aims, content, target group, organization, and research design (see table shown in Table 4).

The aim of ThinkMath intervention programs is to support the learning of early numeracy skills, especially the skills related to an understanding of numerical relations, counting, and basic arithmetic. Our focus is on low-performing children in kindergarten, first and second grade, and our goal is to prevent long-term learning difficulties in mathematics. ThinkMath materials are designed to be used with groups of 4–6 children as supplementary materials to typical classroom instruction. The teacher manual includes 12–15 lesson plans of 35–45 min each. The preliminary idea was to have 2–3 sessions per week, so that the intervention phase would last about 6 weeks (together about 8 sessions of 30 min, making total 540 min), hence suitable for Tier 1 and 2 support. ThinkMath materials are designed to be used by educators and are freely available (https://thinkmathglobal.com/english/materials/).

Instruction follows explicit teaching guidelines (Gersten et al., 2008, 2009) along with several ways to practice the specific skills. Each lesson consists of a teacher-guided activity that models a new numeracy concept or strategy as well as guided and peer activities (e.g., hands-on activities with manipulatives, or card and board games based on the current topic). At the end of the lesson, there is a short, paper-and-pencil individual activity. Another general instructional feature is that early numeracy ideas are represented following the concrete, representational, and abstract levels, thus giving meaning to abstract concepts by using visual representations (e.g., cubes, bundles of sticks, dot cards structured in tens and hundreds). During the development of the intervention materials, we worked closely with educators and investigated the effects of these programs on low-performing children using a quasi-experimental design: we conducted pretest, posttest, and delayed posttest measurements with intervention and control (business-as-usual) groups. The outcome measurements were the ThinkMath-math scales.

TABLE 4 Characters of Educational Intervention Studies

Character		Example of Studies Including the Character
Aims of intervention	Preventive	Toll & Van Luit, 2014
	Remedial	Kucian et al., 2011
Content of intervention	*Math skills practiced*	
	Specific mathematic skills	Siegler & Ramani, 2009
	Multiskills	Clarke et al., 2011
	Basic skills	Aunio, Hautamäki, & Van Luit, 2005
	complex skills	Sharp & Dennis, 2017
	Instructional features	
	Explicit teaching	Toll & Van Luit, 2012, 2014
	Use of concrete manipulatives	Toll & Van Luit, 2012, 2014
	Several ways to practice	Mononen & Aunio, 2014
	Use of computers	Aunio & Mononen, 2017
Target group	Children with low performance	Toll & Van Luit, 2012
	Children at risk for mathematical learning difficulties	Clarke et al., 2016
	Children with mathematical learning difficulties	Kucian et al., 2011
	Preschool and kindergarten children	Dyson, Jordan, & Glutting, 2011
	Primary school students	Opitz et al., 2017
	Middle school students	Zhang, Stecker, Huckabee, & Miller, 2016
Organization	*Size of intervention group*	
	Individual	Superkar et al., 2013
	Small group (3–7 children)	Bryant, Bryant, Gersten, Scammacca, & Chavez, 2008
	Large group	Clarke et al., 2011
	Relation to classroom instruction	

TABLE 4 Characters of Educational Intervention Studies—Cont'd

Character		Example of Studies Including the Character
	Core instruction	Clarke et al., 2011
	Supplementary intervention	Moran, Swanson, Gerber, & Fung, 2014
	Time allocated	see meta-analysis by Dennis et al., 2016
	How long time totally?	
	How many sessions?	
	How long sessions?	
	How many sessions in a week?	
	Intervention agent	
	Teacher	Powell et al., 2015
	Researcher	Dyson et al., 2011
Research design	*Time of measurements*	
	Pretest-posttest	Clarke et al., 2016
	Preimmediate postdelayed posttests	Dyson et al., 2011
	Outcome measures	
	Specific skills practiced	Siegler & Ramani, 2009
	Large set of mathematical skills	Sood & Jitendra, 2011
	Researcher made	Klein, Starkey, Clements, Sarama, & Iyer, 2008
	Standardized test batteries	Praet & Desote, 2014
	Between group comparison	
	Active control group	Fuchs, Fuchs, & Compton, 2012
	Passive control group	Toll & Van Luit, 2014

Studies With ThinkMath Intervention Programs

The second-grade intervention study was done with 88 children (M age 8 y. 2 m.) from four classes in schools located in two southern Finnish cities (Mononen & Aunio, 2014). The intervention program was Improving Mathematics Skills in the Second Grade (Mononen & Aunio, 2012a). The program focuses on the practice number word sequence skills, counting, and conceptual place value knowledge in the 1–1000 range and follows the guidelines of explicit instructions. Children's early numeracy skills were measured with the Assessment of Mathematics Skill in the Second Grade (AMS-2) (Aunio & Mononen, 2012c). It is a paper-and-pencil test and measures (1) skills to recite number words forward and backwards, (2) numerical relational skills associated with base 10 and place value knowledge, (3) addition and subtraction word problems, (4) multidigit addition and subtraction calculations with number symbols, all within a 1–1000 range, and (5) addition and subtraction facts in the 1–20 range (40 items, 2 min' time). Children's reasoning skills were assessed using the Assessment of Thinking Skills in the Second Grade (Hotulainen, Mononen, & Aunio, 2012a). Reading comprehension and fluency skills were measured using a standardized reading test for primary grades (Lindeman, 1998). Early numeracy skills were measured three times: before the intervention, immediately following the intervention, and 3 months after the intervention. The thinking and reading skills were assessed at the first time point.

The criterion for low performance was a score at or below the 25th percentile in the early numeracy test in the beginning of the study. Children were divided in the low-performing intervention group ($n=11$), the low-performing control group ($n=13$), and the typically performing control group ($n=64$ children). The intervention program lasted 6 weeks and included two 45-min intervention sessions per week. The results demonstrated that the children in the low-performing intervention group made significant improvements in early numeracy overall, and addition and subtraction facts but did not show significantly better scores compared to the low-performing control group. Furthermore, neither the intervention children nor the control children were able to perform at the same level as their peers following the intervention. One interesting detail is that after the intervention, children's' fluency in addition and subtraction actually decreased among the intervention group. There was no difference between low-performance children in the control and intervention groups in terms of their reasoning and reading skills. Although there were not many scientifically significant differences between intervention and control children, there was a trend such that when children with low early numeracy skills received extra support, their skills developed. However, then the support ended so did the development of their skills (see also Bailey, this volume). This was especially true with arithmetical fluency skills.

Mononen and Aunio (2016) investigated the impact of ThinkMath intervention for Finnish first graders ($N=151$, M age = 7 y. 2 m.) with low numerical performance. The children were from nine classrooms. This program focused on increasing the counting skills knowledge, including the number sequence and enumeration skills in number range 1–20 (Mononen & Aunio, 2012b). The study used a quasi-experimental design with control groups. The intervention effects were examined using the Assessment of Mathematics

Skills in the First Grade (Aunio & Mononen, 2012b). This group-based paper-and-pencil test includes mathematical tasks in the range from 1 to 100: (1) numerical relational skills (i.e., number comparison), (2) counting skills (i.e., verbal and object counting), and (3) word problem-solving skills (i.e., verbal addition and subtraction problems). Single-digit addition and subtraction facts in the range from 1 to 20 are also assessed (within a 2-min time limit). A sum score for the subscales 1–3 (i.e., a combined scale) was used to identify the low-performing children. In addition, the relations between inductive reasoning (Assessment of Thinking Skills in the First Grade by Hotulainen, Mononen, & Aunio, 2012b), language (reading fluency Allu by Lindeman, 1998; listening comprehension Ytte test by Kajamies, Poskiparta, Annevirta, Dufva, & Vauras, 2003), and early numeracy skills were examined.

The intervention program was administered 12 times over 8 weeks to small groups of children in sessions that lasted about 45 min. We used the same at-risk cutoff as in our previous study (i.e., at or below the 25th percentile in early numeracy). The numerical development of intervention children ($n = 11$) was compared to the development of low-performing ($n = 26$) and average-performing ($n = 114$) children. The results showed that children in the intervention group made significantly greater gains in their early numeracy performance from Time 1 to Time 2, compared with the low-performing control and average-performing children. One important finding was that the children with low performance in early numeracy skills also showed relatively week reasoning and reading skills relative to children with average performance, also found by others (e.g., Geary, Hoard, Nugent, & Bailey, 2012). This means that we need to also think about how to support children's thinking and language skills early on (see also Fuchs et al., this volume).

In our kindergarten intervention study (Mononen, Aunio, & Leijo, in revision), the children in the low-performing group were studied in more detail. The children in this group had early numeracy scores below the 10th percentile (i.e., very low performing, $n = 20$), and children whose early numeracy scores were between the 11th and 25th percentiles (i.e., low performing, $n = 18$). The data were collected with a scale measuring (1) numerical relational skills (i.e., number comparison), (2) number-word-sequence skills, and (3) counting skills (Aunio & Mononen, 2012a). The results showed that the number of intervention children who reached an average level of performance at the posttest stage was higher among the group of children with low performance (67%) versus those with very low performance (35%; Mononen, Tapola, & Aunio, 2015).

Westerholm and Aunio (in press) investigated the effects of ThinkMath intervention for nine kindergarteners for whom Finnish was a second language. The children were included in a one-year, special education program due to their challenges in learning early academic skills in Finnish. Children's early numeracy skills were measured with the Early Numeracy Test (Van Lujt, Van de Rijt, & Aunio, 2006) before, immediately after, and 5 weeks after the intervention ended. The test measures numerical knowledge, including the concepts of comparison, classification, one-to-one correspondence, seriation, the use of number words, structured counting, resultative counting, and the general understanding of numbers.

In this study, we used Mathematical relational and counting skills 1–20 intervention program (Mononen & Aunio, 2012b). Children practiced making comparisons of quantities and numbers using related concepts such as more and less and counting number sequences orally. In addition, counting objects and matching them with number words and number symbols were practiced. Based on the research literature concerning the learning challenges of children for whom the language of instruction is a second language (Arnold, Fisher, Doctoroff, & Dobbs, 2002; Clements & Sarama, 2008; Codding, Chan-Iannetta, Palmer, & Lukito, 2009; Fuchs, Compton, et al., 2008; Fuchs, Fuchs, et al., 2008; Klein et al., 2008; Mercer & Sams, 2006; Starkey et al., 2004), we added motivation (i.e., prizes) and explanations of basic concepts into the program. The intervention program lasted 8 weeks and included 35–75 min sessions twice a week. The intervention effects were significant immediately after the intervention for the early numeracy scores and numerical relational tasks. There were also lasting effects on numerical relational skills 5 weeks after the intervention ended. The ThinkMath intervention program with added motivating features and explicit talk was a useful way to support the early numeracy learning of children who had Finnish as a second language.

LukiMat web service provides two evidence-based computer games in which children can practice their early numeracy skills. The first, Number Race is based on an idea that number skills develop from approximate representations of magnitudes, and with the aid of counting these representations are connected to numbers (Wilson, Dehaene, Pinel, Revkin, & Cohen, 2006; Wilson, Dehaene, Dubois, & Fayol, 2009; Wilson, Revkin, Cohen, Cohen, & Dehaene, 2006). The second game GraphoGame Math was originally designed as a part of the GraphoGame project at the University of Jyväskylä in Finland (Richardson & Lyytinen, 2014). The GraphoGame Math is based on the idea that learning the correspondences between small sets of objects and numbers helps the child to discover the relationships in the number system and arithmetic. The research has demonstrated that both games have positive effects on early numeracy learning of children with learning difficulties in mathematics (Räsänen et al., 2009; Salminen, Koponen, Leskinen, Poikkeus, & Aro, 2015; Salminen, Koponen, Räsänen, & Aro, 2015).

To sum up, our early numeracy intervention programs show promise, but their effectiveness needed to be assessed with larger samples and including randomized designs. We also need to follow-up on how educators are adapting their instruction based on individual needs of the children in their classroom (Fuchs & Fuchs, 2016).

CONCLUSIONS AND FUTURE DIRECTIONS

The work on Finnish evidence-based web services has not only shown that it is possible to develop reliable and useful assessment and intervention instruments for educators, but also that Web services are a very efficient way to deliver such information and materials. As there has been substantial progress in research knowledge related to mathematical learning difficulties in various parts of the world, it would be beneficial to consider to how this valuable knowledge can be provided for all educators so that they can support children's learning the best way.

The experience with designing evidence-based assessment and intervention instruments has also shown that identification of young children with problems in mathematics learning and providing educational support needs to into account the other skills that support mathematics learning, including basic language (see Purpura et al., this volume) and reasoning skills. We need to be able to design our interventions in ways that accommodate language, thinking, or executive function skills (see also Fuchs et al., this volume). This is in line with Bailey et al. (2016) who found out that preexisting differences in children and their home environments explain more of the fadeout effects of interventions than school environmental factors (see also Bailey, this volume).

REFERENCES

Ansari, D. (2015). *No more math wars—an evidence-based, developmental perspective on math education*. Education Canada, September 2015. Retrieved from www.cea-ace.ca.

Arnold, D. H., Fisher, P. H., Doctoroff, G. L., & Dobbs, J. (2002). Accelerating math development in head start classrooms. *Journal of Educational Psychology*, *94*(4), 762–770.

Aunio, P., Hautamäki, J., Sajaniemi, N., & Van Luit, J. E. H. (2009). Early numeracy in low-performing young children. *British Educational Research Journal*, *35*(1), 25–46.

Aunio, P., Hautamäki, J., & Van Luit, J. E. H. (2005). Mathematical-thinking intervention programmes for preschool children with normal and low number sense. *European Journal of Special Needs Education*, *20*, 131–146.

Aunio, P., Korhonen, J., Ragpot, L., Mononen, R., Törmänen, M. & Henning, E. (in revision). Multi-factorial approach to early numeracy—the effects of cognitive skills, language factors and kindergarten attendance on early numeracy performance of South African school beginners.

Aunio, P., & Mononen, R. (2012a). *Assessment of mathematical skills in the Kindergarten*. Unpublished. Helsinki, Finland: University of Helsinki.

Aunio, P., & Mononen, R. (2012b). *Assessment of mathematical skills in the first grade*. Unpublished. Helsinki, Finland: University of Helsinki.

Aunio, P., & Mononen, R. (2012c). *Assessment of mathematical skills in the second grade*. Unpublished. Helsinki, Finland: University of Helsinki.

Aunio, P., & Mononen, R. (2017). The effects of educational computer game on low performing children's early numeracy skills—an intervention study in preschool setting. *European Journal of Special Needs Education*, 1–15.

Aunio, P., Mononen, R., Ragpot, L., & Törmänen, M. (2016). Early numeracy performance of South African school beginners. *South African Journal of Childhood Education*, *6*(1), 1–8. https://doi.org/10.4102/sajce.v6i1.496.

Aunio, P., & Niemivirta, M. (2010). Predicting children's mathematical performance in grade one by early numeracy skills. *Learning and Individual Differences*, *20*, 427–435.

Aunio, P., & Räsänen, P. (2016). Core numerical skills for learning mathematics in children aged five to eight years—a working model for educators. *European Early Childhood Education Research Journal*, *24*(5), 684–704. https://doi.org/10.1080/1350293X.2014.996424.

Bailey, D. H., Nguyen, T., Jenkins, J. M., Domina, T., Clements, D. H., & Sarama, J. S. (2016). Fadeout in an early mathematics intervention: constraining content or pre-existing differences? *Developmental Psychology, 52*(9), 1457–1469.

Baker, S., Gersten, R., & Lee, D. (2002). A synthesis of empirical research on teaching mathematics to low-achieving students. *The Elementary School Journal, 103*, 51–73.

Bryant, D. P., Bryant, B. R., Gersten, R. M., Scammacca, N. N., & Chavez, M. M. (2008). Mathematics intervention for first- and second-grade students with mathematics difficulties: the effects of tier 2 intervention delivered as booster lessons. *Remedial and Special Education, 29*, 20–32.

Clarke, B., Doabler, C. T., Smolkowski, K., Baker, S. K., Fien, H., & Cary, M. S. (2016). Examining the efficacy of a tier 2 kindergarten mathematics intervention. *Journal of Learning Disabilities, 49*(2), 152–165.

Clarke, B., Smolkowski, K., Baker, S., Fien, H., Doabler, C. T., & Chard, D. J. (2011). The impact of comprehensive tier 1 core kindergarten program on the achievement of students at risk in mathematics. *The Elementary School Journal, 111*(4), 561–584.

Clements, D. H., & Sarama, J. (2008). Experimental evaluation of the effects of a research-based preschool mathematics curriculum. *American Educational Research Journal, 45*(2), 443–494.

Codding, R. S., Chan-Iannetta, L., Palmer, M., & Lukito, G. (2009). Examining a classwide application of cover-copy-compare with and without goal setting to enhance mathematics fluency. *School Psychology Quarterly, 24*(3), 173–185.

Dennis, S. M., Sharp, E., Chovanes, J., Thomas, J., Burns, R. M., Custer, B., et al. (2016). A meta-analysis of empirical research on teaching students with mathematics learning difficulties. *Learning Disabilities Research and Practice, 31*(3), 156–168.

Desoete, A., Stock, P., Schepense, A., Baeyens, D., & Roeyers, H. (2009). Classification, seriation, and counting in grades 1, 2, and 3 as two-year longitudinal predictors for low achieving in numerical facility and arithmetical achievement? *Journal of Psychoeducational Assessment, 27*(3), 252–264.

Dyson, N. I., Jordan, N. C., & Glutting, J. (2011). A number sense intervention for low-income kindergartners at risk for mathematics difficulties. *Journal of Learning Disabilities, 46*(2), 166–181.

Fuchs, D., & Fuchs, L. (2016). Responsiveness-to-intervention: a "systems" approach to instructional adaptation. *Theory Into Practice, 55*(3), 225–233. https://doi.org/10.1080/00405841.2016.1184536.

Fuchs, L. S. (2004). The past, present, and future of curriculum-based measurement research. *School Psychology Review, 33*(2), 188–192.

Fuchs, L. S. (2017). Curriculum-based measurement as the emerging alternative: two decades later. *Learning Disabilities Research and Practice, 32*(1), 5–7.

Fuchs, L. S., Compton, D. L., Fuchs, D., Hollenbeck, K. N., Craddock, C. F., & Hamlett, C. L. (2008). Dynamic assessment of algebraic learning in predicting third graders' development of mathematical problem solving. *Journal of Educational Psychology, 100*(4), 829–850.

Fuchs, L. S., Fuchs, D., & Compton, D. L. (2012). Intervention effects for students with comorbid forms of learning disability: understanding the needs of nonresponders. *Journal of Learning Disabilities, 46*(86), 534–548.

Fuchs, L. S., Fuchs, D., Craddock, C., Hollenbeck, K. N., Hamlett, C. L., & Schatschneider, C. (2008). Effects of small-group tutoring with and without validated classroom instruction on at-risk students' math problem solving: are two tiers of prevention better than one? *Journal of Educational Psychology, 100*(3), 491–509.

Geary, D. C. (2013). Learning disabilities in mathematics. Recent advances. In H. L. Swanson, K. R. Harris, & S. Graham (Eds.), *Handbook of learning disabilities* (2nd ed., pp. 239–255). New York: The Guilford Press.

Geary, D. C., Hoard, M. K., Nugent, L., & Bailey, D. H. (2012). Mathematical cognition deficits in children with learning disabilities and persistent low achievement: a five-year prospective study. *Journal of Educational Psychology*, *104*, 206–223.

Geary, D. C., van Marle, K., Chu, F., Rouder, J., Hoard, M. K., & Nugent, L. (2017). Early conceptual understanding of cardinality predicts superior school-entry number system knowledge. *Psychological Science*, 1–15.

Gersten, R., Chard, D., Jayanthi, M., Baker, S. K., Morphy, P., & Flojo, J. (2009). Mathematics instruction for students with learning disabilities: a meta-analysis of instructional components. *Review of Educational Research*, *79*(3), 1202–1242.

Gersten, R., Chard, D. J., Jayanthi, M., Baker, S. K., Morphy, P., & Flojo, J. (2008). *Mathematics instruction for students with learning disabilities or difficulty learning mathematics. A synthesis of the intervention research*. Portsmouth, NH: RMC Research Corporation, Centre on Instruction.

Göbel, S. M., Watson, S. E., Lervåg, A., & Hulme, C. (2014). Children's arithmetic development: it is number knowledge, not the approximate number sense, that counts. *Psychological Science*, *25*(3), 789–798.

Hakkarainen, A., Holopainen, L., & Savolainen, H. (2013). Mathematical and reading difficulties as predictors of school achievement and transition to secondary education. *Scandinavian Journal of Educational Research*, *57*(5), 488–506.

Hellstrand, H., Korhonen, J., Räsänen, P., Linnanmäki, K. & Aunio, P. (in revision). Validating the theoretical model of core numerical skills for learning mathematics in children aged five to eight years.

Hotulainen, R., Mononen, R., & Aunio, P. (2012a). *Assessment of thinking skills in the second grade*. Unpublished. Helsinki, Finland: University of Helsinki.

Hotulainen, R., Mononen, R., & Aunio, P. (2012b). *Assessment of thinking skills in the first grade*. Unpublished. Helsinki, Finland: University of Helsinki.

Jacob, R., & Parkinson, J. (2015). The potential for school-based interventions that target executive function to improve academic achievement: a review. *Review of Educational Research*, *85*(4), 512–552.

Jordan, N. C., Glutting, J., & Ramineni, C. (2010). The importance of number sense to mathematics achievement in first and third grades. *Learning and Individual Differences*, *20*, 82–88. https://doi.org/10.1016/j.lindif.2009.07.004.

Jordan, N. C., Kaplan, D., Oláh, L., & Locuniak, M. N. (2006). Number sense growth in kindergarten: a longitudinal investigation of children at risk for mathematics difficulties. *Child Development*, *77*(1), 153–175. https://doi.org/10.1111/j.1467-8624.2006.00862.x.

Kajamies, A., Poskiparta, E., Annevirta, T., Dufva, M., & Vauras, M. (2003). *YTTE Luetun ja kuullun ymmärtäminen ja lukemisen sujuvuuden arviointi (Assessment of reading and listening comprehension and reading fluency)*. Oppimistutkimuskeskus OTUK, Turku, Finland: University of Turku.

Klein, A., Starkey, P., Clements, D., Sarama, J., & Iyer, R. (2008). Effects of a pre-kindergarten mathematics intervention: a randomized experiment. *Journal of Research on Educational Effectiveness*, *1*(3), 155–178.

Korhonen, J., Linnanmäki, K., & Aunio, P. (2014). Learning difficulties, academic well-being and educational dropout: a person-centred approach. *Learning and Individual Differences*, *31*, 1–10.

Kroesbergen, E. H., & Van Luit, J. E. H. (2003). Mathematics interventions for children with special educational needs. A meta-analysis. *Remedial and Special Education*, *24*(2), 97–114.

Kucian, K., Grond, U., Rotzer, S., Henzi, B., Schönmann, C., et al. (2011). Mental number line training in children with developmental dyscalculia. *NeuroImage*, *57*(3), 782–795.

Kunsch, C. A., Jitendra, A. K., & Sood, S. (2007). The effects of peer-mediated instruction in mathematics for students with learning problems: a research synthesis. *Learning Disabilities Research and Practice*, *22*(1), 1–12.

Laracy, S. D., Hojnoski, R. L., & Dever, B. V. (2016). Assessing the classification accuracy of early numeracy curriculum-based measures using receiver operating characteristic curve analysis. *Assessment for Effective Intervention*, *41*(3), 172–183.

LeFevre, J.-A., Fasr, L., Skwarchuk, S.-L., Smith-Chant, B. L., Bisanz, J., Kamawar, D., et al. (2010). Pathways to mathematics: longitudinal predictors of performance. *Child Development*, *81*(6), 1753–1767. https://doi.org/10.1111/j.1467-8624.2010.01508.x.

Lei, P.-W., Wu, Q., DiPerna, J. C., & Morgan, P. L. (2009). Developing short forms of the EARLI numeracy measures. Comparison of item selection methods. *Educational and Psychological Measurement*, *69*(5), 825–842.

Lindeman, J. (1998). *Ala-asteen lukutesti ALLU. [Reading test for primary grades]*. Turku, Finland: University of Turku.

Lopez-Pedersen, A., Mononen, R., Korhonen, J. & Aunio, P. (in preparation). The validity of ThinkMath screener in Norwegian first graders.

Mazzocco, M. M. M. (2009). Defining and differentiating mathematical learning disabilities and difficulties. In M. M. M. Mazzocco & D. B. Berch (Eds.), *Why is math so hard for some children? The nature and origins of mathematical learning difficulties and disabilities* (pp. 29–48). Baltimore, Maryland: Paul H. Brookes.

Mercer, N., & Sams, C. (2006). Teaching children how to use language to solve maths problems. *Language and Education*, *20*(6), 507–528.

Merkley, R., & Ansari, D. (2016). Why numerical symbols count in the development of mathematical skills: evidence from brain and behavior. *Current Opinion in Behavioral Sciences*, *10*, 14–20.

Miller, S. P., Butler, F. M., & Lee, K. (1998). Validated practices for teaching mathematics to students with learning disabilities: a review of literature. *Focus on Exceptional Children*, *31*(1), 1–24.

Missall, K. N., Mercer, S. H., Martinez, R. S., & Casebeer, D. (2012). Concurrent and longitudinal patterns and trends in performance on early numeracy curriculum-based measures in kindergarten through third grade. *Assessment for Effective Intervention*, *37*, 95–106. https://doi.org/10.1177/1534508411430322.

Mononen, R., & Aunio, P. (2012a). *Improving mathematics skill in the second grade. An intervention program.* Unpublished. Helsinki, Finland: University of Helsinki.

Mononen, R., & Aunio, P. (2012b). *Improving mathematics skill in the first grade. An intervention program.* Unpublished. Helsinki, Finland: University of Helsinki.

Mononen, R., & Aunio, P. (2014). A mathematics intervention for low-performing Finnish second graders: findings from a pilot study. *European Journal of Special Needs Education*, *29*(4), 457–473. https://doi.org/10.1080/08856257.2014.922794.

Mononen, R., & Aunio, P. (2016). Math intervention for low-performing first graders. *African Journal of Childhood Education*, *6*(1), a407.

Mononen, R., Aunio, P., Koponen, T., & Aro, M. (2014). A review of early numeracy interventions: effectiveness and pedagogical implementation. *International Journal of Early Childhood Special Education*, *6*(1), 25–54.

Mononen, R., Aunio, P. & Leijo, S. (in revision). Matemaattisten taitojen harjoittelun vaikutukset matematiikan taidoiltaan heikkojen esiopetusikäisten lasten osaamiseen (The effects of mathematical intervention on low performing children's performance in kindergarten).

Mononen, R., Tapola, A., & Aunio, P. (2015). In *Intensified mathematical support for low performing kindergarteners. Paper presentation in symposium. Conference of European Association for Research on Learning and Instruction, 25–29 August, 2015, Limassol, Cyprus.*

Moran, A. S., Swanson, H. L., Gerber, M. M., & Fung, W. (2014). The effects of paraphrasing interventions on problem solving accuracy for children at risk for math disabilities. *Learning Disabilities Research and Practice*, *29*(3), 97–105.

Morgan, P. L., Farkas, G., & Wu, Q. (2009). Five-year growth trajectories of kindergarten children with learning difficulties in mathematics. *Journal of Learning Disabilities*, *42*, 306–321. https://doi.org/10.1177/0022219408331037.

Muldoon, K., Towse, J., Simms, V., Perra, O., & Menzies, V. (2013). A longitudinal analysis of estimation, counting skills, and mathematical ability across the first school year. *Developmental Psychology*, *49*(2), 250.

National Core Curriculum for Basic Education. (2014/2016). *Finnish national board of education.* Finland: Helsinki.

Opitz, E., Freesemann, O., Prediger, S., Grob, U., Matull, I., & Husmann, S. (2017). Remediation for students with mathematics difficulties: an intervention study in middle schools. *Journal of Learning Disabilities*, *50*(6), 724–736.

Pinto, G., Bigozzi, L., Tarchi, C., Vezzani, C., & Accorti Gamannossi, B. (2016). Predicting reading, spelling, and mathematical skills: a longitudinal study from kindergarten through first grade. *Psychological Reports*, *118*(2), 413–440.

Powell, S. R., Fuchs, L. S., Cirino, P. T., Fuchs, D., Compton, D. L., & Changas, P. C. (2015). Effects of a multitier support system on calculation, word problem and prealgebraic performance among at-risk learners. *Exceptional Children*, *81*(4), 443–470.

Praet, M., & Desote, A. (2014). Enhancing young children's arithmetic skills through non-intensive, computerised kindergarten interventions: a randomised controlled study. *Teaching and Teacher Education*, *39*, 56–65.

Purpura, D. J., Hume, L. E., Sims, D. M., & Lonigan, C. (2011). Early literacy and early numeracy: the value of including early skills in the prediction of numeracy development. *Journal of Experimental Child Psychology*, *110*, 647–658.

Purpura, D. J., Reid, E., Eiland, M. D., & Baroody, A. J. (2015). Using a brief preschool early numeracy skills screener to identify young children with mathematics difficulties. *School Psychology Review*, *44*(1), 41–59.

Purpura, D. J., & Reid, E. E. (2016). Mathematics and language: individual and group differences in mathematical language skills in young children. *Early Childhood Research Quarterly*, *36*, 259–268.

Räsänen, P., Salminen, J., Wilson, A., Aunio, P., & Dehaene, S. (2009). Computer-assisted intervention for early number skills in children with low numeracy: a short review and a randomized controlled trial. *Cognitive Development*, *24*(4), 450–472.

Richardson, U., & Lyytinen, H. (2014). The GraphoGame method: the theoretical and methodological background of the technology-enhanced learning environment for learning to read. *Human Technology: An Interdisciplinary Journal on Humans in ICT Environments*, *10*(1), 39–60.

Salminen, J., Koponen, T., Räsänen, P., & Aro, M. (2015). Preventive support for kindergarteners most at-risk for mathematics difficulties: computer-assisted intervention. *Mathematical Thinking and Learning*, *17*(4), 273–295. https://doi.org/10.1080/10986065.2015.1083837.

Salminen, J. B., Koponen, T. K., Leskinen, M., Poikkeus, A.-M., & Aro, M. (2015). Individual variance in responsiveness to early computerized mathematics intervention. *Learning and Individual Differences*, *43*, 124–131.

Sharp, E., & Dennis, M. S. (2017). Model drawing strategy for fraction word problem solving of fourth-grade students with learning disabilities. *Remedial and Special Education*, *38*(3), 181–192.

Siegler, R. S., & Ramani, G. B. (2009). Playing linear board games—but not circular ones—improves low-income preschoolers' numerical understanding. *Journal of Educational Psychology*, *101*(3), 545–560.

Slavin, R. E., & Lake, C. (2008). Effective programs in elementary mathematics: a best-evidence synthesis. *Review of Educational Research*, *78*(3), 427–515.

Sood, S., & Jitendra, A. K. (2011). An exploratory study of a number sense program to develop kindergarten students' number proficiency. *Journal of Learning Disabilities*, *46*(4), 328–346.

Starkey, P., Klein, A., & Wakeley, A. (2004). Enhancing young children's mathematical knowledge through a pre-kindergarten mathematics intervention. *Early Childhood Research Quarterly*, *19*(1), 99–120.

Superkar, K., Swigart, A. G., Tenison, C., JOlles, D. D., Rosenberg-Lee, M., Fuchs, L., et al. (2013). Nerual predictors of individual differences in response to math tutoring in primary-grade school children. *PNAS*, *110*(20), 8230–8235.

Swanson, H. L., Hoskyn, M., & Lee, C. (1999). *Interventions for students with learning disabilities: A meta-analysis of treatment outcomes*. New York: Guilford.

Toll, S. W. M., & Van Luit, J. E. H. (2012). Early numeracy intervention for low-performing kindergartners. *Journal of Early Intervention*, *34*(4), 243–264.

Toll, S. W. M., & Van Luit, J. E. H. (2014). Effects of remedial numeracy instruction throughout kindergarten starting at different ages. Evidence from a large-scale longitudinal study. *Learning and Instruction*, *33*, 39–49.

Van Lujt, J. E. H., Van de Rijt, B. A. M., & Aunio, P. (2006). *Lukukäsitetesti. (Early Numeracy Test)*. Helsinki, Finland: Psykologien kustannus Oy.

Weiland, C., Wolfe, C. B., Hurwitz, M. D., Clements, D. H., Sarama, J. H., & Yoshikawa, H. (2012). Early mathematics assessment: validation of the short form of a prekindergarten and kindergarten mathematics measure. *Educational Psychology*, *32*(3), 311–333. https://doi.org/10.1080/01443410.2011.654190.

Westerholm, A. & Aunio, P. (in press). Maahanmuuttajataustaisten lasten varhaisten matemaattisten taitojen interventiotutkimus (ThinkMath-intervention for Finnish as a Second Language kindergarteners). *NMI Bulletin*.

Wilson, A., Dehaene, S., Dubois, O., & Fayol, M. (2009). Effects of an adaptive game intervention on accessing number sense in low-socioeconomic-status kindergarten children. *Mind, Brain, and Education*, *3*(4), 224–234.

Wilson, A. J., Dehaene, S., Pinel, P., Revkin, S. K., & Cohen, D. (2006). Principles underlying the design of "the number race" an adaptive computer game for remediation of dyscalculia. *Behavioral and Brain Functions*, *2*(1), 1–14.

Wilson, A. J., Revkin, S. K., Cohen, D., Cohen, L., & Dehaene, S. (2006). An open trial assessment of the number race an adaptive computer game for remediation of dyscalculia. *Behavioral and Brain Functions*, *2*(2), 1–36.

Xin, Y. P., & Jitendra, A. K. (1999). The effects of instruction in solving mathematical word problems for students with learning problems: a meta-analysis. *The Journal of Special Education*, *32*(4), 207–225.

Zhang, D., Stecker, P., Huckabee, S., & Miller, R. (2016). Strategic development for middle school students struggling with fractions: assessment and intervention. *Journal of Learning Disabilities*, *49*(5), 515–531.

Chapter 9

Improving the Mathematical Knowledge of At-Risk Preschool Children: Two Approaches to Intensifying Early Math Intervention

Alice Klein*, Prentice Starkey* and Lydia DeFlorio†
**Center for Early Learning, STEM Program, WestEd, San Francisco, CA, United States*
†College of Education, University of Nevada, Reno, Reno, NV, United States

THE SES-RELATED GAP IN CHILDREN'S EARLY MATHEMATICAL KNOWLEDGE

American students' math performance is well below what is desired and needed for success in the modern world (National Mathematics Advisory Panel, 2008; Provasnik et al., 2012). The challenge of improving this performance has been addressed with new standards to improve mathematics education, and some progress has been made in schools that have instituted reforms (National Governors Association Center for Best Practice & Council of Chief State School Officers, 2010; Rampey, Dion, & Donahue, 2009). The most widely adopted set of math standards, the Common Core State Standards for Mathematics, raises expectations about what children are supposed to learn at every grade level, including kindergarten. This, in turn, requires greater mathematics readiness for children prior to kindergarten entry. At present, children from low-income families lag significantly behind their middle-class peers in mathematics achievement (Aud et al., 2011). Socioeconomic status- (SES-) related gaps are larger than gaps related to other demographic characteristics, with the math gap in kindergarten between children in the lowest and highest quintiles of SES estimated at 1.3 standard deviations (Duncan & Magnuson, 2011; Duncan & Sojourner, 2013).

These SES-related achievement differences emerge in early childhood. The early math gap is conceptually broad, spanning numerical knowledge,

Mathematical Cognition and Learning, Vol. 5. https://doi.org/10.1016/B978-0-12-815952-1.00009-8

informal arithmetic and geometric thinking, and informal measurement skills (Starkey, Klein, & Wakeley, 2004). Differences are present by 2–3 years of age and widen during early childhood, with many low-SES children being almost 1 year behind their middle-class peers by the end of the prekindergarten (pre-k) year (Jordan, Huttenlocher, & Levine, 1992; Starkey & Klein, 2008). Left unaddressed, an abundance of research shows that this gap continues throughout formal schooling (Anunola, Leskinen, Lerkkanen, & Nurmi, 2004; Duncan et al., 2007; Entwisle & Alexander, 1989, 1990; Geary, Hoard, Nugent, & Bailey, 2013; Morgan, Farkas, & Wu, 2009; Rampey, Dion, & Donahue, 2009; Rathbun & West, 2004). This early gap makes children's learning of high mathematics standards, such as the Common Core standards, an ongoing challenge in kindergarten and beyond.

The reason for the early SES-related gap in mathematical knowledge stems, at least in part, from differential levels of support for early mathematical development that children receive in their early learning environments, namely, the home and preschool. Low-income children receive less support in both environments (Starkey & Klein, 2008). Parents' beliefs about early mathematical development and young children's home learning activities are correlated with the extent of children's mathematical knowledge (Barbarin et al., 2008; LeFevre, Polyzoi, Skwarchuk, Fast, & Sowinski, 2010; Melhuish et al., 2008; Skwarchuk, 2009; Thompson, Napoli, & Purpura, 2017). Middle-SES parents, compared to low-income parents, not only have higher expectations regarding the kinds of mathematics skills children should possess at kindergarten entry, but are also more likely to believe that the responsibility for teaching these skills falls on parents, as opposed to preschool teachers (DeFlorio & Beliakoff, 2015; Starkey & Klein, 2008). Middle-SES children engage in a wider range of home learning activities that support the development of mathematical knowledge (DeFlorio & Beliakoff, 2015), and the mathematical content embedded within these activities is generally more advanced (DeFlorio, 2011; Saxe, Guberman, & Gearhart, 1987).

Children from low-income families also receive relatively little support for mathematical development in their preschool classrooms. Preschool teachers report that they spend little time during the school day engaging children in math activities (Graham, Nash, & Paul, 1997; Thornton, Crim, & Hawkins, 2009). These reports are confirmed by direct observations of teaching in preschool classrooms, which have revealed that children spend only 6%–8% of their school day engaged in mathematics activities (Connor, Morrison, & Slominski, 2006; Early et al., 2007; LaParo et al., 2009). Mathematics appears to be a challenging subject for preschool teachers to teach without professional development (PD), so they often focus repeatedly on a few limited topics such as counting (Copley, 2010). The limited amount and low quality of math instruction in preschool classrooms likely result, in part, from teachers' inadequate preparation and knowledge of informal mathematics as it relates to children's early mathematical development (e.g., Isenberg, 2000; Lobman, Ryan, & McLaughlin, 2005).

Further, in a joint position paper, the National Association for the Education of Young Children and the National Council of Teachers of Mathematics specifically attributed the weak quality of preschool mathematics instruction to the general lack of good preschool teacher preparation in mathematics (NAEYC/NCTM, 2002).

The inadequate preparation of preschool teachers in mathematics is compounded by the quality of preschool mathematics curricula in common use. This can clearly be seen in Head Start, the largest public preschool program for economically disadvantaged children in the United States. The most widely used curriculum in Head Start programs is the Creative Curriculum (US Department of Health and Human Services, Administration for Children and Families, 2010). This curriculum has been shown to be ineffective in the area of mathematics in two independent randomized controlled trials (Preschool Curriculum Evaluation Research Consortium, 2008). The second most widely used curriculum in Head Start, High/Scope, is based on Piagetian theory, and therefore excluded mathematics on theoretical grounds (Weikart, Rogers, Adcock, & McClelland, 1970; also see Starkey & Klein, 2008). Given Head Start's reliance on these curricula, the findings of a national randomized controlled trial of the effectiveness of Head Start are not surprising. This study found Head Start to be ineffective in the area of mathematics (US Department of Health and Human Services, Administration for Children and Families, 2010; U.S. Department of Education, Institute of Education Sciences, What Works Clearinghouse, 2015, July). Like Head Start, most state-funded preschool programs also do not use effective mathematics curricula (Dahlin & Squires, 2016; Preschool Curriculum Evaluation Research Consortium, 2008). This combination of inadequate teacher preparation and ineffective mathematics curricula results in a classroom learning environment that is not as rich mathematically as economically disadvantaged children need it to be.

In summary, low-SES preschool children, relative to their middle-class peers, receive less support for mathematics learning in both their home and classroom learning environments. This contributes to the early SES-related gap in informal mathematical knowledge. To address this gap we developed an early math intervention for use with the general population of economically disadvantaged children in public preschool programs.

POTENTIAL FOR EARLY CURRICULAR INTERVENTION TO REDUCE THE MATH GAP

The promise and benefits of early childhood intervention programs have long been acknowledged (e.g., Karoly, Kilburn, & Cannon, 2005; Lazar & Darlington, 1982; Ramey & Campbell, 1984). In the 21st century, intervention research has become more focused on characteristics of early childhood programs, such as early childhood curricula. Some curricula have been rigorously evaluated and found to increase young children's school readiness, including a small number of early math interventions (e.g., Preschool Curriculum Evaluation Research Consortium, 2008; see also Aunio, this volume). We will next discuss one of those effective math interventions.

The *Pre-K Mathematics* Intervention

We began developing and rigorously evaluating components of a curricular intervention, *Pre-K Mathematics*, in the late 1990s and have continued doing so to the present (Starkey et al., 2004; Klein, Starkey, Clements, Sarama, & Iyer, 2008; Starkey & Klein, 2000; Starkey, Klein, DeFlorio, & Swank, 2014). This intervention is based on an assumption that the primary conceptual foundations of children's early mathematical knowledge are the cognitive domains of number and space, domains which are partly structured in infancy and continue developing during early childhood and beyond. Children first develop informal mathematical knowledge, which can be characterized as mathematical knowledge that depends upon the presence or mental representation of sets of concrete objects and a spatial layout. In contrast, formal mathematical knowledge can be characterized as mathematical knowledge that utilizes abstract numerical notation, such as written numerals (1, 2, 3, etc.), arithmetic operation signs (+, −, etc.), and 2-dimensional spatial conventions. Informal mathematical knowledge is theorized to serve as a conceptual foundation for the acquisition of formal mathematical knowledge (e.g., Geary et al., 2018).

Several mathematical competencies, such as principle-based counting and addition and subtraction problem solving, develop to some extent by the time children enter preschool at 3 or 4 years of age (e.g., Geary, 2006; Ginsburg, Klein, & Starkey, 1998; Nunes & Bryant, 2015). The extent of children's mathematical knowledge at the beginning of preschool depends in part on experience, especially mathematically specific inputs and scaffolding provided by caregivers. Differences in these early supports are responsible, at least in part, for the SES-related gap in early mathematical knowledge. The *Pre-K Mathematics* curricular intervention is designed to help children progress from adult-scaffolded (interpsychological) problem solving to independent (intrapsychological) problem solving (e.g., Lantolf & Poehner, 2014; Lantolf, Thorne, & Poehner, 2015; Vygotsky, 1978).

Pre-K Mathematics was designed to support the development of mathematical abilities that have been identified in cognitive studies of young children's emerging understanding of quantity. It provides conceptually broad support for the development of informal mathematical knowledge and the use of numerals, a formal ability, in the presence of concrete sets. It includes activities to support the development of numerical abilities, arithmetic problem solving, spatial abilities and geometric reasoning, pattern knowledge, and nonstandard measurement knowledge (see Table 1 for examples of activities). It has been updated periodically as basic research has progressed. *Pre-K Mathematics* is comprised of adult- (teacher- and parent-) participant math

TABLE 1 Mathematical Domains and Representative Activities Included in *Pre-K Mathematics*

Mathematical Domain	*Pre-K Mathematics* Activity	Mathematical Objective
Number	*Watch Me Count*	Count objects
	Watch Me Make a Set	Produce a cardinal set of objects
	Ladybug Game	Relate sets and numerals
	Let's Play the Number Race	Use a number line
Arithmetic operations	*The Dinosaur Game*	Solve addition and subtraction word problems with objects
	The Sharing Game	Divide sets
	Frogs Sit on the Rocks	Decompose numbers using sets and numerals
Space & geometry	*Watch Me Make a Shape: 2D and 3D*	Construct 2D and 3D forms
	Imagine a Shape	Transform mental images of shapes
Patterns	*Let's Make a Pattern*	Duplicate repeating patterns
Measurement	*Which Stick is Longer?*	Directly compare lengths of objects

activities that enrich the preschool classroom and home learning environments of young children. The intervention consists of small-group math activities with concrete manipulatives for the preschool classroom. Teachers typically conduct small-group math activities twice per week for 20–25 min, with a third day reserved for children needing review or makeups. Teachers also send home English or bilingual (English-Spanish) versions of dyadic (parent-child) math activities on a weekly basis. Activities are organized and sequenced for use across the school year based on mathematical content and developmental research. In recent years the intervention has been aligned at the cluster level with the Common Core math standards for kindergarten (see National Governors Association Center for Best Practice & Council of Chief State School Officers, 2010).

Several key features of the intervention are designed to be sensitive to the learning needs of young children. Each math activity contains a downward (less challenging) extension for children who encounter difficulty at the beginning of the basic (modal) activity and an upward (more challenging) extension for children who complete the basic activity without needing scaffolding. Furthermore, suggestions for scaffolding are provided to support children who experience commonly encountered difficulties with parts of the activity. An instructional goal that teachers adopt is for children to successfully complete the mathematical activities without scaffolding. Teachers are taught to use a progress monitoring instrument that enables them to track individual children's progress and make instructional decisions about when to provide additional support to individual children.

Pedagogical content knowledge, which combines an understanding of content (e.g., the substance of informal mathematics such as sets of concrete objects and lengths, shapes, and spatial positions of concrete objects) and pedagogy (how to teach this substance to young, developing minds), is embedded structurally into each activity (Hill, Ball, & Schilling, 2008; Shulman, 1986). It also is explicitly discussed in *Pre-K Mathematics* teacher workshops. This knowledge is provided because preschool teachers usually receive little preservice training in mathematical development or instruction (Lobman, Ryan, & McLaughlin, 2005).

Effectiveness of the *Pre-K Mathematics* Intervention

The *Pre-K Mathematics* intervention has been rigorously evaluated in a series of randomized trials that have increased in size and diversity of samples and settings (e.g., Klein, Starkey, Clements, Sarama, & Iyer, 2008; Starkey, Klein, DeFlorio, & Swank, 2014). All studies used experimental designs and revealed positive effects of the intervention. Here, we summarize findings from a recent multistate scale-up study. The study was conducted with economically disadvantaged children in public preschool programs. Classrooms were randomly assigned to treatment (*Pre-K Mathematics*) or control (business-as-usual) conditions. Teachers in treatment classrooms received professional development in *Pre-K Mathematics* that enabled them to implement the intervention with high fidelity (i.e., as intended by the developers). Children's mathematical knowledge was assessed at pretest (fall) and posttest (spring) of the pre-k year using the Child Math Assessment (CMA; Klein & Starkey, 2012) and the Test of Early Mathematics Ability, 3rd Edition (TEMA-3; Ginsburg & Baroody, 2003). The intervention was found to have a positive impact on children's mathematical knowledge, with substantively important effect sizes on the CMA and TEMA-3.

Responsiveness of Low-Performing Children

Although the *Pre-K Mathematics* intervention was found to be effective generally with the sample of economically disadvantaged children, a subgroup of children assigned to the treatment condition continued to show low growth in math over the pre-k year. Specifically, treatment children who entered the pre-k year in the lowest quartile (1st Quartile) tended to remain below treatment children who entered in the higher quartiles (2nd–4th Quartiles) at the end

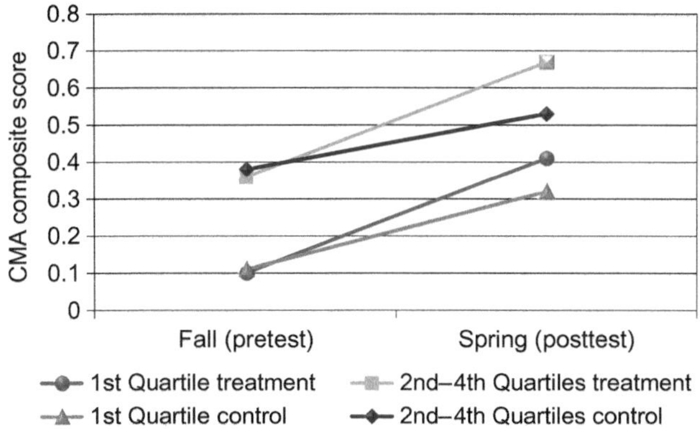

FIG. 1 Child Math Assessment scores of treatment children who received *Pre-K Mathematics* and control children by quartile in fall and spring of the pre-k year.

of the pre-k year, as shown later (Fig. 1). These treatment children in the lowest quartile even performed below control children in the higher quartiles (2nd–4th Quartiles) who did not receive the intervention. Since this subgroup of children demonstrated low response to an effective Tier 1 math intervention, they exited pre-k well below established school readiness benchmarks for mathematics (NAEYC/NCTM, 2002). This puts them at high risk for math difficulties throughout elementary school, as research has found that children entering and exiting kindergarten below the 10th percentile in math have a 70% chance of scoring below the 10th percentile at the end of elementary school (Morgan, Farkas, & Wu, 2009).

Finding solutions for children with a low response to generally effective interventions is a topic of considerable urgency and interest in the special education research and practice community (Intensive Interventions Summit, 2015). Potential solutions include providing more intense interventions or addressing other cognitive or linguistic limitations of low-performing children (e.g., Fuchs & Fuchs, 2014; Fuchs, Fuchs, & Vaughn, 2014). In the studies reported later, we examined both of these potential solutions.

FIRST APPROACH: TUTORIAL INTERVENTIONS IN MATHEMATICS AND ATTENTION

We have argued that there is a need to improve the mathematical knowledge of preschool children who show low response to an effective Tier 1 math intervention and, thus, are at risk for math difficulties in school. One approach for improving the knowledge of these children involves intensifying early math support through a tutorial (Tier 2) intervention. To test this approach, we conducted a randomized experiment that examined the relative effectiveness of two types of tutorial interventions: (1) a tutorial-based math intervention, *Pre-K Mathematics Tutorial*, to supplement the math instruction that at-risk children received in their preschool classrooms; and (2) a tutorial-based math intervention combined with attention training to address the cognitive weaknesses of pre-k children at risk for math difficulties. Both types of tutorial-based math interventions were designed to supplement, not supplant, the Tier 1 math instruction in preschool classrooms. Accordingly, the overarching objective was to test whether increasing the intensity of a math intervention, with or without attention training, would improve the math knowledge of very low-performing pre-k children. In the next section, we summarize the key research underlying these two tutorial interventions and the principal research questions addressed in the study.

The research literature on math interventions for students with math learning difficulties has indicated the importance of increasing the intensity of math instruction through a multitier system that provides explicit, systematic instruction with opportunities for cumulative review (Bryant, Bryant, Gersten, Scammacca, & Chavez, 2008; Fuchs & Fuchs, 2014; Powell et al., 2015). Multitier systems typically involve tutorial, pull-out interventions that supplement Tier 1 instruction in the classroom. However, although this tutorial-based approach to intensifying math instruction has been effective with school-age children at risk for math difficulties, it has not been tested at the pre-k level.

In addition to testing the effect of a tutorial-based math intervention for low-performing pre-k children, the current study also examined whether combining a cognitive intervention with an intensive early math intervention would increase the beneficial effects on young children's math outcomes. We chose to focus on cognitive weaknesses in attention for this at-risk sample, because considerable research has demonstrated that attention abilities in preschool and kindergarten are significant predictors of later mathematics and reading achievement (Duncan et al., 2007; McClelland, Acock, Piccinin, Rhea, & Stallings, 2013). In fact, attention has been viewed as a "hub" cognitive domain, particularly for young children, because it is central to the acquisition of skills across multiple domains (Posner & Rothbart, 2007; Wass, Scerif, & Johnson, 2012). Previous studies have shown that the positive effects of attention training on academic achievement measures are larger for younger and at-risk children (Peng & Miller, 2016; Wass, Scerif, & Johnson, 2012), but there has been little research on the effects of combining both academic and cognitive interventions to improve the academic outcomes of at-risk children. Thus we asked whether this intervention strategy to address the multifaceted nature of learning difficulties would be most effective with very low-performing pre-k children.

We examined two principal research questions in this study. First, does an intensive early math intervention that supplements Tier 1 math instruction with a Tier 2 tutorial intervention significantly improve the math knowledge of very low-performing pre-k children as compared to business-as-usual instruction in pre-k classrooms? Based on the research evidence demonstrating the effectiveness of multitier systems of intervention, we predicted that pre-k children who received a tutorial-based math intervention (with or without attention training) would outperform the control group who received only Tier 1 math instruction in their classrooms. Second, does combining attention training with an intensive early math intervention improve the math knowledge of very low-performing pre-k children as compared to children who receive only the intensive math intervention? This question explored whether there is any synergistic effect of combining both cognitive and academic interventions on children's math outcomes and, if so, we predicted that pre-k children who received the combined math and attention training intervention would outperform the group who received only the intensive math intervention (see also Fuchs et al., this volume).

Study Design and Participants

This study was conducted in public preschool programs in Texas and California. In Texas, one large urban school district participated in the study, while in California three moderate-sized districts from the greater Bay Area were included. All of the participating preschool programs reported that they used a core (Tier 1) math curriculum in their classrooms, and this was independently confirmed in the first year of the study through classroom observations.

Children were selected for participation through a multistep process that entailed determining their age and language eligibility, screening for very low math performance, and then applying specific exclusion criteria. Children with autism spectrum disorders, severe cognitive delays, or significant behavioral difficulties were excluded. However, children with other learning disorders were eligible for participation.

The first step was to obtain date of birth information for all children in participating classrooms. To be included in the study, children were required to be 4 years of age by the start of the tutorial interventions. In addition, children who did not speak either English or Spanish (e.g., a child who only spoke Vietnamese) were excluded from the sample, because they were not able to participate in the cognitive and math assessments. Then, all pre-k children in participating preschool classrooms who were eligible based on age and language criteria were screened at the beginning of the year on a math measure to identify risk for math difficulties in kindergarten.

Math Screening Measure

A math screening measure was designed to identify a group of very low-performing children in the pre-k year who were at risk for low math achievement in kindergarten. Low math achievement was defined as performance in kindergarten below the 25th percentile on the TEMA-3. The specific tasks on the screening measure were drawn from the CMA through a series of logistic regression analyses to determine which CMA tasks in pre-k best predict low performance (<25th percentile) on the TEMA-3 in kindergarten.

Three CMA tasks, comprised of 11 individual items, were found to be significant predictors: (1) Object Counting, (2) Addition/Subtraction with Hidden Objects, and (3) Shape Recognition. These 11 items comprised the math screener. A cut-point (four items) was set to balance false positives and false negatives in the screening process. Pre-k children who received a total score of four items or less on the math screener were likely to score below the 25th percentile on the TEMA-3. These children were considered at risk for low math achievement in kindergarten and, therefore, eligible to participate in this tutorial intervention study.

Child Sample

A total of 541 children who met the math screening criterion were included in the study at pretest, 264 in California (CA) and 277 in Texas (TX). All children who participated in the study were from economically disadvantaged families and were income eligible to attend public preschool programs in their respective states. Overall, the children in the sample were 4.5 years of age at pretest, 47% were girls, and 48% were English Language Learners (i.e., greater proficiency in Spanish than in English for the purposes of both assessment and tutoring). Although approximately half of this at-risk sample were English Language Learners (ELL), two-thirds came from CA sites and only one-third from TX sites. This state difference in English language proficiency may be a contributing factor to the overall higher performance of children in TX that we discuss later.

It is worth noting that the math screener did identify a very low-performing sample of pre-k children for this study. At the pretest, all children were assessed on the TEMA-3. We found that 86% of the sample scored below the 10th percentile, and 75% scored at or below the 6th percentile. Thus the very low level of math knowledge of this sample at the beginning of the pre-k year underscores their risk for math difficulties in school and provides further justification for intensifying an early mathematics intervention for this group of children.

Study Design

A randomized experiment was used to compare the effects of two types of tutorial-based math interventions (math tutoring by itself or combined with attention training) with a control condition. Blocking on classroom, children who met the screening criteria were randomly assigned to one of three conditions: (1) *Math+Attention* condition, in which children received both the tutorial math intervention and attention training ($n = 181$); (2) *Math Only* condition, in which children received the tutorial math intervention as well as a control activity (listening to books on tape) for the same amount of time as the attention training ($n = 180$); and (3) *Control* condition, in which children received business-as-usual Tier 1 math instruction from their classroom teacher, but did not receive either the tutorial math intervention or attention training ($n = 180$). Children in all three conditions continued to receive their Tier 1 math instruction from their classroom teachers. Accordingly, by randomly assigning children to conditions *within* classrooms, the study design controlled for different amounts of Tier 1 math instruction across conditions.

Trained tutors implemented the math and attention training interventions outside of the classroom to maximize children's ability to focus on learning and to prevent treatment diffusion to control children in the preschool classroom. The tutorial math intervention was delivered to pairs of children from the same classroom, one from the Math + Attention condition and one from the Math Only condition, 4 days a week for 15–20 min per day. The attention training intervention was delivered individually to children in the Math + Attention condition in one, 8-min session per week (generally on Fridays, when math the Tier 2 math activities were not implemented). In addition, children in the Math Only condition listened to books on tape 1 day per week for 8 min to control for the additional time that children in the Math + Attention condition spent with the tutor during attention training.

The Math Intervention: Pre-K Mathematics Tutorial

A tutor-delivered math intervention, *Pre-K Mathematics Tutorial (PKMT)*, was designed to provide intensive instruction on a core set of math concepts and skills to help very low-performing pre-k children build a strong foundation of informal mathematical knowledge. This intervention was adapted from the *Pre-K Mathematics* intervention described previously. The *PKMT* intervention is comprised of 20 math activities that use concrete materials to engage pre-k children in an active learning process. A new math activity was introduced each week, and four review weeks were scheduled over the course of the tutorial intervention for a total duration of 24 weeks.

The content of the *PKMT* math activities focuses primarily on Number and Operations, with a secondary emphasis on Space and Geometry. The math activities are designed to be sensitive to the learning needs of low-performing pre-k children. They are scripted to enable tutors to provide explicit instruction (e.g., modeling how to solve a problem) and to scaffold children when they encounter difficulty. Moreover, each activity includes a downward extension for children who are not ready for the specific skill being taught and an upward extension for children who succeed easily on the activity.

A unique feature of the *PKMT* intervention is its focus on developing a core set of math concepts and skills that children should know by the pre-k year. For example, tutors worked with pairs of children on a foundational activity, *Which Animal Has More?* depicted later (Fig. 2). The objective of this math activity was to help children learn to compare two small sets of objects (stars) and identify which set had more. Since the very low-performing sample who met the screening criteria for this study did not have this core math knowledge, the first half of the tutorial math intervention started with foundational math

FIG. 2 Tutor conducting a foundational math activity, *Which Animal Has More*, with a small group of children during a *Pre-K Mathematics Tutorial* session.

activities (e.g., verbal counting, object counting with small sets) and then proceeded to more advanced pre-k math activities (e.g., matching sets of objects to numerals) in the second half of the intervention. Thus the *PKMT* intervention incorporates many of the key characteristics of Tier 2 interventions that have been shown to be effective with school-age children: (1) intensifying math instruction in addition to the Tier 1 math curriculum in the classroom, (2) providing explicit instruction with modeling and scaffolding as needed, (3) following a carefully sequenced curriculum plan that includes cumulative review of fundamental concepts and skills, and (4) emphasizing teaching to mastery and monitoring children's progress toward mastery on each math activity (see also Fuchs et al., this volume). All these characteristics were essential for implementing an effective tutorial math intervention with very low-performing pre-k children.

The Attention Intervention

Children assigned to the Math+Attention condition received an attention-training intervention in addition to the *PKMT* intervention. The attention intervention was developed by McCandliss and colleagues (Rueda, Rothbart, McCandliss, Saccomanno, & Posner, 2005) and consisted of computerized, attention training games in Vigilance and Conflict. We chose this specific attention intervention, because it had prior evidence of improving attention outcomes for preschool children (Rueda, Checa, & Combita, 2012) and it utilized adaptive training programs. In adaptive training, the difficulty level of successive trials on each game was adjusted to the individual child's response on previous trials; as a child's performance improved on the games, the trials became faster (Vigilance) or involved more complex rules (Conflict). Children participated in an attention training session once a week, with 8 sessions each of Vigilance and Conflict games, for a total duration of 16 weeks.

The Vigilance game required children to sustain their attention and respond to a "go" signal to feed different animals their favorite food. For example, Francis the frog loves to eat flies that are trapped inside a jar. When the fly buzzes in the jar and then exits the jar, the child must press the spacebar on the laptop as quickly as possible for the frog to "catch" the fly. The Conflict game required children to choose between two objects to feed a monster based on rules that changed over the trials of the game. For example, Kiki the monster eats red flowers and red trucks during the day (signaled by a sun), but at night (signaled by a moon) Kiki eats trucks regardless of their color. This game was designed to engage executive attention processes, because it required children to inhibit previous responses, ignore task-irrelevant information, and switch flexibly between rules across trials.

Measures and Assessment Procedures

Two measures of early mathematical knowledge, the CMA and TEMA-3, were used to address the research questions related to the impact of the interventions on children's math development. Children were assessed on both measures at pretest and at posttest of the pre-k year. Attention, language, and other cognitive measures were examined in this study, but those measures are not described or reported in this paper (see Barnes et al., 2016).

The CMA measures preschool children's informal mathematical knowledge across a broad range of concepts and skills, including number, arithmetic operations, geometry and spatial thinking, nonstandard measurement, and patterns. It is comprised of nine tasks, with multiple items per task, and the range of difficulty is appropriate for children 4 and 5 years of age. The CMA is administered individually to children in one 20-min session, and the instrument is available in both English and Spanish to enable assessors to test children in their dominant language. The psychometric properties of the CMA are very good for this age group of children. Test-retest reliability over a 2-week interval is .91, and internal consistency (coefficient alpha) is .92. Furthermore, the concurrent validity of the CMA with the TEMA-3 is .74 ($P < .01$).

The TEMA-3 measures children's informal and formal mathematical knowledge in the areas of number and arithmetic operations. It is designed for use with children from 3 to 8 years of age, and it provides raw scores as well as norms in this age range. The TEMA-3 is administered individually to preschool children in one session of 20–30 min. As reported in the TEMA-3 manual, test-retest reliability ranges from .82 to .93, and concurrent validity with other math measures ranges from .54 to .91. It should be noted that since the TEMA-3 does not include items that measure spatial, geometric, or measurement concepts, it is not as well aligned with the *PKMT* intervention as the CMA. Thus we would not expect it to be as sensitive to growth in math knowledge for intervention children as we will discuss in the findings later. Nevertheless, the TEMA-3 was included as an additional math measure in this study, because scores on this instrument can be represented as percentiles and then used to establish cut-points of risk for math difficulties in this sample of children.

Findings

Is the Tutorial-Based Math Intervention Effective?

Our first question asked whether very low-performing children who received a tutorial-based math intervention (*PKMT*) plus Tier 1 math instruction in their classrooms would show greater improvement in their math knowledge than children in the control group. To answer this question, we conducted three-level, hierarchical linear analyses (child nested within classroom and classroom nested within school) to estimate the effects of the math intervention groups on the two math outcome measures, CMA and TEMA-3. Children's pretest scores on the outcome measures and age at testing were included as controls, and State (CA, TX) was included as a fixed effect. Although analyses of the CMA used total proportion-correct scores and those of the TEMA-3 used total raw scores, the distributional properties (e.g., skewness, ceiling effects) of both outcome measures did not differ significantly from normal.

We hypothesized that children who received the *PKMT* intervention (both the Math+Attention and the Math Only groups) would show more improvement in their math knowledge at the end of the pre-k year than children in the control group who received only Tier 1 math instruction and that is exactly what we found. Posttest scores of both math intervention groups on the CMA, a broad measure of informal mathematical knowledge, were significantly higher than the scores of the control group across states ($P < .001$). In individual comparisons with the Control group, both the Math+Attention group (effect size, ES = .43) and the Math Only group (ES = .60) demonstrated greater math knowledge on the CMA at the end of the pre-k year than the Control group shown in the following figure (Fig. 3). In addition, there was a main effect for State, with CMA scores in TX higher, on average, than scores in CA. The same pattern of effects was found on the TEMA-3 where the effect of State was even more evident.

The effects of the math intervention groups on the TEMA-3, a measure of informal and formal numerical knowledge, were significant, but smaller. As indicated in our description of the TEMA-3 previously, this measure is not as well aligned with the *PKMT* intervention and, therefore, we did not expect it to capture the full impact of the tutorial-based math intervention. Moreover, the main effect for State, showing that TEMA-3 posttest scores

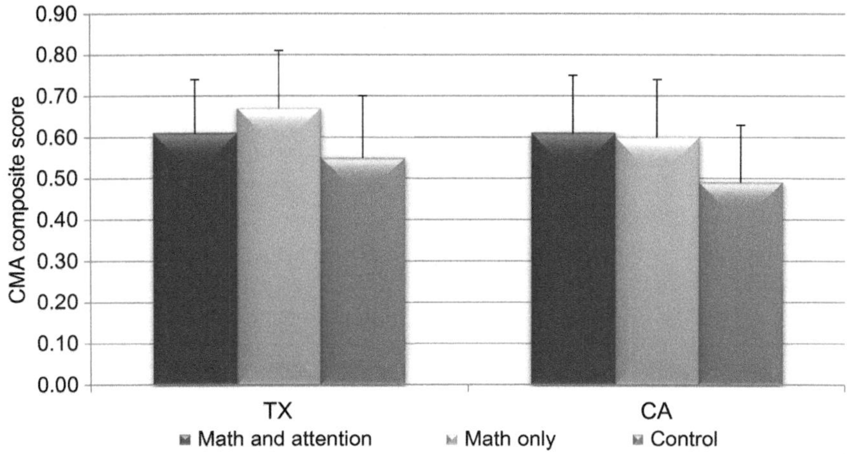

FIG. 3 Child Math Assessment scores at the end of the pre-k year for low-performing children in the Math and Attention, Math Only, and Control groups in Texas and California. *Bars* represent the standard deviations.

were higher overall in TX than in CA, must be interpreted in the context of a significant State by Condition interaction ($P = .02$). Posttest scores of the Math Only group on the TEMA-3 were significantly higher than the Control group in both states ($P < .01$, ES = .23), but the treatment effect for the Math + Attention group was only significant in CA. That is, the difference between Math + Attention and Control groups in CA was greater than the difference between the Math + Attention and Control groups in TX. We interpret this interaction in terms of how the business-as-usual control condition differed between states, as noted later.

Does Attention Training Have a Facilitative Effect on Math Outcomes?

Our second question asked whether combining attention training with an intensive early math intervention (*PKMT* intervention) would facilitate improvement of math knowledge in the very low-performing sample as compared to receiving the intensive early math intervention alone. Three-level, hierarchical linear analyses were also used to address this question. We predicted that if the attention training intervention had a facilitative effect on children's math outcomes, posttest scores on the CMA or TEMA-3 would be higher for the Math + Attention group than the Math Only group. However, there were no significant differences between the two math intervention groups on either the CMA or TEMA-3. Thus there was no evidence that attention training combined with the *PKMT* intervention had a synergistic effect on math outcomes for this very low-performing pre-k sample.

Summary of Findings From Tutorial Intervention Study

This is the first study to test a tutorial-based approach to intensifying early math intervention for pre-k children at risk for math difficulties in school. The study found that very low-performing children who received the tutorial math intervention (*PKMT*) in addition to their Tier 1 math instruction showed significantly greater improvement in their math knowledge than children in the control group who received only Tier 1 math instruction. Furthermore, the tutorial math intervention was effective regardless of whether it was combined with the attention training intervention; there was no facilitative effect of attention training on children's math outcomes. It is important to note, however, that a subgroup of intervention children in both states (18% in TX and 30% in CA) continued to show low response even to this intensive math intervention. We will discuss this subgroup of children at the end of our chapter.

The second significant finding concerns the intensity and quality of the Tier 1 math instruction as a component of an effective tutorial math intervention. The study found a main effect for State as well as for Condition in the analyses of the *PKMT* intervention; posttest scores on the CMA and TEMA-3 were higher in TX than in CA. This finding can be attributed to the difference in intensity of the Tier 1 math programs being implemented in the two states. Classroom observations revealed that teachers in TX classrooms implemented a numeracy-focused Tier 1 program on a regular schedule and spent about twice as many minutes on math instruction per day than teachers in CA. Nevertheless, very low-performing children in TX who received the *PKMT* intervention (with or without attention training) still demonstrated significant improvement in their mathematical knowledge over the pre-k year. Next, we examine an approach for intensifying early math intervention by implementing 2 consecutive years of an effective Tier 1 math intervention.

SECOND APPROACH: INTENSIFICATION BY PROVIDING 2 YEARS OF TIER 1 MATH INTERVENTION

A second approach for improving the mathematical knowledge of pre-k children who show low response to effective Tier 1 math interventions is to increase the duration of the Tier 1 math intervention. Many children enter public preschool at age 3 and could, therefore, begin receiving an effective math intervention at that point. Our cross-cultural research revealed that an SES-related math gap was already present at age 3 in China as well as in the United States (Starkey & Klein, 2008). In China, however, 3-, 4-, and 5-year-old children enrolled in "kindergarten" received a broad, standards-based Tier 1 math curriculum that was tailored developmentally for each age group. Chinese "kindergarten" comprises three age levels, corresponding to the pre-pre-k and pre-k years of preschool and the kindergarten year in the United States.

We found that this SES math gap closed in China by the end of the second year of kindergarten. In contrast, in the United States, where we observed no evidence of systematic Tier 1 math curricula being used in preschool, the SES math gap became more pronounced over the same 2-year period. This set of cross-cultural findings suggest that starting earlier and intervening longer might produce needed growth in the mathematical knowledge of low-SES children, including those entering preschool with very low math knowledge.

The current study sought to answer two principal questions. First, do 2 consecutive years of Tier 1 math intervention, beginning at age 3, improve the mathematical knowledge of children from economically disadvantaged backgrounds? Second, does this 2-year intervention improve the mathematical knowledge of a subgroup of very low-performing children in this sample who scored below the 10th percentile on the TEMA-3 at the beginning of preschool? This low-performing subgroup is arguably comparable to the children examined in the tutorial math intervention study discussed previously. Recall that a large majority (86%) of the children in that study scored below the 10th percentile on the TEMA-3 at the beginning of the pre-k year. At the end of the chapter, we will compare the relative effectiveness of these two different approaches to intensifying early math intervention for this at-risk subgroup of very low-performing children.

Study Design and Participants

Forty-two classrooms located at 30 Head Start sites participated in this study. The classrooms were randomly selected from a larger group of classrooms and then randomly assigned to either a treatment or control condition. Teachers of classrooms assigned to the treatment condition were trained to implement Tier 1 mathematics curricula in their classrooms with the same children over 2 consecutive years of preschool, and teachers of classrooms assigned to the control condition continued with business-as-usual practices over the course of the same two years. If more than one classroom from a Head Start site was randomly selected for participation, the classrooms were assigned to the same condition. This was necessary to prevent treatment diffusion, sometimes also referred to as contamination, which can occur when treatment and control participants communicate with each other and some aspects of the intervention are shared by the treatment teacher with the control teacher.

Child Sample

Within each classroom, 8–10 children were randomly selected from a pool of all eligible children for participation in the study. Children were eligible for participation if they (1) would be age eligible for kindergarten 2 years from the study start date; (2) did not have any known, significant developmental delays that would potentially impact their ability to participate in the intervention and assessments; and (3) had parental consent. To the extent possible, we also attempted to select equal numbers of boys and girls. This process resulted in a sample of 354 children across the treatment and control conditions. Overall, the children were 3.4 years of age at pretest, and 55% were girls. The ethnic composition of the sample was as follows: 60% were Latino, 19% African American, 5% Asian American/Pacific Islander, 4% Caucasian, and 12% multiethnic.

Study Design

The findings reported here are part of a larger randomized experiment that was conducted to investigate the impact of varying the duration of Tier 1 math interventions on children from low-income families. Sixty-three classrooms were randomly assigned to one of three conditions: (1) 2-year math intervention, (2) 1-year math intervention, and (3) no math intervention (business-as-usual control condition). We will be reporting the findings from the full study elsewhere (Klein, Starkey, & DeFlorio, 2013; Starkey, Klein, DeFlorio, & Swank, 2013). In the present chapter, we compare the impact of the 2-year preschool math intervention ($n = 175$) with the control condition ($n = 179$) on the mathematical development of economically disadvantaged children, including a subgroup of very low performers within this sample.

The Math Intervention: Pre-Pre-K Mathematics and Pre-K Mathematics

It is a common practice for American preschool teachers to use the same general curriculum with 3- and 4-year-old children. Contrary to this practice, we developed and implemented an age-specific Tier 1 mathematics intervention for 3- and 4-year-olds to facilitate young children's rapidly developing mathematical knowledge. Teachers who were randomly assigned to the 2-year math intervention condition implemented *Pre-Pre-K Mathematics*, a Tier 1 mathematics curriculum for 3-year-olds, during the first year of the study, and then implemented *Pre-K Mathematics,* which was described previously as a Tier 1 mathematics curriculum for 4-year-olds, during the second year of the study. *Pre-Pre-K Mathematics* was designed to provide a foundation for learning *Pre-K Mathematics*.

Both curricula cover the same mathematical domains (e.g., number, arithmetic operations, spatial sense and geometry, patterns, and measurement), but the activities for 3-year-olds are developmentally simpler. For example, 3-year-olds learn to construct sets of up to five objects, whereas 4-year-olds construct sets up to 10 objects. In a prior intervention study, we found it feasible for preschool teachers to conduct math activities with a small group of 3-year-olds when a second teacher was present to manage children who were not currently participating in the small group (Landry et al., 2014). *Pre-Pre-K Mathematics,* like *Pre-K Mathematics*, is comprised of a sequenced set of classroom math activities and corresponding home activities for parents to play with their children. However, since children's mathematical knowledge is less developed at age three, *Pre-Pre-K Mathematics* includes approximately half the number of activities as *Pre-K Mathematics* and exposes children to each activity twice as often (e.g., introducing a new math activity every 2 weeks) over the school year. Furthermore, the math activities for 3-year-olds were designed to be conducted in less time, approximately 10–15 min per small group, than the activities for 4-year-olds.

In each year of the study, treatment teachers participated in 7 full days of workshops and received biweekly on-site coaching from our professional development staff to learn how to implement each curriculum activity. During each on-site coaching session, the coaches completed a fidelity assessment during the first small-group math activity that the teacher conducted. A fidelity assessment measures the degree to which the teacher is implementing the curriculum activities as intended by the developer. Fidelity was evaluated on five dimensions of curricular implementation deemed to be essential by the curriculum developers: (1) following the schedule of activities in the curriculum plan, (2) preparing all materials needed to conduct the activity, (3) delivering all parts of the small-group math activity, (4), providing developmental adjustments (e.g., scaffolding, downward extensions, upward extensions) as needed by individual children in the group, and (5) keeping written records of individual children's performance on the math activity. Prior to implementation, the developers of the intervention established that an overall score of 0.90 or greater (possible range, 0–1) on the fidelity measure was considered high. The fidelity of implementation score for the treatment teachers over the 2 years of the study was 0.98. This indicates that the professional development support was sufficient and both early math interventions were implemented well.

The amount of exposure to the curriculum that children received, referred to as curriculum dosage, was obtained from teachers' Assessment Record Sheets (ARS). Teachers were asked to use the ARS forms to record individual children's participation and performance on the math curriculum activities. In the first year of the study, the implementation goal for *Pre-Pre-K Mathematics* was to expose each child to each of the 13 math activities at least four times or 52 planned doses per year. The average number of doses received by children over the first year was 47.2 (91% of the planned dosage). This average was slightly below the planned dosage primarily due to child absences that teachers were not able to make up during review periods.

In the second year of the study, the implementation goal for *Pre-K Mathematics* was to expose each child to each of the 24 math activities at least two times or 48 planned doses per year. The average number of doses children received was 45.18 (94% of the planned dosage). Again, child absences contributed to the average number being slightly below the planned dosage.

In addition to curriculum dosage, children's mastery of the math activities was calculated for each intervention year from the teachers' ARS forms. Mastery was defined as the child's ability to succeed on the basic activity without scaffolding or explicit support by the teacher. Note that when children did not master a given activity within the planned number of doses (four in the pre-pre-k year or two in the pre-k year), teachers provided additional doses during review periods as time allowed. In year 1, treatment children mastered an average of 9.85 (76%) of 13 math activities, and in year 2, they mastered an average of 19.5 (81%) of 24 activities. The high levels of math mastery observed both years support the view that children were learning the math concepts targeted by the intervention.

Information on the use of the home activities was obtained from Parent Feedback Forms (PFF) completed by parents. Parents were asked to use the PFF to record whether they had received these math activities from their child's teacher and, if so, how often they used the activities with their child. Over the first year of the study, parents reported, on average, using 83% (10 of 12) of the *Pre-Pre-K Mathematics* home activities. Over the second year, parents reported using 85% (11 of 13) of the *Pre-K Mathematics* home activities.

Measures and Assessment Procedures

Children's mathematical knowledge was assessed at the beginning (fall) and end (spring) of each year using both the CMA and TEMA-3 measures. To minimize the risk of floor effects, the CMA-DE, a downward extension of the CMA that is developmentally appropriate for 3-year-old children, was administered in the fall and spring of the pre-pre-k year. The CMA-DE comprises the following 9 tasks: subitizing (quickly apprehend and state the number of objects in small sets without counting), verbal counting, object counting (count sets of objects and state cardinal value), constructing equivalent sets, adding and subtracting with small sets of hidden objects, dividing objects between two recipients, shape naming, duplicating pattern units (e.g., A-B or A-A-B-B), and measuring the length and size of two objects by direct comparison. The CMA-DE is administered to children in one 15-min session and is scored using the same procedures as the CMA.

All child assessments were administered individually by a trained project research assistant in a quiet space outside of the classroom at their schools over two sessions. Approximately half of the children received the CMA (or CMA-DE) in the first session, and the other half received the TEMA-3 first. Emerging bilingual or Spanish-only speaking children were assessed by a bilingual (English-Spanish) research assistant. Children's responses were accepted in either language and scored conceptually.

Findings

Do 2 Consecutive Years of Tier 1 Math Intervention Improve the Mathematical Knowledge of Children From Economically Disadvantaged Backgrounds?

Our first question addresses the extent to which economically disadvantaged children benefit from receiving the *Pre-Pre-K Mathematics* and *Pre-K Mathematics* interventions, beginning at age 3. Children who met the classroom dosage criterion (75%) for both years were included in the treatment-on-the-treated (TOT) analyses reported later. For treatment children, this meant that they received at least 75% of *Pre-Pre-K Mathematics* activities and at least 75% of *Pre-K Mathematics* activities. For control children, this meant that they had at least 75% preschool attendance for each of the 2 years of the study. CMA composite scores and TEMA-3 raw scores were the outcome measures in separate analyses. Controlling for children's pretest score at preschool entry and age at testing, math scores were significantly higher for children in the treatment group than in the control group on both the CMA and TEMA-3, with effect sizes of 1.10 and 0.65, respectively. Thus the 2-year intervention clearly improved the mathematical knowledge of economically disadvantaged children by the end of pre-k.

Does This 2-Year Intervention Improve the Mathematical Knowledge of Very Low-Performing Children?

Our second question concerns whether the subgroup of very low-performing children in this sample also benefit from receiving the *Pre-Pre-K Mathematics* and *Pre-K Mathematics* interventions consecutively. Specifically, we examined whether treatment children who entered preschool below the 10th percentile on the TEMA-3 demonstrated greater math knowledge at the end of pre-k as compared to control children who also entered preschool below the 10th percentile. This very low-performing subgroup included 58 (of the 175) children in the treatment condition and 45 (of the 179) children in the control condition. Controlling for children's pretest scores at preschool entry and age at testing, math scores at the end of pre-k were significantly higher for children in the treatment group than in the control group on both the CMA and TEMA-3, ES = 0.92 and 0.67, respectively (Fig. 4). Furthermore, the two-year intervention appeared to alter the math trajectory for many children in this low-performing subgroup, as measured by their TEMA-3 percentile scores. For the treatment children who scored below the 10th percentile at preschool entry, only 26% remained below the 10th percentile at the end of pre-k, and 21% scored above the 50th percentile. In contrast, for the control children who scored below the 10th percentile at preschool entry, 42% remained below the 10th percentile at the end of the pre-k year, and only 9% scored above the 50th percentile.

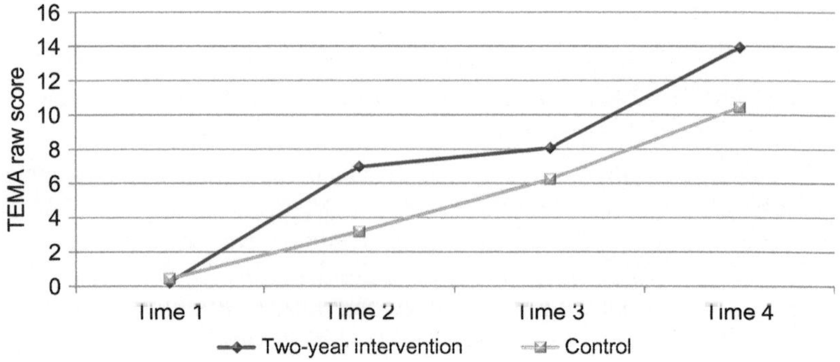

FIG. 4 TEMA scores for all low-performing children in the Two-Year Tier 1 Math Intervention Treatment and Control groups at Time 1 (beginning of pre-pre-k year), Time 2 (end of pre-pre-k year), Time 3 (beginning of pre-k year), and Time 4 (end of pre-k year).

Taken together, these findings indicate that the 2-year math intervention improved the mathematical knowledge of both economically disadvantaged preschool children and the subgroup of very low-performing children. It should be noted, however, that 26% of the low-performing treatment children remained below the 10% percentile at the end of pre-k, despite having received 2 years of an effective Tier 1 mathematics intervention. Some implications of these findings will be discussed in the final section of this chapter.

SUMMARY OF FINDINGS

It has previously been found that the SES-related gap in children's early mathematical knowledge can be reduced significantly by providing a high quality, Tier 1 math intervention during the pre-k year (Clements, Sarama, & DiBiase, 2004; Klein, Starkey, Clements, Sarama, & Iyer, 2008). A subgroup of very low-performing children, however, demonstrated low response to a generally effective Tier 1 math intervention, *Pre-K Mathematics*. We reported two studies that tested different approaches to intensifying early math intervention for this at-risk sample of pre-k children: a tutorial-based Tier 2 math intervention, and a 2-year Tier 1 math intervention.

Effectiveness of the First Intervention Approach

We addressed two questions in our study about the effectiveness of a tutorial-based math intervention. One question was whether a Tier 2 math intervention that supplements Tier 1 math instruction improved the math outcomes of very low-performing pre-k children. The second question was whether combining attention training with a Tier 2 math intervention facilitates the math learning of these at-risk children. We found that this tutorial intervention approach was highly effective. Low-performing children who received the tutorial math intervention showed significantly greater improvement in their math knowledge at the end of the pre-k year than the control group who received only Tier 1 math instruction. However, there was no evidence that attention training had an additional facilitative effect on the math outcomes for this at-risk group of children.

Effectiveness of the Second Intervention Approach

We addressed two additional questions in our study about the effectiveness of extending an early math intervention for 2 years. One question was whether 2 consecutive years of Tier 1 math intervention, beginning at age 3, improved the mathematical knowledge of economically disadvantaged preschool children. The second question focused on the effectiveness of this intervention for the subgroup of very low-performing children who scored below the 10th percentile on the TEMA-3 at preschool entry. The findings clearly indicated that this intervention approach was effective for both groups of children. The mathematical knowledge of all treatment children, compared to control children, improved significantly by the end of the pre-k year. Furthermore, intensifying early math support by providing 2 years of an effective Tier 1 math intervention had a significant impact on the math outcomes of this very low-performing subgroup of children. At the end of the 2-year intervention, there were significant differences between the math scores of treatment and control children who entered preschool below the 10th percentile. Moreover, there were no significant differences in math scores between treatment children who entered preschool below the 10th percentile and those who entered above the 10th percentile. In contrast, control children who entered preschool below the 10th percentile continued to perform below their higher-performing peers at the end of the pre-k year.

Comparison of the Two Intervention Approaches for Very Low-Performing Children

The final analysis compared the relative effectiveness of these two approaches to intensifying early math intervention for pre-k children who typically show low response to an effective Tier 1 math intervention. Since this subgroup of children are at risk for math difficulties in elementary school, there are important policy implications of determining the most effective intervention approach. We compared CMA scores at the end of the pre-k year for low-performing children in the 2-year math intervention study with low-performing children in the California sample of the tutorial math intervention study (Fig. 5). Findings revealed that treatment children demonstrated significantly greater math knowledge than control children in both studies, and their posttest scores on the CMA were virtually equivalent. In addition, a similar percentage of children did not

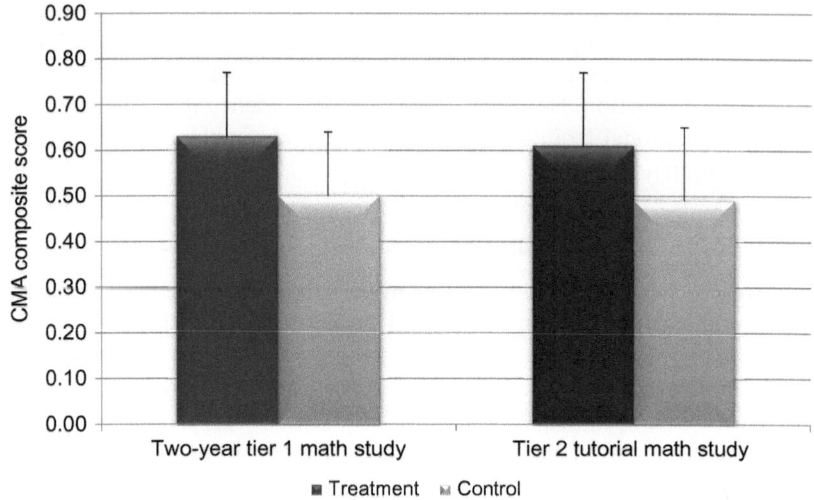

FIG. 5 Child Math Assessment scores at the end of the pre-k year for all low-performing children in the Two-Year Tier 1 Math study and the Tier 2 Tutorial Math study in California. *Bars* represent the standard deviations.

respond to either of these intervention approaches, with 30% not responding to the tutorial-based math intervention and 26% not responding to the 2-year math intervention. Thus the two approaches were similar in effectiveness for this sample of very low-performing preschool children.

CONCLUSIONS AND FUTURE DIRECTIONS

We drew three main conclusions from these studies. The first is that most low-performing preschool children respond to intensified support in mathematics. Second, there is a residual subgroup of very low-performing children who do not respond to either Tier 1 or Tier 2 approaches to intensifying math support; research is needed to understand why. And finally, public preschool programs should provide high-quality, intensive math support in the form of effective Tier 1 and Tier 2 interventions, but they will need new policy initiatives and resources to do this.

Conclusion 1. Most Low-Performing Children Respond to Intensified Support in Mathematics

We tested two approaches that intensified the math support that very low performing children received during the preschool years. Children benefitted both when a Tier 2 math intervention was used to intensify Tier 1 math in the pre-k year and when 2 years of Tier 1 math intervention was used in the pre-pre-k year and pre-k years. The relative effectiveness of these two approaches appears to be roughly similar. The early math gap can be reduced in a large majority of children from low-income families by providing high-quality, intensive math support through public preschool programs.

Conclusion 2. A Residual Subgroup of Low-Performing Children Showed Low Response to Intensified Math Support; Research is Needed to Understand Why

Research is needed to identify and comprehensively assess young children who do not respond to intensified early math intervention. This research could shed light on the genesis of mathematical learning disabilities and determine whether for some subgroups of children there are other cognitive factors such as working memory or aspects of executive functioning that are involved. The percentage of American children who fall into this subgroup can be roughly estimated to be at least 7% percent. This calculation is based on estimates that 26%–30% of our samples did not respond and that approximately 1 in 4 American children under the age of 6 live in poor families (Jiang, Granja, & Koball, 2017). Moreover, this estimate fails to take into consideration some unknown percentage of children from more affluent families who need, but may not respond to, intensified math support. Thus research is also called for to determine the percentage of young children from middle-class backgrounds who show low response to effective math interventions. The research we suggest could establish whether some at-risk preschool children will require special accommodation in math, such as Tier 3 math instruction.

Conclusion 3. Public Preschool Programs Should Provide High-Quality, Intensive Math Support, But New Policies and Resources Are Needed

Intervention science has found that a large majority of economically disadvantaged children benefit from early math intervention. At present, state and federal policies for public preschool programs has not utilized this research. These programs are not required to provide high quality, evidence-based mathematics enrichment. We recommend (1) policies that require public preschool programs to provide effective enrichment in mathematics, and (2) adequate funding earmarked for this purpose. In the absence of such policies, however, we recommend that early childhood administrators determine what resources they would need to implement high quality, effective mathematics curricula. Our work shows that an effective Tier 1 math curriculum in the pre-k year is an important first step for programs to take, but it will not be sufficient for some children. Programs that enroll large numbers of 3-year-olds (e.g., Head Start and the Illinois state preschool program) could provide additional enrichment by differentiating Tier 1 math support by age, thereby providing 2 years of math enrichment. Programs that enroll few or no 3-year-olds (e.g., the Georgia state preschool program) could offer additional enrichment by providing both Tier 1 and Tier 2 math instruction during the pre-k year. Again, however, resources are needed to deliver the professional development and tools needed for teachers and administrators to implement interventions effectively and on a sustained basis, and for parents to support their children effectively at home.

In closing, it is good news that for a large majority of economically disadvantaged children, early intervention can significantly improve mathematical knowledge and, hence, readiness for elementary school mathematics. These children do not learn differently than their middle-class peers. They just have fewer opportunities to learn mathematics due to their early learning environments. School readiness has been a cornerstone of public preschool programs, such as Head Start, since their inception. Basic developmental and intervention research has shown what mathematical concepts and abilities should be supported in preschool classrooms and how this can be done effectively. The application of this research to improve children's mathematical learning environments prior to kindergarten now depends on policy makers and program providers.

ACKNOWLEDGMENTS

The research reported in this chapter was supported by the Institute of Education Sciences, US Department of Education through Grant No. R305K050004 and R305A080697 to WestEd (P. Starkey & A. Klein), and Grant No. R324A120410 to UT, Austin (M. Barnes, PD). The opinions expressed are those of the authors and do not represent views of the US Department of Education.

REFERENCES

Anunola, K., Leskinen, E., Lerkkanen, M. K., & Nurmi, J. E. (2004). Developmental dynamics of math performance from preschool to grade 2. *Journal of Educational Psychology, 96*, 699–713. https://doi.org/10.1037/0022-0663.96.4.699.

Aud, S., Hussar, W., Kena, G., Bianco, K., Frohlich, L., Kemp, J., et al. (2011). *The condition of education 2011 (NCES 2011-033)*. U.S. Department of Education, National Center for Education Statistics, Washington, DC: U.S. Government Printing Office.

Barbarin, O. A., Early, D., Clifford, R., Bryant, D., Frome, P., Burchinal, M., et al. (2008). Parental conceptions of school readiness: relation to ethnicity, socioeconomic status, and children's skills. *Early Education & Development, 19*, 671–701. https://doi.org/10.1080/10409280802375257.

Barnes, M. A., Klein, A., Swank, P., Starkey, P., McCandliss, B., Flynn, K., et al. (2016). Effects of tutorial interventions in mathematics and attention for low-performing preschool children. *Journal of Research on Educational Effectiveness, 9*(4), 577–606. https://doi.org/10.1080/19345747.2016.1191575.

Bryant, D. P., Bryant, B. R., Gersten, R., Scammacca, N., & Chavez, M. M. (2008). Mathematics intervention for first- and second-grade students with mathematics difficulties: the effects of tier 2 intervention delivered as booster lessons. *Remedial and Special Education, 29*, 20–32. https://doi.org/10.1177/0741932507309712.

Clements, D. H., Sarama, J., & DiBiase, A. M. (Eds.), (2004). *Engaging young children in mathematics: Standards for early childhood mathematics education*. Mahwah, NJ: Erlbaum.

Connor, C. M., Morrison, F. J., & Slominski, L. (2006). Preschool instruction and children's emergent literacy growth. *Journal of Educational Psychology*, *98*, 665–689. https://doi.org/10.1037/0022-0663.98.4.665.

Copley, J. (2010). *The young child and mathematics* (2nd ed.). Washington, DC: NAEYC.

Dahlin, M. S., & Squires, J. (2016). *State pre-k approved curricula (NAECS-SDE/CEELO listserv summary)*. New Brunswick, NJ: Center on Enhancing Early Learning Outcomes. Retrieved from http://ceelo.org/wp-content/uploads/2016/05/ceelo_listserv_summary_approved_curriculum_web.pdf.

DeFlorio, L. L. (2011). *The influence of the home learning environment on preschool children's informal mathematical development: Variation by age and socioeconomic status*. (Doctoral dissertation). Retrieved from: https://escholarship.org/uc/item/4534f34v.

DeFlorio, L., & Beliakoff, A. (2015). Socioeconomic status and preschoolers' mathematical knowledge: the contribution of home activities and parent beliefs. *Early Education and Development*, *26*, 319–341. https://doi.org/10.1080/10409289.2015.968239.

Duncan, G. J., Dowsett, C. J., Claessens, A., Magnuson, K., Huston, A. C., Klebanov, P., et al. (2007). School readiness and later achievement. *Developmental Psychology*, *43*, 1428–1446. https://doi.org/10.1037/0012-1649.43.6.1428.

Duncan, G. J., & Magnuson, I. K. (2011). The nature and impact of early achievement skills, attention skills, and behavior problems. In G. J. Duncan & R. J. Murname (Eds.), *Whither opportunity: Rising inequality, schools, and children's life chances* (pp. 47–69). New York: Russell Sage Foundation.

Duncan, G. J., & Sojourner, A. J. (2013). Can intensive early childhood intervention programs eliminate income-based cognitive and achievement gaps? *Journal of Human Resources*, *45*, 945–968.

Early, D. M., Maxwell, K. L., Burchinal, M., Alva, S., Bender, R. H., Bryant, D., et al. (2007). Teachers' education, classroom quality, and young children's academic skills: results from seven studies of preschool programs. *Child Development*, *78*, 558–580. https://doi.org/10.1111/j.1467-8624.2007.01014.x/epdf.

Entwisle, D. R., & Alexander, K. L. (1989). Early schooling as a "critical period" phenomenon. *Research in Sociology of Education and Socialization*, *8*, 27–55.

Entwisle, D. R., & Alexander, K. L. (1990). Beginning school competence: minority and majority comparisons. *Child Development*, *61*, 454–471.

Fuchs, D., & Fuchs, L. S. (2014). Rethinking service delivery for students with significant learning problems: developing and intensive instruction. *Remedial and Special Education*, *36*(2), 105–111. https://doi.org/10.1177/0741932514558337.

Fuchs, D., Fuchs, L. S., & Vaughn, S. (2014). What is intensive intervention and why is it important? *Teaching Exceptional Children*, *46*(4), 13–18.

Geary, D. C. (2006). Development of mathematical understanding. In D. Kuhl, R. S. Siegler, & W. Damon (Eds.), *Handbook of child psychology: Vol. 2. Cognition, perception, and language* (pp. 777–810). New York: John Wiley & Sons.

Geary, D. C., Hoard, M. K., Nugent, L., & Bailey, H. D. (2013). Adolescents' functional numeracy is predicted by their school entry number system knowledge. *PLoS ONE*, *8*(1), e54651.

Geary, D. C., van Marle, K., Chu, F., Rouder, J., Hoard, M. K., & Nugent, L. (2018). Early conceptual understanding of cardinality predicts superior school-entry number system knowledge. *Psychological Science*, *29*(2), 191–205. https://doi.org/10.1177/0956797617729817.

Ginsburg, H., & Baroody, A. (2003). *Test of early mathematics ability* (3rd ed.). Austin, TX: Pro-Ed.

Ginsburg, H. P., Klein, A., & Starkey, P. (1998). The development of children's mathematical thinking: connecting research with practice. In W. Damon, I. Siegel, & A. Renninger (Eds.), *Handbook of child psychology: Child psychology in practice* (pp. 401–476). Hoboken, NJ: John Wiley.

Graham, T. A., Nash, C., & Paul, K. (1997). Young children's exposure to mathematics: the child care context. *Early Childhood Education Journal*, *25*, 31–38.

Hill, H. C., Ball, D. L., & Schilling, S. G. (2008). Unpacking pedagogical content knowledge: conceptualizing and measuring teachers' topic- specific knowledge of students. *Journal for Research in Mathematics Education*, *39*(4), 372–400. Retrieved from http://www.jstor.org/stable/40539304.

Intensive Interventions Summit. (2015). OSEP-IES, March 23–24, 2015, Washington, D.C.

Isenberg, J. P. (2000). The state of the art in early childhood professional preparation. In D. Horm-Wingerd & M. Hyson (Eds.), *New teachers for a new century: The future of early childhood professional preparation* (pp. 17–58). Washington, DC: U.S. Department of Education.

Jiang, Y., Granja, M. R., & Koball, H. (2017). *Basic facts about low-income children*. National Center for Children in Poverty. Retrieved from http://www.nccp.org/publications/pdf/text_1172.pdf.

Jordan, N. C., Huttenlocher, J., & Levine, S. C. (1992). Differential calculation abilities in young children from middle- and low-income families. *Developmental Psychology*, *28*, 644–653. https://doi.org/10.1037/0012-1649.28.4.644.

Karoly, L. A., Kilburn, R. M., & Cannon, J. S. (2005). *Proven benefits of early childhood interventions*. Santa Monica, CA: RAND Corporation Research Brief Series. Retrieved from http://www.rand.org/pubs/research_briefs/RB9145.

Klein, A., & Starkey, P. (2012). *Psychometric study of the child math assessment*. Final report on scaling up the implementation of a pre-kindergarten mathematics intervention in public preschool programs (IES Grant R305K050004).

Klein, A., Starkey, P., Clements, D., Sarama, J., & Iyer, R. (2008). Effects of a pre-kindergarten mathematics intervention: a randomized experiment. *Journal of Research on Educational Effectiveness*, *1*, 155–178. https://doi.org/10.1080/19345740802114533.

Klein, A., Starkey, P., & DeFlorio, L. (2013). *Changing the developmental trajectory in early math through a two-year preschool math intervention*. Washington, DC: Paper presented at the Fall 2013 meeting of the Society for Research on Educational Effectiveness.

Landry, S. H., Zucker, T. A., Taylor, H. B., Swank, P. R., Williams, J. M., Assel, M., et al. (2014). Enhancing early child care quality and learning for toddlers at risk: the responsive early childhood program. *Developmental Psychology*, *50*, 526–541.

Lantolf, J. P., & Poehner, M. (2014). *Sociocultural theory and the pedagogical imperative in L2 education: Vygotskian praxis and the research/practice divide*. New York: Routledge.

Lantolf, J. P., Thorne, S. L., & Poehner, M. E. (2015). Sociocultural theory and second language development. In B. Van Patten & J. Williams (Eds.), *Theories in second language acquisition: An introduction* (pp. 207–226). New York: Routledge.

LaParo, K. M., Hamre, B. K., Locasale-Crouch, J., Pianta, R. C., Bryant, D., Early, D., et al. (2009). Quality in kindergarten classrooms: observational evidence for the need to increase children's learning opportunities in early childhood classrooms. *Early Childhood Research Quarterly*, *20*, 657–692. https://doi.org/10.1080/10409280802541965.

Lazar, I., & Darlington, R. (1982). Lasting effects of early education: a report from the Consortium for longitudinal studies. *Monographs of the Society for Research in Child Development, 47*, 2–3. Serial No. 195.

LeFevre, J.-A., Polyzoi, E., Skwarchuk, S.-L., Fast, L., & Sowinski, C. (2010). Do home numeracy and literacy practices of Greek and Canadian parents predict the numeracy skills of kindergarten children? *International Journal of Early Years Education, 18*, 55–70. https://doi.org/10.1080/09669761003693926.

Lobman, C., Ryan, S., & McLaughlin, J. (2005). Reconstructing teacher education to prepare qualified preschool teachers: lessons from New Jersey. *Early Childhood Research & Practice, 7*, 1–15. Retrieved from https://eric.ed.gov/?id=EJ1084844.

McClelland, M. M., Acock, A. C., Piccinin, A., Rhea, S. A., & Stallings, M. C. (2013). Relations between preschool attention span-persistence and age 25 educational outcomes. *Early Childhood Research Quarterly, 28*, 314–324.

Melhuish, E. C., Phan, M. B., Sylva, K., Sammons, P., Siraj-Blatchford, I., & Taggart, B. (2008). Effects of the home learning environment and preschool center experience upon literacy and numeracy development in early primary school. *Journal of Social Issues, 64*, 95–114. https://doi.org/10.1111/j.1540-4560.2008.00550.x.

Morgan, P. L., Farkas, G., & Wu, Q. (2009). Five year growth trajectories of kindergarten children with learning difficulties in mathematics. *Journal of Learning Disabilities, 42*, 306–321. https://doi.org/10.1177/0022219408331037.

National Association for the Education of Young Children and National Council of Teachers of Mathematics. (2002). *Early childhood mathematics: Promoting good beginnings.* Washington, D.C.: National Association for the Education of Young Children.

National Governors Association Center for Best Practice, & Council of Chief State School Officers. (2010). *Common core state standards for mathematics.* Washington, DC: National Governors Association Center for Best Practices, Council of Chief State School. Retrieved from http://www.corestandards.org/Math/.

National Mathematics Advisory Panel. (2008). *Foundations for success: The final report of the National Mathematics Advisory Panel.* Washington, DC: U.S. Department of Education.

Nunes, T., & Bryant, P. (2015). The development of mathematical reasoning. In L. Liben, U. Muller, & R. Learner (Eds.), *Handbook of child psychology and developmental science* (pp. 715–762). Hoboken, NJ: John Wiley & Sons.

Peng, P., & Miller, A. C. (2016). Does attention training work? A selective meta-analysis to explore the effects of attention training and moderators. *Learning and Individual Differences, 45*, 77–87. Retrieved from http://www.sciencedirect.com.udel.idm.oclc.org/science/article/pii/S1041608015300157.

Posner, M. I., & Rothbart, M. K. (2007). Research on attention networks as a model for the integration of psychological science. *Annual Review of Psychology, 58*, 1–23. Retrieved from https://doi.org/10.1146/annurev.psych.58.110405.085516.

Powell, S. R., Fuchs, L. S., Cirino, P. T., Fuchs, D., Compton, D. L., & Changas, P. C. (2015). Effects of a multitier support system on calculation, word problem, and prealgebraic performance among at-risk learners. *Exceptional Children, 81*(4), 443–470. https://doi.org/10.1177/0014402914563702.

Preschool Curriculum Evaluation Research Consortium. (2008). Literacy express and DLM early childhood express supplemented with open court reading. Pre-K Florida State University. In *Effects of preschool curriculum programs on school readiness* (pp. 117–130). Washington, DC: U.S. Department of Education, Institute of Education Sciences, National Center for Education Research.

Provasnik, S., Kastberg, D., Ferraro, D., Lemanski, N., Roey, S., & Jenkins, F. (2012). *Highlights from TIMSS 2011: Mathematics and science achievement of U.S. fourth- and eighth-grade students in an international context (NCES 2013-009).* Washington, DC: National Center for Education Statistics, Institute of Education Sciences, U.S. Department of Education.

Ramey, C., & Campbell, F. (1984). Preventive education for high-risk children: cognitive consequences of the Carolina Abecedarian Project. *American Journal of Mental Deficiency, 88*, 512–523.

Rampey, B. D., Dion, G. S., & Donahue, P. L. (2009). *NAEP 2008 trends in academic progress (NCES 2009-479).* Washington, DC: National Center for Education Statistics, Institute of Education Sciences, U.S. Department of Education.

Rathbun, A., & West, J. (2004). *From kindergarten through third grade: Children's beginning school experience.* National Center for Education Statistics. Retrieved from http://nces.ed.gov/pubsearch/pubsinfo.asp?pubid=2-4-7.

Rueda, M. R., Checa, P., & Combita, L. M. (2012). Enhanced efficiency of the executive attention network after training in preschool children: immediate changes and effects after two months. *Developmental Cognitive Neuroscience, 2*, 192–204.

Rueda, R., Rothbart, M., McCandliss, B., Saccomanno, L., & Posner, M. (2005). Training, maturation, and genetic influences on the development of executive attention. *Proceedings of the National Academy of Sciences of the United States of America, 102*, 14931–14936.

Saxe, G. B., Guberman, S. R., & Gearhart, M. (1987). Social processes in early number development. *Monographs of the Society for Research in Child Development, 52*, 2. Serial No. 216.

Shulman, L. S. (1986). Those who understand: a conception of teacher knowledge. *American Educator, 10*(1), 4–14. https://doi.org/10.3102/0013189X015002004.

Skwarchuk, S. L. (2009). How do parents support children's preschool numeracy experiences at home? *Early Childhood Education Journal, 37*, 189–197. https://doi.org/10.1007/s10643-009-0340-1.

Starkey, P., & Klein, A. (2000). Fostering parental support for children's mathematical development: an intervention with head start families. *Early Education and Development, 11*, 659–680. https://doi.org/10.1207/s15566935eed1105_7.

Starkey, P., & Klein, A. (2008). Sociocultural influences on young children's mathematical knowledge. In O. N. Saracho & B. Spodek (Eds.), *Contemporary perspectives on mathematics in early childhood education* (pp. 253–276). Charlotte, NC: Information Age Publishing.

Starkey, P., Klein, A., DeFlorio, L., & Swank, P. (2013). *Closing the SES-related gap in early mathematical knowledge.* Seattle, WA: Paper presented at the 2013 biennial meeting of the Society for Research in Child Development.

Starkey, P., Klein, A., DeFlorio, L., & Swank, P. (2014). Promoting math readiness through a sustainable mathematics intervention. In K. L. Bierman & M. Boivin (Eds.), *Promoting school readiness and early learning: The implications of developmental research for practice.* New York: Guilford.

Starkey, P., Klein, A., & Wakeley, A. (2004). Enhancing young children's mathematical knowledge through a pre-kindergarten mathematics intervention. *Early Childhood Research Quarterly, 19*, 99–120. https://doi.org/10.1016/j.ecresq.2004.01.002.

Thompson, R. J., Napoli, A. R., & Purpura, D. J. (2017). Age-related differences in the relation between the home numeracy environment and numeracy skills. *Infant and Child Development, 26*, 1–13.

Thornton, J. S., Crim, C. L., & Hawkins, J. (2009). The impact of an ongoing professional development program on prekindergarten teachers' mathematics practices. *Journal of Early Childhood Teacher Education, 30*(2), 150–161. https://doi.org/10.1080/10901020902885745.

U.S. Department of Education, Institute of Education Sciences, What Works Clearinghouse. (2015, July). *Early childhood education intervention report: Head start*. Retrieved from http://whatworks.ed.gov.

US Department of Health and Human Services, Administration for Children and Families. (2010). *Head start impact study*. Final report. Washington, DC. Retrieved from www.acf.hhs.gov/programs/opre/hs/impact_study/.

Vygotsky, L. S. (1978). *Mind in society: The development of higher psychological processes*. Cambridge, MA: Harvard University Press.

Wass, S. V., Scerif, G., & Johnson, M. H. (2012). Training attentional control and working memory–is younger, better? *Developmental Review*, *32*(4), 360–387. https://doi.org/10.1016/j.dr.2012.07.001.

Weikart, D. P., Rogers, L., Adcock, C., & McClelland, D. (1970). *The cognitively oriented curriculum: A framework for preschool teachers*. Washington, DC: U.S. Department of Health, Education, and Welfare. Retrieved from: https://files.eric.ed.gov/fulltext/ED044535.pdf.

Chapter 10

The Use of Analogies in Mathematics Instruction: Affordances and Challenges

Xenia Vamvakoussi
Department of Early Childhood Education, University of Ioannina, Ioannina, Greece

> *Looking in a natural history museum at the skeletons of various mammals, you may find them all frightening. If this is all the similarity you can find between them, you do not see much analogy. Yet you may perceive a wonderfully suggestive analogy if you consider the hand of a man, the paw of a cat, the foreleg of a horse, the fin of a whale, and the wing of a bat, these organs so differently used, as composed of similar parts similarly related to each other*
>
> (Pólya, 1954, p. 13).

From very early on in life, people need to make sense of unfamiliar situations, solve novel problems, and learn about new domains. How do they cope with these challenges? One mechanism that has been acknowledged by philosophers, mathematicians, scientists, and, of course, by psychologists, as instrumental for invention as well as for learning is reasoning by analogy.

In the first section of this chapter I highlight the important role of analogical reasoning in invention and learning in the context of mathematics, and I discuss the challenges of using analogies in instruction.

ANALOGY AND ANALOGICAL REASONING IN MATHEMATICS

The word *analogy* comes from the Greek word αναλογία. One of the meanings of this word in Greek is *proportion*, that is, the relation of equivalence between two ratios. Although proportion is indeed an analogy, the term *analogy* has a broader meaning. If one looks up the word *analogy* in English dictionaries, it is typically associated with *similarity*,[1] *correspondence*, and *comparison*. All of these words are relevant to analogy and to the processes of analogical reasoning.

1. http://dictionary.cambridge.org/dictionary/english/analogy; https://en.oxforddictionaries.com/definition/analogy

Mathematical Cognition and Learning, Vol. 5. https://doi.org/10.1016/B978-0-12-815952-1.00010-4

Analogy is indeed a special kind of similarity (Holyoak, 2005; Pólya, 1954). Analogical reasoning relies on the comparison between two systems that can belong to the same domain or to similar or to different domains and are perceived as similar in some respects (English, 1997; Gentner et al., 1997; Vosniadou, 1989). The one deemed more familiar to the individual is usually termed "the source" or "the base," while the less familiar one is termed "the target." Analogical reasoning involves mapping between the source and the target (i.e., finding correspondences between the two systems). Pólya's (1954) example presented previously illustrates nicely that, although perceiving of similarity is a necessary first step to establish an analogy, not all kinds of similarity are equally useful. An effective match requires perceiving of relational rather than superficial similarities. If such a match is accomplished, then knowledge of the source can be employed productively to draw inferences about the target.

Analogical reasoning is deemed a form of inductive reasoning (Holyoak, 2005; Pólya, 1954) and also an important component of relational reasoning, defined by Alexander, Jablansky, Singer, and Dumas (2016, p. 37) as "the ability to detect meaningful patterns within streams of information." As such, analogical reasoning is considered a centerpiece of higher order thinking and acknowledged as aid for invention as well as for learning. For instance, Pólya claims that the use of analogies is indispensable in practically every discovery in elementary as well as in advanced mathematics, substantiating this claim with several historical examples. Taking an even stronger position, Núñez and Lakoff (2005) argue that practically all mathematical ideas rely on cross-domain mappings from one conceptual domain to another termed *conceptual metaphors*, which can be considered as a species of analogy (Bowdle & Gentner, 2005). A prominent example of cross-domain mapping in the history of mathematics is the one between arithmetic and geometry (Dantzig, 2005), in essence between discrete and continuous quantities. A product of this long-term cross-domain mapping is the (real) number line, an extensively used representation of numbers in school settings that will be discussed further in subsequent sections.

Because analogical reasoning employs prior knowledge of the source to construct new knowledge about the target, it is considered an important mechanism for learning and transfer (Brown, 1989; Vosniadou, 1989). In particular, cross-domain mapping is considered a "bootstrapping process" that supports learning when what is to be learned transcends what is already known not merely in a quantitative, but also in some qualitative way (Carey, 2004; Smith, Solomon, & Carey, 2005). It is also considered a mechanism for conceptual restructuring, since in the process of comparing the two domains, either or both may be reorganized to improve the match (Gentner et al., 1997; Gentner & Wolff, 2000; Vosniadou, 1989). A prominent example of cross-domain mapping in the development and restructuring of number concepts is the one between physical quantity and number (Resnick & Singer, 1993;

Siegler, 2016; Smith et al., 2005). For instance, Smith et al. (2005) found that elementary students' understanding of number as infinitely divisible (i.e., that numbers can be divided repeatedly, without ever "getting to zero") was strongly related to their understanding of physical quantities as infinitely divisible (i.e., that matter would continue to exist and occupy space with repeated divisions).

The need for analogical reasoning is prevalent in mathematics learning through instruction. Analogical reasoning is required, for example, when students are presented with worked-out examples of problems (Quilici & Mayer, 1996); when they deal with spatial representations such as maps (Yuan, Uttal, & Gentner, 2017); when teachers draw analogies between mathematical and nonmathematical situations (e.g., "an equation is like a scale"); or between two mathematical situations, as when an arithmetical expression is used as an analog of an algebraic expression (see Richland, Holyoak, & Stigler, 2004, for examples of analogies produced by teachers in the mathematics classroom). And, importantly, analogical reasoning lies at the heart of the use of external representations to construct meaning for abstract mathematical ideas. For example, in the first years of instruction concrete representations are used as a source for the learning of natural number concepts (English, 1997). Consider also that the number line is grounded on the analogy "numbers are points on the line."

Analogical reasoning is a fundamental component of human cognition: Even very young children spontaneously construct analogies, possibly since infancy (English, 2004; Goswami, 2004). In the context of instruction, negative transfer (i.e., the inappropriate transfer of prior knowledge in domains where it is not valid) can be attributed to spontaneous reasoning by analogy. For example, students make inferences about rational numbers via reasoning by analogy to natural numbers (Siegler, Thompson, & Schneider, 2011). This may have some positive results, but it often yields transferring to rational numbers attributes and properties that are particular to natural numbers (Ni & Zhou, 2005; Vamvakoussi, Christou, & Vosniadou, 2018; Van Hoof, Vamvakoussi, Van Dooren, & Verschaffel, 2017), which is certainly not what instruction intended.

However, there is great deal of evidence that when people are presented with analogies and are expected to use them, they often fail (Duit, 1991; Dunbar, 2001). It appears that there is a great difference between an analogy drawn on the individual's own conceptual resources—be it an expert mathematician who engages intentionally in analogical reasoning, or a child that spontaneously draws an analogy in naturalistic settings—and an analogy that is in some sense imposed by someone else.

This phenomenon becomes quite problematic when it comes to instructional analogies. Indeed, instructional analogies are used in the classroom or in the textbooks in the hopes that students will identify similarities among concepts, procedures, problems, and solution methods. However, students often miss the intended connections between what teachers perceive as analog situations. For example, students face great difficulty in identifying word problems that share the same mathematical structure and, instead, deem similar two problems that have the same surface features (e.g., they both refer to fruits; Richland & McDonough, 2010).

There are several reasons why the use of analogies in instruction does not always yield the expected results (Clement, 2008; Vamvakoussi, 2017): Sometimes the analogies used are simply not appropriate. Other times, however, the source is not as familiar to students as expected; or students may interpret it on the basis of their initial ideas that are not consistent with the intended scientific ones. Typically, the relational similarities between the source and the target are salient for teachers, but not transparent to students. As Richland et al. (2004) note, it might be the case that students do not actually engage in analogical reasoning when presented with teacher-led analogies.

It appears that for students to be able to use instructional analogies productively, substantial support is required. There is agreement among the researchers who have offered principles and guidelines for the use of instructional analogies that the selection of the source is important, that the students should be made and kept aware of the intended connections, and that the connections between the source and the target should be made explicit and clarified (e.g., English, 1997; Richland, Zur, & Holyoak, 2007). In addition, Hatano and Sakakibara (2004) stress the importance of enhancing students' knowledge of the source domain, inducing students' understanding of the relation between its elements.

A specific teaching strategy, proposed by Clement and colleagues (e.g., Brown & Clement, 1989; Clement, 1993, 2008) is the use of "bridging analogies." This strategy takes into consideration students' initial ideas of a situation and bridges their initial understanding with the intended scientific idea using an anchoring situation and one or more bridging situations. An anchoring situation is meant to be consistent with students' initial ideas but also compatible with the intended scientific idea; it is thus expected to trigger an intuition that can lead to the understanding of the target situation. This teaching approach was originally developed to address students' conceptual difficulties in science. The bridging analogies strategy has been implemented, for example, in the context of acting–reacting forces, a science topic that is very challenging for students. For instance, when a book lies on a table, students typically neglect the force exerted by the table on the book. Clement and colleagues designed interventions based on an anchoring example (i.e., a book on a spring), followed by bridging examples (e.g., a book on foam rubber), gradually reaching the target situation (i.e., the book on the table).

In the following I will elaborate on the empirical background and the rationale of a study that used the analogy "numbers are points on the line," and a bridging analogy ("the number line is like a rubber line") to facilitate students grasping the density property of rational numbers.

WHERE DOES THE ANALOGY "NUMBERS ARE POINTS ON THE LINE" COME FROM?

The number line is grounded on the analogy "number are points on the number line"—in a more typical notation, number: number system:: point: line (Herbst, 1997). As already mentioned, it is a product of the cross-domain mapping between discrete and continuous quantities, in particular between number and the Euclidean line (Dantzig, 2005). The comparison between number and the line was initiated already in ancient Greek mathematics, triggered by the question whether any length could be measured by a number. However, an explicit and satisfying mapping was not accomplished before the 19th century (Bell, 2005; Dantzig, 2005). This is because the meaning of the terms "number," "line," "point," and the relations among them had to be constructed, expanded, revised, and reconstructed again several times over the course of centuries. It is worth considering that in ancient Greek mathematics (and for many centuries on) number was viewed as a collection of units, thus discrete in nature. In contrast, the Euclidean line was deemed inherently continuous in the sense of an entity "unbroken" or "uninterrupted" that has no gaps (Bell, 2005). This conception of the Euclidean line is close to what Núñez and Lakoff (1998) call "holistic line," which can be described as the trace produced by a moving point. The relation between line and points has been puzzling mathematicians for centuries. In particular, the idea that points are constituent elements of the line (i.e., that infinitely many points can be assembled to produce the line) has been highly controversial.

During this long-term comparison, number and the line acted interchangeably as source and target domain and gradually more clarified ideas had been transferred from one domain to the other. In this process, the concept of number as well as of the line underwent radical changes of meaning: The category of number, initially including only the natural numbers, changed for rational and irrational numbers to be admitted as members of the category. Associating numbers via measurement first with segment lengths, and then with points as locations indicating the end point of the segment on the number line, had been instrumental in this respect. Points were reconceptualized as locations on the number line and as limiting positions in an infinite process applied on line segments (Dantzig, 2005; Martin, 1998). The Euclidean line was reconceptualized as a set of points and again as an isomorphism of the real numbers.

This long-term cross-domain mapping reached a crucial point in the 19th century when Cantor and Dedekind, in their efforts to define real numbers rigorously, systematically employed the idea that there is an order preserving, one-to-one correspondence between the points on a line and the real numbers, which is known today as the Cantor-Dedekind axiom. This allowed for a rigorous definition of real numbers (i.e., the arithmetical continuum). Then, the line was defined precisely so that there is an order-preserving, one-to-one correspondence between the points of the line and the real numbers. With this new definition at hand, the Euclidean line transformed into the Cartesian line, and the relation between numbers and points became valid by definition (Dantzig, 2005; Martin, 1998).

Thus the analogy "numbers are points on the number line" is a very elaborated analogy that took centuries to develop and become what Pólya (1954) characterizes as a "clarified analogy": two systems are analogous, if they agree in clearly definable relations of their respective parts; and if relations within the two domains are governed by the same laws. In this case, the analogy number: number system:: point: line (Herbst, 1997) relies on one-to-one correspondence between the points on a line and the real numbers, and the relation in question is an ordering relation (Dantzig, 2005; Martin, 1998): For any two real numbers one can always tell which one is the greater. Similarly, for any two points, one can always tell which one is to the right. In addition, the greater number corresponds to a point that is further to the right. Furthermore, this ordering relation is dense, for numbers as well as for points: Between any two elements, one can always insert a third one.

This (very rough) historical account is presented not only as an example of cross-domain mapping in the discipline of mathematics, but also as a reminder that some analogies used (explicitly or implicitly) in mathematics instruction may be too elaborate to be readily accessible to students. The analogy "numbers are points on the line" entails background assumptions about the nature of the entities involved and their relations that have far-reaching implications and are not transparent to students.

In the following I will focus on students' understanding of the dense ordering in an arithmetical context (density of numbers) and in a geometrical context (density of points), and I will present evidence from a series of studies illustrating the challenges that secondary students face with this counterintuitive notion.

TAPPING INTO STUDENTS' UNDERSTANDING OF THE DENSE ORDERING OF RATIONAL NUMBERS

Similar to real numbers, but unlike natural numbers, rational numbers are densely ordered, that is, between any two rational numbers, there is always a third one. This is known as the "betweenness property" of rational numbers and implies that between any two rational numbers there are infinitely many others (hereafter, "infinity of intermediates"); and that no number has a (unique) successor (hereafter, the "no successor" principle). This property is also shared by the points of the line that correspond to rational numbers. In fact, the idea of density emerged in geometrical context as an attribute of the Euclidean line (Bell, 2005).

Understanding the dense ordering of rational numbers is very challenging for students. In a series of studies we investigated secondary students' (7th to 11th graders) understanding of density, gathering data in a variety of ways:

individual interviews (Vamvakoussi & Vosniadou, 2004); open and forced-choice questionnaires (Vamvakoussi, Christou, & Van Dooren, 2011; Vamvakoussi & Vosniadou, 2007, 2010); and in instructional settings, where students were asked to collaboratively use, evaluate, compare, and construct representations of numbers and the number line (Vamvakoussi, Kargiotakis, Kollias, Mamalougos, & Vosniadou, 2003, 2004).

We thus had rich data on students' understandings of density and also on their strategies when dealing with tasks targeting density. We found that students were very likely to say that there is a finite number of intermediates between two given rational numbers, consistent with prior research with students from primary to tertiary education (Giannakoulias, Souyoul, & Zachariades, 2007; Hannula, Pehkonen, Maijala, & Soro, 2006; Hartnett & Gelman, 1998; Merenluoto & Lehtinen, 2002).

In addition, we found that students' judgments of the number of numbers in an interval were heavily affected by the type of numbers used as interval end points (i.e., whether they were natural or nonnatural numbers, and whether they were represented as decimals or as fractions). More specifically, students were more prone to accept the infinity of intermediates between two natural numbers than two nonnatural numbers; and between two decimals, than between two fractions. In fact, a considerable percentage of students across studies were very consistent in treating natural numbers, decimals, and fractions differently with respect to the number of their intermediate numbers. Students were also reluctant to accept that there may be decimals between fractions and vice versa. Even when they accepted the infinity of intermediates in an interval, they would say that these have the same representational form as the interval end points (i.e., infinitely many decimals between decimals, and infinitely many fractions between fractions).

The previous findings suggested two major issues with students' understanding of density, in particular of the "infinity of intermediates" aspect: Students transfer the property of discreteness from natural to rational numbers. In contrast, they do not immediately transfer this property from one arithmetical context to the other, indicating that they view (the positive) rational numbers as a collection of unrelated sets (natural numbers, decimals, fractions) that are allowed to behave differently with respect to order (discrete/dense). To substantiate the claim that students face great difficulty perceiving the rational as well as the real numbers set as unified systems of numbers, Fig. 1 presents the model built by two 9th graders for real numbers (see also Kilpatrick, Swafford, & Findell, 2001).

Thinking in terms of instruction, we were presented with a threefold problem: What would be an appropriate way to support secondary school students (a) to conceptualize numbers as individual entities, invariant under different forms of representation, (b) perceive rational numbers as a unified system of numbers, and (c) rerepresent them as a dense array? We reasoned that the number line could be instrumental in this respect.

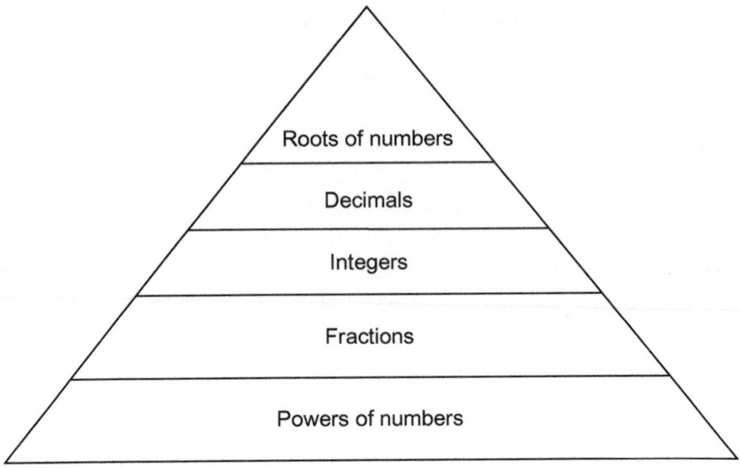

FIG. 1 A model for real numbers produced collaboratively by 2 9th graders.

COULD THE NUMBER LINE SUPPORT STUDENTS' UNDERSTANDING OF DENSITY?

The number line is a powerful representation of numbers. The analogy "numbers are points on the line" has the potential to help students conceptualize the rational (and real) numbers as a unified system of numbers (Kilpatrick et al., 2001; Sirotic & Zazkis, 2007; for a similar argument, see Siegler et al., 2011). In addition, the numbers-to-points correspondence could help students understand that the decimal and the fractional form of rational numbers are interchangeable representations rather than different kinds of numbers, and thus facilitate more sophisticated understandings of density (Vamvakoussi & Vosniadou, 2010). The number line can and has been used to support students' understanding of density, in particular of the betweenness property (Herbst, 1997). Furthermore, it could be argued that density in a geometrical context (i.e., density of points) could be more accessible to students, because points on the line are a homogeneous aggregate (unlike students' conceptualizations of numbers); and because the "carrier" of points, namely, the line is intuitively perceived as continuous.

However, as discussed in detail previously, the number line is a very elaborate representation relying on sophisticated, abstract conceptualizations of all the mathematical concepts involved (numbers, points, line) and also of the relations among them. Thus students' interpretations of the number line may not actually be helpful. Indeed from early on, students are exposed in instruction to experiences with the "number line" in the form of "number tracks" (i.e., a sequence of numbers typically placed in a series of adjacent squares). This may result in students' conceptualizations of the number line as a series of "stepping stones" with an empty space in between (English, 1993). Another common conceptualization of the number line as a ruler

(Doritou & Gray, 2009) may also convey the idea that the amount of numbers in an interval is finite. In addition, the Euclidean line as well as the points may be confused with their physical representations, triggering pragmatic considerations that stand in the way of students' grasping notions related to infinity (Tirosh & Stavy, 1996). In particular, conceptions of points as material spots are at odds with the infinite amount of points on a line segment and may underlie the idea of a segment as "a necklace of beads" (Sbaragli, 2006), as well as the belief that longer segments have more points (Fischbein, 1987). In fact, the idea of the line as a set of points is compatible with the idea of a series of discrete points, even if one avoids the pitfall of conceiving points as material spots.

Based on these considerations, we conducted two studies to investigate the following two research questions:

(a) Does the number line have an effect on students' thinking about the number of numbers between two given rational numbers?
(b) Is the idea of density more accessible to students in a geometrical ("infinitely many *points* in a straight line segment") or in a numerical context ("infinitely many *numbers* between two numbers")?

We piloted the tasks used in these studies with 15 10th graders who participated in individual interviews. Part of the qualitative data presented in the two subsequent sections comes from this pilot study and have been reported in Vamvakoussi (2010).

Does the Number Line Have an Effect on Students' Reasoning About Density?

In the first study (Vamvakoussi & Vosniadou, 2007) we tested whether the number line would have an effect on students' responses in tasks targeting the "infinitely many intermediates" aspect of density. The participants were 164 9th graders (about 15 years of age) and 137 11th graders (about 17 years of age), a total of 301 secondary students. We designed two types of questionnaires, an open-ended (QT1) and a forced-choice one (QT2). Both types of questionnaires had two parts with three items each, targeting the number of numbers between two rational numbers in decimal or in fractional form. In the second part of the questionnaires, however, the given numbers were presented on the number line.

The students were equally divided between the QT1 and the QT2 condition. Half of them received the items with the number line (hereafter, NL-items) first, whereas the other half received first the items without the number line (hereafter, N-items). In both cases the first part of the questionnaire was withdrawn before the second part was administered.

We found no significant effect of the order of the presentation of the number line on students' performance: Within both age groups and regardless of the type of the questionnaire they received, students who received the NL items first did not differ significantly from the students who received the N-items first.

We then compared students' performance in the NL items to their performance in the N-items. We found that the number line had no significant effect on 11th graders' performance, for both types of questionnaire. However, we found a significant effect on the 9th graders' performance, in the open-ended as well as in the forced-choice condition: The younger students performed better in the NL-items than in the N-items.

Looking closer at students' responses, we found that 52 (31.7%) of the 9th graders performed better in the NL-items. For about half of them, this increase was due to the fact that they had moved from "there is no other number" to "there is a finite number of intermediates" responses. We also found that a considerable percent of 9th graders (21.9%) performed worse in the NL-items. This was due to the fact that they gave less "infinitely many intermediates" answers in the NL-items than in the N-items. The majority of students (59.8% of the 9th graders and 54.7% of the 11th graders) had the same performance in the N- and the NL-items.

Qualitative data may help clarify these findings. The students in the following examples were 10th graders who were first asked how many numbers there are between two pseudo-successive rational numbers (such as 0.005 and 0.006, or 3/5 and 4/5). Then they were asked to place these numbers on the number line and to answer the same question again.

Student A and Student B initially answered that there is no other number in between. Both students allowed for a distance between the given numbers when they located them on the number line. Student A answered again that there are no numbers in between. When specifically prompted, she acknowledged the space between the given numbers but insisted on her initial answer: "No, there are no numbers in between. There is space, but there are no numbers." In contrast, Student B revised his initial answer, explicitly referring to the ruler: "There are ... I don't remember how they're called...Thousandths? You know these little lines ... Like there are on the ruler."

In contrast to Students A and B, Student C initially answered that there are infinitely many intermediates. When he placed the numbers on the number line, however, he was no longer confident in his answer. He stated that "there are a few ... ten or something ... maybe one hundred ... I don't know." When explicitly confronted with the inconsistency in his responses, he appeared confused and said: "I don't know ... There is only so much space here."

On the other hand, we also had examples of students who, although asked about density is a numerical context, spontaneously invoked the number line and reasoned that there are infinitely many intermediates. Such an example is presented in Bempeni and Vamvakoussi (2015): When a 9th grade female (Student D) was asked how many numbers there are between 3/5 and 4/5, she answered: "If we locate them on the number-line, there is definitely a gap in between. In this gap, there are infinitely many numbers" (p. 26).

Consider also Student E, a 10th grader who was asked whether it is possible to know how many numbers there are between two real numbers a, b (a < b). The student answered that there are infinitely many intermediates, no matter which number a and b stand for. She explained that 'if we place a and b on the number line, no matter where, this segment, big or small, always consists of infinitely many numbers.'

Based on these examples, some plausible explanations about the findings of this study (Vamvakoussi & Vosniadou, 2007) could be offered: Conceptions of the number line as a physical object, particularly as a ruler, could lead students to infer that there are some more numbers between two pseudo-successive ones, due to the salient space between the two numbers (like Student B and presumably the students in our sample who moved from "no intermediate" to "a finite number of intermediates"). The very same conception, however, could impose some pragmatic constraints, hindering students' understanding of the infinity of intermediates (like Student C, and possibly some of the students in our sample whose performance decreased in the presence of the number line). On the other hand, for other students the number line does not make much difference, albeit for different reasons: For some (like Student A), the space between two numbers represented on the number line can be "empty"; for others, the infinity of numbers is understood in both contexts, as for Students D and E (or understood in one context first, and then transferred to the other).

Is Density More Accessible to Students in a Geometrical Rather Than in an Arithmetical Context?

In the study reported as Experiment 1 in Vamvakoussi and Vosniadou (2012) we systematically compared students' understanding of density in a purely arithmetical context, on the number line, and in a geometrical context (density of points).

The participants of the study (secondary students from 7th to 11th grade) were administered a questionnaire consisting of three item blocks with five items each. The items of the first block (hereafter, N-items) targeted the notion of density in numerical context. Students were asked how many numbers there are between two rational numbers, varying the interval endpoints (i.e., integers, decimals, fractions).

The items of the second block (hereafter, NL-items) were similar to the N-items, but the given numbers were presented on the number line. The items of the third block (hereafter, P-block) asked the number of points on a straight line segment, varying the length and the direction of the segment. The following multiple-choice responses were offered to the students, and they were the same for all blocks: (a) there is no intermediate, (b) there is a finite number of intermediates, and (c) there are infinitely many intermediates, the intermediates being numbers in the first two blocks and points in the third.

Finally, a sixteenth item in the form of a thought experiment was used to tap into students' understanding of the "no successor" aspect of density in a geometrical context. The students were asked to imagine that they had the possibility to see a segment AB under infinite magnification and to choose the drawing showing how they would "see" the point A and its neighboring points. Three options were offered, in the form of drawings. As shown in Fig. 2, the first drawing showed the segment growing in length and also in width, similar to a physical object; in the second, the segment remained continuous, increasing in length, but not in width; in the third the segment was presented as a "necklace of beads." Students were also offered the alternative to present a drawing of their own.

The results showed that students performed significantly better in the P-items, compared to the N-items and also to the NL-items, that is, students were more prone to answer that there are infinitely many points on a line segment than between two numbers, with or without the number line. It should be noted, however, that only 24 (10.5%) students consistently gave the "infinity of intermediates" answer across all P-items.

We again found that the number line had no significant effect on students' performance, and that for some students it actually had an adverse effect. For example, 58 (25.3%) students answered that there are infinitely many numbers between two decimals with three decimals digits; the same students answered that there is a finite number of intermediates, when two similar decimals were presented on the number line.

In the thought experiment, the majority of students (90, 39.3%) chose the third option (i.e., the segment as a "necklace of beads," see Fig. 3). Seventy-six students (33.2%) chose the first option (i.e., the segment as a physical object). Looking at the students who were consistent in responding that there are infinitely many points on a line segment (24, 10.5%) we found that only four of them had chosen the second option.

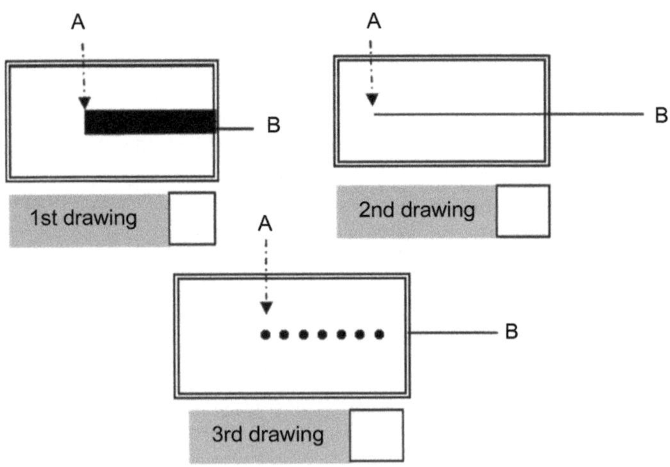

FIG. 2 How does a line segment look under infinite magnification? Options offered to students.

- **Anna says:** The successor of 3 is the number 3.0000...01. The three dots mean that there are many-many zeros.
  - Do you agree with Anna? **Yes** ☐ **No** ☐
  - **Explain your answer:**

- **Dimitris says:** The numbers 3/7 and 4/7 correspond to two successive points on the number line. If I had an imaginary microscope so I could see these points and their nearby points from really close, I would see the following picture:

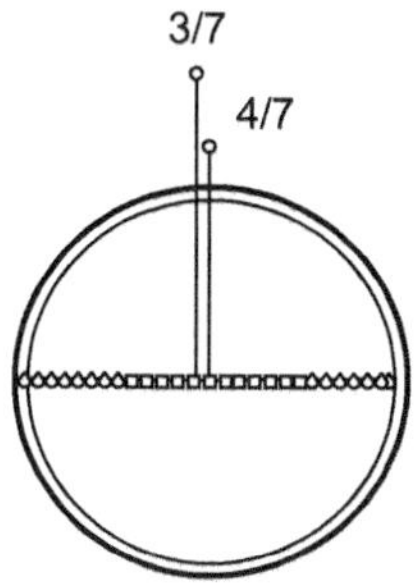

  - Do you agree with Dimitris? **Yes** ☐ **No** ☐
  - **Explain your answer:**

FIG. 3 Examples of "no successor" items.

To sum up, the findings of these two studies (Vamvakoussi & Vosniadou, 2007, 2012) indicated, first, that the number line has a very limited (positive) effect on students' judgments about the infinity of intermediates. Conceptions of the line as a "number track," as a ruler or, in a more general fashion, as a physical object can explain this finding (Doritou & Gray, 2009; English, 1993; Fischbein, 1987; Tirosh & Stavy, 1996). Such conceptions are also consistent with the idea that a line segment grows in length as well as in width under magnification.

Second, the infinity of intermediates may be more accessible for students in a geometrical than in an arithmetical context, since students were more prone to accept the infinity of points between two points on the line than between two numbers. This, however, should not be taken to imply that students have a sophisticated understanding of this notion. The great majority of students answered at least once that there is a finite number of intermediate points to the P-points, presumably because they were susceptible to variations of the segment length. Conceptions of points as material spots, and possibly, the use of the term "infinitely many" to actually refer to a very large amount, may explain this finding. Consider, for example, that conceptions of the line segment as carrying a very large amount of tiny little spots, similar to grains of sand, are consistent with this finding.

Third, it appears that whatever understanding students have for the amount of points on the line segment is not transferred to numbers, even when these numbers are presented on the number line. This suggests that the numbers-to-points correspondence is not evoked by students.

Finally, our findings suggested that the "no successor" aspect of density is more difficult for students than the "infinity of intermediates" aspect. This is because students who were consistently on the "infinite" side in the P-items of Study 2 still conceptualized the line segment as a "necklace of beads" (Sbaragli, 2006). This finding deserves more attention. Indeed, there is evidence that the "no successor" aspect of density is particularly challenging for students, up to the tertiary level: Students have been found to perform better in tasks targeting the "infinity of intermediates" aspect of density than the "no successor" aspect (Giannakoulias et al., 2007; Hannula et al., 2006; Merenluoto & Lehtinen, 2002). One way to explain this phenomenon is to attribute it to the interpretation of the term "infinitely many" as a "very large amount." This might not be the case, however, for some students, like Student E, the 10th grader presented in the previous section as an example of rather sophisticated understanding of the infinity of intermediates in an interval. Consider Student E's response to the task asking if it is possible to specify the first (smallest) value that a real variable x takes, given that $x > 10$.

> Student E: I would say 11, but there are more numbers between 10 and 11: 10.5, 10.1, 10.01, 10.001 and so on.
> Researcher: So what do you think; is there such a number?
> Student E: There is, but I do not know precisely which. It is what we were talking about before, there are infinitely many numbers in between, and so we may not be able to define the one immediately after 10.
> Researcher: So, let me draw a line and place 10 here. You say that 11 is not immediately after 10, since there are infinitely many others in between. But are you saying that there is some number immediately next to 10?
> Student E: Yes, it is the one that is immediately after 10, the immediately next point, the successive point!

For Student E, "infinitely many" means more than "a very large amount." Considering her strategy to produce examples of numbers between 10 and 11 ("…10.1, 10.01, 10.001 and so on"), it probably means "an unending amount" of numbers, in the sense that there is always one more to be found. This does not imply, however, the "no successor" principle, as is clear from Student E's response in the context of numbers as well as points.

In a more general fashion, recursive processes like the one evoked by Student E lie at the heart of students' approaches to the tasks targeting the infinity of intermediates. Consider that one can always find (some) more fractions between two given fractions (e.g., 3/5 and 4/5) by converting them to equivalent forms (e.g., 6/10 and 8/10). Similarly, one can always find (some) more decimals between two decimals (e.g., 0.5 and 0.6) also by converting them to equivalent forms (e.g., 0.50 and 0.60). Some students take the next step and realize that such processes can be repeated again and again. Consider the example of Student F, a female 10th grader who realizes this possibility "on the spot":

> Student F: There is no number between 2.3 and 2.4 (hesitates). Actually, I could convert them into 2.30 and 2.40. But this does not have an end!
> Researcher: What do you mean?
> Student F: I could add another digit, and another, and another. And this can go on forever!

The same strategy can be used to tackle a task on the "no successor" principle. Consider the answer provided (in written) by a pair of 9th graders when asked whether it is possible to specify which number is closest to 2.3 (Vamvakoussi et al., 2003, 2004):

> We cannot give a precise answer to this question, but we can say that the number we are looking for is approximately
> 2.3000000000000000000000000000000000000000000000000000
> 000000000000000000000000000000000000000000000000000000
> 000000000000000000000001

It appears that the use of recursive processes can lead students to infer the infinity of intermediates, but not necessarily the "no successor" principle. Nevertheless, recursive processes, consisting of discrete steps (and thus appealing to the notion of potential infinity), are a fundamental way of approaching concepts of infinity, and are accessible to students, even to elementary school students (e.g., Fischbein, 1987; Hartnett & Gelman, 1998; Lakoff & Núñez, 2000). We exploited this idea for our intervention study.

Using the "Numbers are Points" Analogy and the "Rubber Line" Bridging Analogy

Building on the findings of our previous studies, we designed an experimental intervention (reported as Experiment 2 in Vamvakoussi & Vosniadou, 2012). We aimed at testing whether purposeful use of analogies could support secondary students' understanding of density, in particular the "no successor" aspect of this notion. We used the analogy "numbers are points on the line" treating the line as the source. This is because our findings had indicated that the infinity of points on a segment was more accessible to students than the infinity of numbers in an interval.

However, we were aware that students' understanding of the density property in the source domain was quite limited and had to be enhanced (Hatano & Sakakibara, 2004). To this end, we used the bridging analogies teaching strategy proposed by Clement and colleagues (e.g., Brown & Clement, 1989; Clement, 1993, 2008). Aiming at decreasing the gap between students' initial ideas and the sophisticated idea of points as a dense array, we devised the "rubber line" bridging analogy, that is, the line was presented as an imaginary rubber band that never breaks, no matter how much it is stretched.

We reasoned that the "rubber line" bridging analogy had the potential to help students grasp the idea of density, particularly its "no successor" aspect, because it builds on students' experiences with a real-world object (i.e., the rubber band), and it is also consistent with students' experiences with physical representations of the number line. On the other hand, it is associated, via the imaginary property of being unbreakable, with a recursive process that, as discussed previously, is a way of approaching concepts of infinity that is accessible to students. Finally, this recursive process produces a sequence of segments rather than a sequence of points that can be deemed discrete.

We designed a text that provided explicit information about the infinity of numbers in a specific interval, made explicit reference to the numbers-to-points correspondence, presented examples of the intermediate numbers, and used the "rubber line" bridging analogy to convey the idea that points (and numbers) can never be found one immediately after the other (Table 1).

The "rubber line" text (T_3 in Table 1) was experimentally tested against two other texts (T_1 and T_2 in Table 1) which contained the same information, except for the rubber line. Compared to T_1, T_2 contained in addition figures illustrating the examples of the intermediate numbers on the number line.

We hypothesized that all texts would have a positive effect on students' judgments about the amount of numbers in an interval, but that students exposed to "rubber line" text would perform better in items related to the "no successor" aspect of density and would be more apt to provide explanations for their answers.

Six classes of 8th and 11th graders (one experimental class per grade), 149 students in total, participated in the study. They received a pretest consisting of 3 N-items, 3 P-items, and 3 NL-items, similar to the ones used in our previous studies. Then they were administered one of the texts. And finally, they received a posttest containing all the pretest tasks as well as 5 additional tasks that examined students' abilities to deal with the no-successor aspect of density (see Fig. 3 for examples of the "no successor items).

The results showed that both age groups performed significantly better in the "infinity items" after the intervention, and that there was no significant effect of the text type. Thus all groups profited from the explicit information about the infinity of numbers presented in the common part of the texts. The added value of the "rubber line" appeared in the "no successor" items. Specifically, within each age group, the experimental students performed significantly higher than their peers in the "no successor items," were more consistent in providing a correct answer across these items, and they were also more consistent in providing explanations for their answers. Interestingly, about half of the experimental students explicitly mentioned the "rubber line" or referred to a recursive process to justify their answers, as illustrated by the following examples (Vamvakoussi & Vosniadou, 2012, p. 279):

TABLE 1 Text Types Used in Study 3

Description	Text	Text Type
Infinity of numbers	Have you ever wondered how many numbers there are between 0 and 1; Well, between these two numbers there are infinitely many numbers. And by "infinitely many," we do not mean a lot of numbers, for example thousands, millions, or even billions of numbers. We mean an unending number of numbers.	T_1, T_2, T_3
Numbers-to-points correspondence:	Let us explain this: Think of the real number line. As you may remember, this line consists of infinitely many points and there is one number corresponding to each of these points. Every real number can be placed on the number line.	T_1, T_2, T_3
Examples of intermediate numbers.	Now think of 0 and 1 placed on the number line. In between there is a space, where more numbers can be placed, such as the decimals 0.1, 0.21, 0.333, 0.456, 0.67893, 0.555555, etc. There is no end to how many such numbers you can find. Thus, there are infinitely many numbers between 0 and 1.	T_1, T_2, T_3
Figures illustrating the examples on the number line		T_2
The "rubber line"	How is it possible for all these numbers to be placed in the interval between 0 and 1 on the number line? Are there enough points available? The mathematical number line is a strange object. You can imagine it as a rubber band that never breaks, no matter how much you stretch it. Place numbers between 0 and 1, until it looks like you have used all the available points. If you stretch the rubber band, then you will find out that between the points that looked as if there were the one next to the other, there are more available points, corresponding to more numbers. This process can be repeated infinitely many times—don't forget that your imaginary rubber band never breaks! So, there are infinitely many points between the points corresponding to 0 and 1—therefore, there are infinitely many numbers between 0 and 1.	T_3

Example 1: Between 2.001 and 2.002 there are infinitely many points. Actually, if we think of the number line, that we can stretch (mentally) as much as we like, then we understand that it is not possible to make a picture showing infinitely many tiny spots between 2.001 and 2.002.
Example 2: I don't agree, because you keep stretching the line and you find more points, infinitely many points.
Example 3: If we had an even better microscope, with better zoom capability, we would find out that between these points there are more points—and this process would never end, therefore the points would be infinitely many.

CONCLUSIONS AND FUTURE DIRECTIONS

In this chapter, I presented an overview of studies of our investigations of secondary students' understanding of the density property in an arithmetical and a geometrical context, and examined the potential instructional value of the analogy "numbers are points on the line" that underlies the number line, a representation of numbers that students regularly encounter in mathematics instruction.

In many respects, this analogy is a deviant case: It refers to fundamental mathematical concepts (i.e., numbers, points, line, and also infinity) and its full *clarification* in the sense suggested by Pólya (1954) requires very abstract and sophisticated conceptualizations of these concepts and the relations among them. As discussed, it took centuries before this analogy was established and clarified within the discipline of mathematics. Nevertheless, this analogy is used, implicitly or explicitly, in school mathematics. And for the purposes of this chapter it is worth examining a deviant case of analogy with a view to accentuate the challenges and affordances of the use of instructional analogies in general. To this end, I will revisit the principles and guidelines for fruitful use of instructional analogies discussed in the introduction, highlighting their relevance to the experimental intervention summarized in the previous section.

This intervention invested on the cross-domain mapping between numbers and the line with a view to fostering students' understanding of the dense ordering of rational numbers. Based on our previous studies, we took into consideration students' initial understandings of this notion in an arithmetical as well as in a geometrical context. This was useful in two respects: It allowed us to select the line as a source, because the density of points appeared to be more accessible to students than the density of numbers; and it also revealed that students' understandings within the source were still quite limited. Limited understanding of the source is one important reason why the use of instructional analogies may not yield the expected results (Clement, 2008; English, 1997; Hatano & Sakakibara, 2004). In such cases, students' understandings of the source need to be enhanced (Hatano & Sakakibara, 2004). Inspired by Clement and colleagues' (e.g., Brown & Clement, 1989; Clement, 1993, 2008) "bridging analogies" teaching strategy, we devised the "rubber line" as a means to decrease the gap between students' initial conceptions of line segments, and the conceptualization of the segment as a dense array of points.

Another reason why instructional analogies may not be effective is that students miss the intended connections between the source and the target (English, 1997; Richland et al., 2007). We thus embedded the "rubber line" bridging analogy in a text that reminded students of the numbers-to-points correspondence and provided cues for the transfer of the property of density from the context of points to the context of numbers (see Table 1: "So, there are infinitely many points between the points corresponding to 0 and 1—*therefore*, there are infinitely many numbers between 0 and 1").

The "rubber line" text had a positive impact on students' understanding of the "no successor" principle, which is a particularly challenging aspect of the density property. This is a quite promising result, considering that the intervention was short and based merely on information presented in a text. This, of course, should not be taken to imply that such an intervention could lead to deep and long-lasting changes to students' conceptions of numbers or points. Rather, its value lies in indicating that instructional analogies can foster students' understandings of even highly counter-intuitive mathematical ideas, under the condition that substantial and thoughtful support is also provided.

REFERENCES

Alexander, P. A., Jablansky, S., Singer, L. M., & Dumas, D. (2016). Relational reasoning: what we know and why it matters. *Policy Insights from the Behavioral and Brain Sciences*, *3*(1), 36–44. https://doi.org/10.1177/2372732215622029.

Bell, J. L. (2005). *Continuity and infinitesimals. In E. N. Zalta (Ed.), The Stanford encyclopedia of philosophy*. Retrieved from http://plato.stanford.edu/archives/win2008/entries/continuity.

Bempeni, M., & Vamvakoussi, X. (2015). Individual differences in students' knowing and learning about fractions: evidence from an in-depth qualitative study. *Frontline Learning Research*, *3*(1), 17–34.

Bowdle, B. F., & Gentner, D. (2005). The career of metaphor. *Psychological Review*, *112*(1), 193–216. https://doi.org/10.1037/0033-295X.112.1.193.

Brown, A. L. (1989). Analogical learning and transfer: what develops? In S. Vosniadou & A. Ortony (Eds.), *Similarity and analogical reasoning* (pp. 369–412). Cambridge, MA: Cambridge University Press.

Brown, D. E., & Clement, J. (1989). Overcoming misconceptions via analogical reasoning: abstract transfer versus explanatory model construction. *Instructional Science*, *18*, 237–261. https://doi.org/10.1007/BF00118013.

Carey, S. (2004). Bootstrapping and the origin of concepts. *Daedalus*, *133*, 59–68. https://doi.org/10.1162/001152604772746701.

Clement, J. (1993). Using bridging analogies and anchoring intuitions to deal with students' preconceptions in physics. *Journal of Research in Science Teaching*, *30*, 1241–1257.

Clement, J. (2008). The role of explanatory models in teaching for conceptual change. In S. Vosniadou (Ed.), *International handbook of research on conceptual change* (1st ed., pp. 417–452). Mahwah, NJ: Lawrence Erlbaum Associates.

Dantzig, T. (2005). *Number: The language of science* (4th ed.). New York: Pi Press.

Doritou, M., & Gray, E. (2009). In *Teachers' subject knowledge: the number line representation. Paper presented at 6th conference of the European Society for Research in Mathematics Education (CERME 6), Lyon, France.*

Duit, R. (1991). On the role of analogies and metaphors in learning science. *Science Education, 75*, 649–672. https://doi.org/10.1002/sce.3730750606.

Dunbar, K. (2001). The analogical paradox: why analogy is so easy in naturalistic settings yet so difficult in the psychological laboratory. In D. Gentner, K. J. Holyoak, & B. N. Kokinov (Eds.), *The analogical mind: Perspectives from cognitive science* (pp. 313–3340). Cambridge, MA: The MIT Press.

English, L. (1993). Reasoning by analogy in constructing mathematical ideas. In G. Bell, B. Wright, N. Leeson, & J. Geake (Eds.), *Challenges in mathematics education: Constraints on construction* (pp. 213–222). Lismore: Mathematics Education Group of Australasia.

English, L. (2004). Mathematical and analogical reasoning in early childhood. In L. English (Ed.), *Mathematical and analogical reasoning of young learners* (pp. 1–22). Mahwah, NJ: Lawrence Erlbaum Associates.

English, L. D. (1997). Analogies, metaphors, and images: vehicles for mathematical reasoning. In L. D. English (Ed.), *Mathematical reasoning: Analogies, metaphors, and images* (pp. 3–18). Mahwah, NJ: Erlbaum.

Fischbein, E. (1987). *Intuition in science and mathematics.* Dordrecht: Kluwer Academic Publishers.

Gentner, D., Brem, S., Ferguson, R. W., Markman, A. B., Levidow, B. B., Wolff, P., et al. (1997). Analogical reasoning and conceptual change: a case study of Johannes Kepler. *The Journal of the Learning Sciences, 6*(1), 3–40. https://doi.org/10.1207/s15327809jls0601_2.

Gentner, D., & Wolff, P. (2000). Metaphor and knowledge change. In E. Dietrich & A. Markman (Eds.), *Conceptual change in humans and machines* (pp. 295–342). Mahwah, NJ: Lawrence Erlbaum Associates.

Giannakoulias, E., Souyoul, A., & Zachariades, T. (2007). In D. Pitta-Pantazi & G. Philippou (Eds.), *Students' thinking about fundamental real numbers properties. Proceedings of the fifth congress of the European Society for Research in Mathematics Education* (pp. 425–426). Cyprus: ERME, Department of Education, University of Cyprus.

Goswami, U. (2004). Analogical reasoning and mathematical development. In L. English (Ed.), *Mathematical and analogical reasoning of young learners* (pp. 169–186). Mahwah, NJ: Lawrence Erlbaum Associates.

Hannula, M. S., Pehkonen, E., Maijala, H., & Soro, R. (2006). Levels of students' understanding on infinity. *Teaching Mathematics and Computer Science, 4*(2), 317–337. https://doi.org/10.5485/tmcs.2006.0129.

Hartnett, P., & Gelman, R. (1998). Early understandings of numbers: paths or barriers to the construction of new understandings? *Learning and Instruction, 8*, 341–374. https://doi.org/10.1016/S0959-4752(97)00026-1.

Hatano, G., & Sakakibara, T. (2004). Toward a cognitive-sociocultural psychology of mathematical and analogical development. In L. English (Ed.), *Mathematical and analogical reasoning of young learners* (pp. 187–200). Mahwah, NJ: Lawrence Erlbaum Associates.

Herbst, P. (1997). The number-line metaphor in the discourse of a textbook series. *For the Learning of Mathematics, 17*(3), 36–45.

Holyoak, K. J. (2005). Analogy. In K. J. Holyoak & R. G. Morrison (Eds.), *The Cambridge handbook of thinking and reasoning* (pp. 117–142). New York: Cambridge University Press.

Kilpatrick, J., Swafford, J., & Findell, B. (2001). *Helping children learn mathematics.* Washington, DC: National Academy Press. Adding + it up.

Lakoff, G., & Núñez, R. (2000). *Where mathematics comes from: How the embodied mind brings mathematics into being.* New York: Basic Books.

Martin, G. E. (1998). *The foundations of geometry and the non-Euclidean plane.* New York: Springer-Verlag.

Merenluoto, K., & Lehtinen, E. (2002). Conceptual change in mathematics: understanding the real numbers. In M. Limon & L. Mason (Eds.), *Reconsidering conceptual change: Issues in theory and practice* (pp. 233–258). Dordrecht: Kluwer.

Ni, Y., & Zhou, Y.-D. (2005). Teaching and learning fraction and rational numbers: the origins and implications of whole number bias. *Educational Psychologist, 40,* 27–52. https://doi.org/10.1207/s15326985ep4001_3.

Núñez, R., & Lakoff, G. (1998). What did Weierstrass really define? The cognitive structure of natural and epsilon-delta continuity. *Mathematical Cognition, 4*(2), 85–101. https://doi.org/10.1080/135467998387343.

Núñez, R., Lakoff, G., & Campbell, J. I. D. (2005). The cognitive foundations of mathematics: the role of conceptual metaphor. In *Handbook of mathematical cognition* (pp. 109–124). New York: Psychology Press.

Pólya, G. (1954). *Mathematics and plausible reasoning: Induction and analogy in mathematics.* New Jersey: Princeton University Press.

Quilici, J. L., & Mayer, R. E. (1996). Role of examples in how students learn to categorize statistics word problems. *Journal of Educational Psychology, 88*(1), 144. https://doi.org/10.1037//0022-0663.88.1.144.

Resnick, L., & Singer, J. (1993). Protoquantitative origins of ratio reasoning. In T. Carpenter, E. Fennema, & T. Romberg (Eds.), *Rational numbers: An integration of research* (pp. 107–130). Hillsdale, NJ: Erlbaum.

Richland, L. E., Holyoak, K. J., & Stigler, J. W. (2004). Analogy use in eighth-grade mathematics classrooms. *Cognition and Instruction, 22*(1), 37–60. https://doi.org/10.1207/s1532690xci2201_2.

Richland, L. E., & McDonough, I. M. (2010). Learning by analogy: discriminating between potential analogs. *Contemporary Educational Psychology, 35*(1), 28–43.

Richland, L. E., Zur, O., & Holyoak, K. J. (2007). Cognitive supports for analogies in the mathematics classroom. *Science, 316,* 1128–1129. https://doi.org/10.1126/science.1142103.

Sbaragli, S. (2006). Primary school teachers' beliefs and change of beliefs on mathematical infinity. *Mediterranean Journal for Research in Mathematics Education, 5*(2), 49–76.

Siegler, R. S. (2016). Magnitude knowledge: the common core of numerical development. *Developmental Science, 19,* 341–361. https://doi.org/10.1111/desc.12395.

Siegler, R. S., Thompson, C. A., & Schneider, M. (2011). An integrated theory of whole number and fractions development. *Cognitive Psychology, 62*(4), 273–296. https://doi.org/10.1016/j.cogpsych.2011.03.001.

Sirotic, N., & Zazkis, R. (2007). Irrational numbers on the number line—where are they? *International Journal of Mathematical Education in Science and Technology, 38*(4), 477–488. https://doi.org/10.1080/00207390601151828.

Smith, C. L., Solomon, G. E. A., & Carey, S. (2005). Never getting to zero: elementary school students' understanding of the infinite divisibility of number and matter. *Cognitive Psychology, 51,* 101–140. https://doi.org/10.1016/j.cogpsych.2005.03.001.

Tirosh, D., & Stavy, R. (1996). Intuitive rules in science and mathematics: the case of 'everything can be divided by two'. *International Journal of Science Education, 18*, 669–683. https://doi.org/10.1080/0950069960180603.

Vamvakoussi, X. (2010). The 'numbers are points on the line' analogy: does it have an instructional value. In L. Verschaffel, E. De Corte, T. de Jong, & J. Elen (Eds.), *Use of external representations in reasoning and problem solving: Analysis and improvement* (pp. 209–224). Routledge: New Perspectives on Learning and Instruction Series.

Vamvakoussi, X. (2017). Using analogies to facilitate conceptual change in mathematics learning. *ZDM Mathematics Education, 49*, 497–507. https://doi.org/10.1007/s11858-017-0857-5.

Vamvakoussi, X., Christou, K. P., & Van Dooren, W. (2011). What fills the gap between the discrete and the dense? Greek and Flemish students' understanding of density. *Learning and Instruction, 21*, 676–685. https://doi.org/10.1016/j.learninstruc.2011.03.005.

Vamvakoussi, X., Christou, K. P., & Vosniadou, S. (2018). Bridging psychological and educational research on rational number knowledge. *Journal of Numerical Cognition, 4*(1), 84–106. https://doi.org/10.5964/jnc.v4i1.82.

Vamvakoussi, X., Kargiotakis, G., Kollias, Mamalougos, N. G., & Vosniadou, S. (2003). In B. Wasson, S. Ludvigsen, & U. Hoppe (Eds.), *Collaborative modelling of rational numbers.* (pp. 103–107). *Designing for change in networked learning environments—proceedings of the international conference on computer support for collaborative learning* (pp. 103–107). Dordrecht: Kluwer.

Vamvakoussi, X., Kargiotakis, G., Kollias, V., Mamalougos, N., & Vosniadou, S. (2004). In M. Gregoriadou, A. Raptis, S. Vosniadou, & C. Kynigos (Eds.), *Understanding the structure of the set of rational numbers: a CSCL environment to facilitate conceptual change. Proceedings of the 4th Hellenic conference with international participation of the Hellenic Scientific Association of Information and Communications Technologies in Education* (pp. 407–409). Vol. 2, Athens: Ekdoseis Neon Technologion.

Vamvakoussi, X., & Vosniadou, S. (2004). Understanding the structure of the set of rational numbers: a conceptual change approach. *Learning and Instruction, 14*, 453–467. https://doi.org/10.1016/j.learninstruc.2004.06.013.

Vamvakoussi, X., & Vosniadou, S. (2007). How many numbers in an interval? Presuppositions, synthetic models and the effect of the number line. In S. Vosniadou, A. Baltas, & X. Vamvakoussi (Eds.), *Reframing the conceptual change approach in learning and instruction* (pp. 267–283). Oxford, UK: Elsevier.

Vamvakoussi, X., & Vosniadou, S. (2010). How many decimals are there between two fractions? Aspects of secondary school students' understanding of rational numbers and their notation. *Cognition and Instruction, 28*(2), 181–209. https://doi.org/10.1080/07370001003676603.

Vamvakoussi, X., & Vosniadou, S. (2012). Bridging the gap between the dense and the discrete: the number line and the "rubber line" bridging analogy. *Mathematical Thinking and Learning, 14*, 265–284. https://doi.org/10.1080/10986065.2012.717378.

Van Hoof, J., Vamvakoussi, X., Van Dooren, W., & Verschaffel, L. (2017). In D. C. Geary, D. B. Berch, R. J. Ochsendorf, & K. Mann Koepke (Eds.), *Mathematical Cognition and Learning Series. Acquisition of complex arithmetic skills and higher-order mathematics concepts* (Vol. 3, pp. 101–123). Cambridge, MA: Academic Press, Elsevier.

Vosniadou, S. (1989). Analogical reasoning as a mechanism in knowledge acquisition: a developmental perspective. In S. Vosniadou & A. Ortony (Eds.), *Similarity and analogical reasoning* (pp. 413–436). Cambridge, MA: Cambridge University Press.

Yuan, L., Uttal, D., & Gentner, D. (2017). Analogical processes in children's understanding of spatial representations. *Developmental Psychology, 53*(6), 1098–1114. https://doi.org/10.1037/dev0000302.

Chapter 11

The Role of Visual Representations in Mathematical Word Problems

Asha K. Jitendra* **and John Woodward**[†]
[*]University of California, Riverside, Graduate School of Education, Riverside, CA, United States
[†]School of Education, University of Puget Sound, Tacoma, WA, United States

Popular anecdotal accounts of successful mathematicians and scientists describe how they use more than verbal and symbolic methods when they reason about ideas in their disciplines. Much has been made of the role of visual reasoning in their thinking (Sfard, 1994; Stylianou, 2002). Einstein's use of visual imagery as the basis for his special theory of relativity is a common, if not overused example of visualization at the intersection of physics and mathematics. Tweney (2009) provides a less well known, but nonetheless fascinating account of Faraday's use of drawings to understand the fundamental properties of electricity and magnetism. These drawings were an important foundation for Maxwell's subsequent landmark equations.

More recently, some have tried to document the role of visual reasoning in a more systematic, empirical manner. Aldous (2007), for example, conducted extensive interviews with five expert problem solvers in science and mathematics in conjunction with a quantitative analysis of over 400 individuals in these fields. It is noteworthy that he then grounded the results of these analyses in cognitive psychology and neuroscience rather than what he describes as a long history of vague terms such as *incubation* and *illumination*.

Aldous (2007) argues that advanced mathematical and scientific thinking involves an interplay between visual-spatial and verbal reasoning. Drawing on Dehaene, Spelke, Pinel, Stanescu, and Tsivkin (1999), he claims that the brain circuit for approximate arithmetic is found in bilateral areas of the parietal lobes where mental transformations and visual-spatial processing occurs. This link is speculative given the qualitative nature of his research, but it is nonetheless a major departure from past frameworks that were difficult to validate through traditional, empirical methods.

Mathematical Cognition and Learning, Vol. 5. https://doi.org/10.1016/B978-0-12-815952-1.00011-6

Aldous's (2007) link to neurocognitive research is also important because it alludes to a spate of new research on the role of visual thinking in mathematics as it unfolds in human development as well as memory. Although this literature is beyond the scope of this chapter, a preverbal sense of numerosity in infants and children is well documented (Izard, Sann, Spelke, & Streri, 2009; McCrink & Wynn, 2004). Even an infant's visual awareness captures changes in quantity when these occur under controlled experimental conditions (Wynn, 1992). There is also a considerable body of research on the role of the visuospatial sketchpad in working memory (Baddeley & Hitch, 1974). A number of studies suggest that the visuospatial sketchpad is not only an important predictor of some aspects of mathematical performance (Lefevre et al., 2010; Li & Geary, 2017; Meyer, Salimpoor, Wu, Geary, & Menon, 2010), but that it also helps distinguish levels of mathematical performance from giftedness to mathematical disabilities (Dark & Benbow, 1990; Raghubar, Barnes, & Hecht, 2010). This literature is increasingly critical to our understanding of mathematics, because it captures the *internal* architecture of human learning.

A second and more relevant takeaway from Aldous's (2007) research for this chapter is *external* in nature. Specifically, Aldous recommends that mathematics and science students be encouraged to use visual-spatial strategies for solving complex problems. Aldous supports both a greater use of visual imagery in imagination as well as the use of drawings, diagrams, and models. This latter proposal is in line with a number of long-standing, K-12 mathematics policy documents. For example, the NCTM Standards (National Council of Teachers of Mathematics, 2000) and the Common Core State Standards for Mathematics (CCSS-M) (National Governors Association Center for Best Practices, Council of Chief State School Officers, 2010) urge K-12 teachers to show students how visual representations can be productive for understanding various concepts as well as a promising strategy for solving word problems.

This chapter will begin by reviewing the literature on visual representations and their effect on student success in word problems solving. This is particularly important because policy documents such as the ones mentioned immediately above tend not to describe in sufficient detail how teachers, or curriculum developers for that matter, can help students interpret, create, or use external representations such as pictures or diagrams in the problem-solving process. The chapter will conclude with an extended discussion of how a specific kind of visual representation has been used successfully to teach proportional thinking to middle school students.

STUDENTS' NATURAL USE OF VISUAL REPRESENTATIONS IN MATHEMATICS

A reasonable starting point for examining the extent to which students create external visual representations is in the context of everyday problem solving. Contexts where students learn mathematics in a traditionally, symbolic intensive manner and where there is little formal instruction in visual representation provides insight into what role a picture, diagram, or drawing plays in the way students solve word problems. An in-depth review of two sets of research—one conducted at the third- and fourth-grade level and the other at the tenth-grade level—demonstrate the natural tendencies for students to use visual imagery and representations as they solve word problems.

Woodward and his colleagues (Baxter, Woodward, & Olson, 2001; Woodward, 2001; Woodward & Baxter, 1997; Woodward, Monroe, & Baxter, 2001) conducted a series of studies of National Science Foundation (NSF)-funded mathematics curricula used in third- and fourth-grade classrooms. These curricula reflected many of the recently developed NCTM standards (National Council of Teachers of Mathematics, 1989). In the first of a series of studies Woodward and Baxter (1997) examined students' unprompted use of visual representations to solve a variety of whole number word problems in the fall and again in the spring. Third-grade students were individually interviewed, and they were given a calculator, ruler, paper and pencil, poker chips, and place value blocks as problem-solving tools. Assessors documented how problems were solved through observation as well as by asking at the end of each problem, "Tell me how you got that answer." Fig. 1 shows the spring version of one of the word problems used in the study.

A total of 205 students participated in this study, 104 at two intervention schools using the NSF-funded curricula and 101 third graders at the comparison school using a traditional mathematics curriculum. Of the 410 responses (205 in the fall and 205 in the spring), only 2% (nine students) created some kind of visual representation using pencil and paper available during each individual assessment. Of the nine students who drew some kind of representation, all of them were pictorial in nature. The overwhelming majority of students in this study used paper and pencil solely for symbolic calculations.

Qualitative observations of the intervention classrooms were conducted throughout the school year in conjunction with these assessments (Baxter et al., 2001). The five 3rd-grade teachers were either in their first or second year of the NSF-funded curriculum, and they were consistently concerned about coverage (i.e., finishing all of the lessons in the curriculum before the end of the school year). Field notes show no cases where teachers devoted any significant

Your friend Devon is in the other third-grade class. Kelly just got this problem for homework and has no idea how to do the problem. How would you solve the problem?

Homework problem: It is 654 footsteps from the parking lot to Thundering Waterfalls. You leave your car in the parking lot and walk 249 footsteps before you remember that you left your camera in the car. You walk back to the car, get your camera, and walk to the waterfalls. What distance did you walk?

FIG. 1 Multistep word problem.

amount of time to showing students how to create visual representations such as pictures or diagrams in conjunction with word problems found in the curriculum. Word problem solving appeared in student workbooks, and teachers tended to review answers and clarify student errors verbally. These observations reinforced the view that even though students in the NSF curriculum spent more time discussing mathematics and even math word problems, relatively little time was spent on the use of visual representations as a strategy for solving word problems. In fact, the distinct feature of the NSF curriculum used in these classrooms was a significant emphasis on alternative algorithms for operations on whole numbers as well as a wide range of math games that supported an informal use and understanding of concepts.

Subsequent research capitalized on annual statewide assessments at the fourth-grade level (Woodward, 2001; Woodward et al., 2001). The mathematics portions of these assessments contained multiple-choice problems along with performance assessments where students were asked to explain their answers to word problems using, "words, numbers, or pictures." As preparation for the spring statewide test, seven 4th-grade teachers in the experimental group used performance assessments at the end of each unit of instruction in their NSF math curriculum. Performance assessment tasks were varied and consisted of multistep word problems involving whole numbers, decimals, logic, geometry, and measurement. These assessments embodied the structure and prompts found on the statewide test, particularly instruction to students to explain their thinking using words, numbers, or pictures.

In this study, teachers taught students how to apply rubrics to each end of unit, performance assessment by presenting samples of experimenter-developed student responses to parallel versions of these problems. These experimenter-developed responses were presented to the class in random order of quality, thus allowing the teacher to conduct a classroom discussion of each response and how it might be scored on the rubric. For those responses deemed low in quality, the class discussed how the response could be improved. Field notes of these classroom discussions indicated that teachers and students focused on all dimensions of each experimenter-developed response (i.e., words, numbers, and pictures), not just the quality of a visual representation that may have been used in the example. In that regard, teachers tended not to expound on the use of drawings or emphasize the virtues of a simple diagram over an elaborate picture. Fig. 2 presents how two students in the spring of the study responded to the waterfall problem presented in Fig. 1.

These responses are indicative of the nature of student responses over the course of the year. Each one reflects a mix of pictures, numerical calculations, and written explanations. The number of students who drew pictures reflected a substantive change from the original study (Woodward & Baxter, 1997) discussed previously. For that reason, it is worth noting that the prompt to explain their thinking using a picture generally resulted in just that—a decorative picture. Michael's drawing in Fig. 2 is indicative of the vast majority

Michael

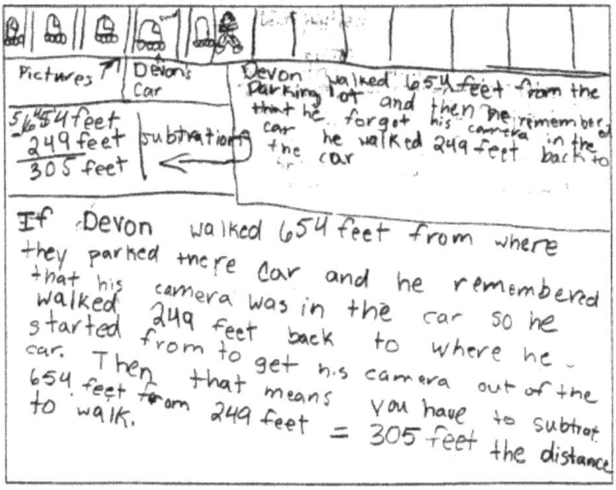

Janet

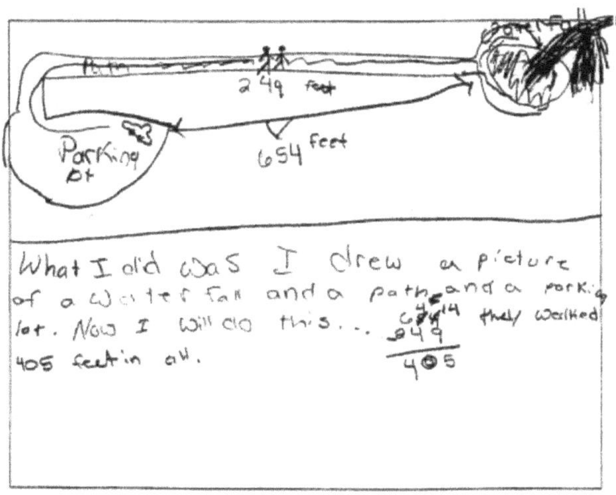

FIG. 2 Pictorial representations of word problems.

of student responses over the course of seven performance assessments administered throughout the year. Considerable space on the page is devoted to a direct, pictorial translation of the problem (e.g., cars, parking lots, waterfalls). Janet's drawing exemplifies a small minority of cases where there is more emphasis on linear relationships that capture the distance walked initially, then back to the car, and finally to the waterfall.

Detailed analyses of all of the performance assessments over the yearlong study showed that 63% of students drew some kind of visual representation as part of their answer. However, only 18% of those drawings (11% overall) created representations that could be considered diagrammatic representations of the problem. In other words, 89% of students who chose to create a visual representation did so by drawing a picture that typically included the kinds of superfluous information found in Fig. 2. The research also found that achieving the correctly calculated answer was instrumental to success on the assessment. Succinct written explanations rather than drawings or diagrams characterized the best responses to the assessments.

At one level, these results are not surprising from a developmental perspective. Children have difficulty constructing mental models of word problems, particularly nonroutine problems (Reuter, Schnotz, & Rasch, 2015). Furthermore, there is a propensity for students to create visual representations containing irrelevant, decorative features that do little to help students identify the core quantitative relationships found in the problem (Hegarty & Kozhevnikov, 1999). It should be noted that these kinds of drawings were not modeled for students, nor were they a feature of the curricular materials. Unfortunately, these kinds of images are the least effective strategy for solving word problems correctly (Diezmann & English, 2001; van Garderen & Montague, 2003).

One might anticipate that students in high school would demonstrate a more sophisticated use of visual representations in their problem solving if for no other reason than the fact that these students have been exposed to an array of images over their school experience. Place value blocks, number lines, Cartesian graphs, and any number of geometric images presented in two-dimensional formats are common features of textbooks and classroom instruction.

Campbell, Collis, and Watson (1995) carefully examined how tenth-grade students used image schemata—imagery that bridges abstract logical structures with concrete images and experiences—in the context of three different kinds of word problems. The three problems varied in topical content (i.e., fractions, geometry, basic linear algebra), and all of them could be solved using a variety of visual or nonvisual methods. Students were administered pretests for reasoning ability in mathematical operations as well as one for vividness of visual imagery. These assessments yielded four types of learners: high operations-high visual, high operations-low visual, low operations-high visual, and low operations-low visual. Overall results of the study indicated that individual differences in visual imagery had no direct effect on problem-solving success, while ability in mathematical operations did.

Of considerable interest in this study was performance on the geometry problem where students were asked to determine the number of faces on a 3 cm by 3 cm cube that had been dipped in red paint and was subsequently cut into 27 smaller, 1 cm by 1 cm cubes. Of the three tasks involving fractions, geometry, and basic linear algebra, the geometry problem was most amenable to imagery, and not surprisingly almost all students stated that they formed some kind of mental image when they solved (or tried to solve) it.

In fact, 84% of students drew one or more diagrams of the cube as part of their method for solving the problem. However, correct performance on this task was almost indistinguishable for high visual-high operations ability (42%), low visual-high operational ability (46%), and high visual-low operations ability students (46%). Low visual-low operations ability students averaged 17% correct in solving this problem. The researchers argued that the nature of the problem availed itself to a kind of concrete image that would have been useful for most students.

Overall, Campbell et al. (1995) concluded that the usefulness of a visual image or representation is directly related to its abstractness (i.e., the extent to which it is tied to and represents the underlying structural relationships in the problem). The linear algebra problem in this study is instructive for that reason. It was the only problem of the three that was created to reflect what students were learning in their curriculum at the time. A graph of a function is also the kind of abstract image that is well suited to the underlying structure of the algebra problem. Nonetheless, differential responses were illuminating because only students with high operations ability—irrespective of their ability for visual vividness—were significantly more likely to solve the problem correctly. Almost half of the high operation students solved the problem correctly as opposed to 15% or less of low operations ability students. There was no significant effect for vividness of visual imagery.

Finally, a tendency to use concrete images in the problems was complemented by verbal elaborations that reinforced these kinds of visual representations. Many students explained that they expanded upon the problem by adding details from their everyday experiences. These additional details on fractions and algebra problems occurred across operations and visual vividness ability levels. At times, these nonmathematical details helped, but at other times, they misled students and led to incorrect solutions.

Campbell et al.'s (1995) study in conjunction with the research on third- and fourth-graders described previously has important implications for understanding the role of visual representations in math problem solving. Without specific instruction—what some researchers call "diagrammatic literacy"—many students are challenged to produce the kind of abstract image schemata (see Johnson, 1987) that transcends a particular example of an image and in turn, helps students understand the essential logical propositions and quantitative relationships in a problem. Otherwise, there is a natural tendency to create concrete and often decorative pictures that have no substantial benefit for problem solving (Elia & Philippou, 2004; van Garderen & Montague, 2003).

A second observation from this research is that visual imagery alone may be insufficient for success. Students still require some kind of operational competence to solve the problem. This may be more than a matter of making the final computations to find an answer. Rather, it is the ability to read problems and identify critical relationships as well as unknown quantities (Hegarty, Mayer, & Monk, 1995; Kintsch & Greeno, 1985; Verschaffel, Greer, & De Corte, 2000).

EMBEDDING VISUAL REPRESENTATIONS IN TEXT

A sensible way to augment word problem solving would be to attach or embed visual representations along with the word problem for students to see as they solve it. After all, curriculum developers are in a position to control the quality of the imagery associated with problem solving rather than hoping that students will create their own high quality representations. Arguably, well-constructed visual representations could serve as an added scaffold for understanding and solving word problems.

Dewolf, Van Dooren, Cimen, and Verschaffel (2014) examined how illustrations depicting the setting or action contained in the problem could foster student success. For example, one problem in their study involved four boards that were 2.5 m each. The question asked how many one-meter sections could be cut out of these boards. This nonroutine problem was more complex than traditional or routine word problems because of the residual half meter on each board was potentially confusing (e.g., what to do with the extra .5 m of board after each board was sawed into 1-m lengths). The associated illustration for this problem showed an adult sawing a board with a pile of boards in the background. Contrary to expectations, these illustrations of "real world settings" had no significant effect on problem-solving performance.

Much of the literature on visual representations would suggest that some kind of well-structured diagram would be more likely to sponsor student success. For example, Hegarty and Kozhevnikov (1999) argue that schematic spatial diagrams help capture the key structural relationships found in word problems and most importantly, they negatively correlate with the kinds of pictorial, decorative representations described earlier. Diezmann and English (2001) provide specific guidance on the kinds of schematic diagrams well suited to word problems. They include lines or networks, matrices, trees or hierarchy, and part-whole diagrams (see Fig. 3).

Booth and Koedinger (2012) embedded a subset of these diagrams into word problems in their study of middle school students. The diagrams used in the experimental condition (diagram plus word problem) were rectangular configurations as well as simple pictographs. The rectangular configurations were akin to what are often called "tape diagrams," where each rectangle in the row of adjoining rectangles represents a quantity. As with the Dewolf et al. (2014) study, experimental students were not taught the meaning of these representations. They were simply assessed on their ability to interpret them in conjunction with the word problem. Booth and Koedinger found no general benefit for including a diagram with the word problem, and academically low-achieving students had varied responses to the diagrams. There was also a developmental effect where sixth graders tended to be confused by the diagrams, but eighth-grade students often found them beneficial. The greatest overall advantage of diagrams presented under these conditions (i.e., embedded along with word problems but not taught) was for older, high ability students.

Line or network diagram

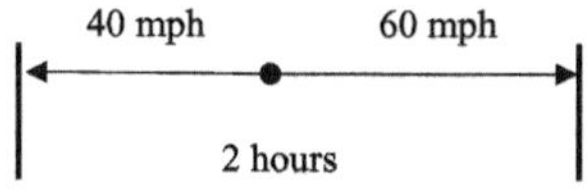

Two cars start out from town in opposite directions traveling at different speeds. The first car is traveling 40 mph and the second car is traveling 60 mph. How far are they apart after 2 hours?

Matrix diagram

	Horror	Action	Comedy
Thomas		x	
Ramona			√
Inez	x	√	x

Thomas, Ramona, and Inez are discussing their favorite kinds of films. Each one of them likes a different kind of movie. Thomas doesn't like movies where there are a lot of fights or car chases. Ramona laughs a lot in her favorite movies. Inez doesn't like scary movies or funny movies. Which movies do Thomas, Ramona, and Inez like the best?

Hierarchy diagram

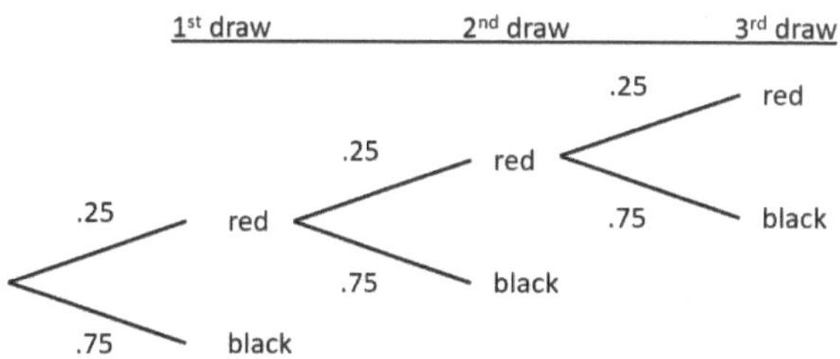

A deck of cards has 40 cards. 30 of the cards are black, and 10 of the cards are red. The deck is shuffled until the cards are in random order. You pick out cards one at a time. Each time you return the card, and shuffle again. What are the chances of drawing a red card, a red card, and then a black card?

Part whole diagram

Cheryl goes to the junkyard to look for a used tire for her vehicle. She counts 30 tires on the cars and motorcycles in the junkyard. There are twice as many cars as there are motorcycles. How many cars and motorcycles are there in the junkyard?

FIG. 3 Schematic diagrams.

Pantziara, Gagatsis, and Elia (2009) obtained similar results in their research on visual representations. Once again, the intervention drew upon classic schematic diagrams described in Diezmann and English (2001): networks, hierarchies, and tables. Sixth-grade students were given nonroutine word problems without diagrams in an initial assessment. One week later, they were administered parallel versions of the same problems with appropriately associated diagrams. Students were not taught how to interpret these diagrams before the administration of the second assessment. Pantziara et al.

found no overall significant differences in performance on the two tests even though one network and two hierarchy diagram problems yielded statistically significant differences in problem-by-problem analyses of the data. Further analysis indicated that diagrams made problems easier for some students and more difficult for others. The researchers attributed the failure of diagrams to the fact that diagrammatic literacy was not a feature of their daily classroom instruction.

Finally, Reuter et al. (2015) examined the effects of varied levels of assistance (or "prestructuring") of diagrammatic representations in their study of fourth-grade students. Students solved word problems in three different sessions: a pretest, posttest, and transfer test with no teacher instruction prior to or between assessments. Problems on the pretest contained no diagrams, while diagrams of various levels of assistance or completeness appeared in conjunction with the word problems for the experimental students on the post and transfer tests. Researchers anticipated that more complete diagrams would enhance a student's ability to understand and successfully solve the word problem. Contrary to expectations, no level of diagrammatic assistance yielded significant effects on the post or transfer assessments.

TEACHING STUDENTS TO CREATE OR COMPLETE DIAGRAMS AS THEY SOLVE PROBLEMS

While the studies described previously show no effects for embedded visual representations, they are counterbalanced by a wide range of research indicating that the kinds of schematic diagrams recommended in the literature are effective when accompanied by systematic instruction and student use with feedback (Diezmann & English, 2001; Hegarty & Kozhevnikov, 1999). Studies range from fifth graders constructing representations collaboratively for percent problems (Terwel, van Oers, van Dijk, & van den Eeden, 2009) to college students studying probability (Beitzel & Staley, 2015; Garcia-Retamero & Hoffrage, 2013). Particularly strong effects for diagrams can be found with academically low-achieving and special education students (Fang, Herron, Zhou, Hartsell, & Mohn, 2015; Jitendra et al., 2013; Jitendra, Dupuis, Star, & Rogriguez, 2016; Jitendra, Harwell, Dupuis, & Karl, 2017; Jitendra, Harwell, Karl, Simonson, & Slater, 2017; Kribbs & Rogowsky, 2016; van Garderen, 2007).

Yet this support for teaching students how to use diagrams as they solve word problems is tempered by *pedagogical* as well as *cognitive* concerns. Pedagogically, Fagnant and Vlassis (2013) note that there is a significant trade-off between problem-solving instruction that provides readymade diagrams that students replicate or complete with conditions where they are encouraged to produce diagrams on their own. In the latter condition, a classroom of students may use a variety of diagrams to solve the same problem. Fagnant and Vlassis argue that student-generated diagrams ultimately lead to greater mathematical flexibility because students have to process more carefully the underlying quantitative relationships in the problem and construct an appropriate diagram that makes sense to them or, in the case of Terwel et al. (2009), a group of students working together collaboratively. This claim is consistent with research on problem solving in the intermediate and middle grades that recommends students be taught a variety of solution methods including well-structured visual models (Woodward et al., 2012).

The limitations of readymade diagrams are apparent when problems shift from routine to complex, multistep, or nonroutine problems. Fig. 4 shows the kind of diagram used in the Fang et al. (2015) study in conjunction with two types of word problems. The first is similar to the kind of addition and subtraction problems used in their research, while the second word problem contains the kind of irrelevant information found in nonroutine problems not included in their study. Fang et al. instructed students how to use the diagram by putting the large number found in the word problem in the first box and then determine what was unknown. If the total was unknown, the problem was addition, while subtraction was inferred to be the correct operation when one of the smaller numbers was unknown or if the difference between two numbers was called for in the problem. Students wrote the operation in the circle and the appropriate label in the parentheses.

This kind of highly structured approach would be undercut by routine addition problems where the total is known (e.g., "Liz has 7 books. She got some more books off of the shelf. Now she has 12 books. How many books did she get off of the shelf?"). Even more challenging to students would be the kind of nonroutine problem shown in Fig. 4. The books on fish are irrelevant to the problem, yet the total number of books on fish is the largest number in the problem. This is an extreme example of the limitations of readymade diagrams, but it is a cautionary example of how a restricted range

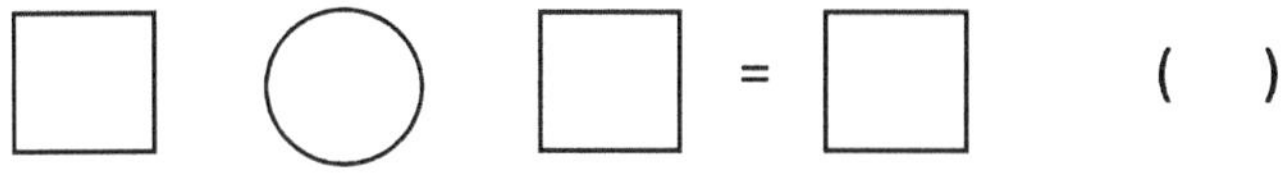

Routine problem: Liz had 7 books. She bought 4 more books. How many books did she have in all?

Non-routine problem: Liz is in the school library because she wants to check out books on lizards. There are 17 books on the shelf about lizards, and 25 books on fish. She picks 3 of the books on lizards to check out. How many books on lizards are on the shelf when Liz leaves the library?

FIG. 4 Readymade diagrams and two types of word problems.

of word problems can give a false impression of problem-solving competence even when the domain is limited to addition and subtraction word problems.

Yet Fagnant and Vlassis's (2013) recommendation that students create their own representations has its own problems. Beyond the student propensity to draw decorative pictures, the array of possible schematic diagrams that students could create is considerable. The general-purpose diagrams shown in Fig. 3 are neither exhaustive nor are they equally useful. Tape diagrams, for example, are well suited to operations on whole numbers and fractions as well as ratio problems (see Murata, 2008). Line diagrams, especially as number lines, are highly effective with operations on whole numbers, fractions, and algebra (Klein, Beishuizen, & Treffers, 1998; Siegler, Fazio, Bailey, & Zhou, 2013; Siegler, Thompson, & Schneider, 2011; Woodward et al., 2012). Within the domain of ratios and proportions alone, the CCSS-M (National Governors Association Center for Best Practices, Council of Chief State School Officers, 2010) recommends that students use tape diagrams, tables, and double number lines. Matrix and hierarchy diagrams tend to have less of a role in K-12 problem solving, and hierarchy/tree diagrams tended to be limited to topics such as probability (Beitzel & Staley, 2015; Garcia-Retamero & Hoffrage, 2013). However, this may be more a function of the kinds of word problems taught in schools. Matrix and hierarchy could be used more widely with increased attention to logic and puzzle problems.

Thus teaching a sufficient variety of representations that do not overwhelm students is its own pedagogical problem. As Reuter et al. (2015) note, students need a kind of "diagram literacy" training where they are shown how to use different representations and then given feedback once they have used them. This entails a considerable investment by teachers into a type of instruction that while advocated by standards and mathematics education research is typically not found in schools. One way to resolve this issue, perhaps, is by concentrating on just a few, high utility representations or making a deeper commitment to one kind of representation such as the empty number line's use in the Netherlands (Klein et al., 1998; van den Heuvel-Panhuizen, 2008).

There is also a *cognitive* issue that arises around students learning to create multiple types of diagrams when they learn to solve certain kinds of word problems (e.g., operations on whole numbers, ratios and proportions, algebra, probability). Booth and Koedinger (2012) speculated that one reason why diagrams did not yield significant results in their study was because the foreign nature of the diagrams increased the students' cognitive load. In addition to reading and interpreting the word problem, students had to make sense of diagrams that were unfamiliar to them. Beitzel and Staley (2015) reported similar concerns when students are asked to generate their own representations. However, Beitzel and Staley also suggested that the distinct benefit of a familiar representation used repeatedly is that it can reduce cognitive load on working memory by offloading some of the task to the diagram. This, in turn, allowed the learner to focus more on reading and interpreting the problem. These suggestions are consistent with Ainsworth's (2006) conceptual framework for learning from multiple representations.

There is also an alternative explanation for Booth and Koedinger's (2012) findings, which suggests that students actually devoted relatively *few* cognitive resources to understanding the diagrams. Some research indicates that students attend *less* to completed diagrams than to text (Schuler, Scheiter, Rummer, & Gerjets, 2012; Schwonke, Berthold, & Renkl, 2009). This observation applies more to diagrams that are embedded for students to interpret than cases where students are asked to either construct their own diagrams or complete ones that are readymade.

INTEGRATING VISUAL REPRESENTATIONS WITH TEXT

The focus to this point in the chapter has been largely on how visual representations can enhance word problem solving. Little has been discussed regarding how these representations play a role in a wider context, either in classroom instruction or curricular materials. For example, the role of visual representations is becoming increasingly important in technology-based learning. Representations such as diagrams may be accompanied by text, narration, video clips, animation, and interactive feedback in the case of computer tutorials. All of this, in turn, has sponsored a number of frameworks for the optimal use of visual representations in complex, multimedia environments (e.g., Ainsworth, 2006; Mayer, 2009; Nathan, Kintsch, & Young, 1992; Reed, 2006; Schnotz & Bannert, 2003). There are components of this literature that pertain to, if not restate observations already made about the integration of visualization representations with traditional word problems as these problems appear to learners in a printed or electronic format. What follows is a brief review of traditional models of how students read and then effectively construct mental representations of word problems. These representations, in turn, serve as a framework for successfully solving the problem. In the latter part of this section we will describe how schematic diagrams can be incorporated into this process, and in doing so, help modulate cognitive load.

Traditional accounts of the difficulties that student experience in solving word problems involve their inability to read beyond the "surface features" of the story (Quilici & Mayer, 1996; Reed, 1999; Weaver & Kintsch, 1992). Consequently, many students are distracted by a word problem's story content and do not detect that two problems with different story contents actually contain the same underlying structure. For example, two word problems where the story content in one involves changes in the price of gasoline and in the other involves variations in annual inches of rainfall may both have an underlying percent increase or decrease structure.

An essential step in modeling a word problem beyond its surface structure is to translate the text into a set of propositions, which create the foundation for a mental model of the problem (Kintsch & Greeno, 1985). Hegarty and her colleagues (Hegarty et al., 1995; Hegarty & Kozhevnikov, 1999) underscore the role visual-spatial representations play in this abstraction process. Successful problem solvers in Hegarty and Kozhevnikov (1999) tend to create schematic representations of the problem, whereas unsuccessful students are prone to pictorial representations. Hence, the issue around success in problem solving may have less to do with visualization in general as described in Campbell et al.'s (1995) research than a specific type of visualization (i.e., schematic representation).

Hegarty et al. (1995) present a detailed account of successful strategies used to solve word problems in their study of college students. It is important to note that by using this kind of sample, they controlled for subjects who lacked basic decoding or reading comprehension skills. By tracking eye movements as subjects read and solved word problems, the researchers were able to construct a model of word problem-solving competence. While successful as well as unsuccessful problem solvers read text in a nonlinear manner (i.e., they look back over text they have already read), their overall strategies fundamentally differed. Successful problem solvers extract propositions by reading and rereading the text and in doing so construct an increasingly rich model of the problem based on numbers, relational terms, and variable names. The eventual object-based model, which is visual-spatial in nature, captures coherently the relationship between key variables or quantities as well as the unknown variable. This model is even retained in working memory as a way of monitoring the solution process. The amount of cognitive processing associated with this kind of textual analysis enables successful problem solvers to adapt to different types of word problems. This is critical given the current word problem-solving literature on routine, nonroutine, inconsistent, and multistep problems.

Unsuccessful problem solvers, in contrast, reduce working memory demands by only looking for numbers and relational terms in a word problem. This obviates the need to examine or reexamine other critical portions of the text, and it leads to a more direct translation of the text to computations and an answer. Unfortunately, the propositional network in the student's model is impoverished, and the relationships between variables or quantities tend to be oversimplified. The very nature of this method of reading and converting text into a mental model makes it unlikely that these students are sensitive to a change in the type or structure of a word problem. Nor are students who use a direct translation strategy likely to change their strategy with more practice.

More recent investigations of word problem solving suggest that the challenges students face is greater than constructing a mental model of the problem. Even though Boonen and his colleagues' research (Boonen, de Koning, Jolles, & van der Schoot, 2016; Boonen, van der Schoot, van Wesel, De Vries, & Jolles, 2013; Boonen, van Wesel, Jolles, & van der Schoot, 2014) reinforces the role of spatial abilities as well as schematic representations in successful problem solving, finer grain reading comprehension skills are also important. Comprehension issues arise when the semantic complexity of the problem increases. Relational terms such as "more than" appear to be easier for students to convert into subtraction than "less than" into the operation of addition (Boonen et al., 2016). These subtle, semantic-linguistic dimensions of word problems appear to challenge students who would otherwise be classified as successful problem solvers.

There are at least two important implications from this research that pertain to math problem solving in schools today. First, Boonen's research (Boonen et al., 2014; Boonen et al., 2016) reminds us that unlike theoretical models of problem solving (Kintsch & Greeno, 1985) or work conducted with college students (Hegarty et al., 1995), there are a number of students in our K-12 schools who are poor in math problem solving *and* reading comprehension. Second, a good deal of problem-solving research is conducted with typical textbook problems that are four sentences or less in length. Longer problems, such as "real world" problems advocated in CCSS-M (National Governors Association Center for Best Practices, Council of Chief State School Officers, 2010) are likely to contain more details and a greater amount of irrelevant information so as to provide an authentic context for the problem solving. These features invoke not only concerns for reading comprehension, but demands made on working memory.

Sweller's highly influential theory of cognitive load (De Jong, 2010; Sweller, 1994) offers an added perspective on the challenges that students face when they are uncertain how to translate a word problem into an appropriate mental model. Excessive cognitive load arises for many reasons, not the least of which have to do with general weaknesses in working memory in academically low-achieving students and those with math disabilities (see Geary, Hoard, Byrd-Craven, & DeSoto, 2004; Swanson & Beebe-Frankenberger, 2004). For capable students such as the ones described previously (e.g., Hegarty et al., 1995), the use of direct translation strategies appears to reflect a learner's effort to reduce cognitive load by attending to only limited features of the problem (i.e., numbers and relational terms). Finally, Boonen et al. (2016) suggest that subtle levels of semantic complexity may create excessive cognitive load in students who would otherwise be deemed successful problem solvers.

Cognitive load theory claims that excessive demands on working memory can be controlled by attending to intrinsic as well as extraneous factors. A method for addressing intrinsic load factors would be to modify word problems in terms of their semantic structure or by removing extraneous information. Yet these kinds of alterations are contrary to the everyday structure of problems as well as mathematics education policy that encourages students to engage in solving more complex, "real world" problems. Thus intrinsic load factors may be beyond a teacher's or curriculum designer's control because of these kinds of policies and their associated assessments.

Furthermore, a narrow range of word problems as shown in the Fang et al. (2015) study can ill prepare students for even word problems that contain a reasonable amount of distracting information (see Fig. 4).

A more profitable instructional approach is to attend to extraneous load factors where curricular materials and/or instruction are designed to enhance schema construction. This could entail teaching students how to construct mental models of problems where they look beyond just numbers and relational terms to variable names, unknown quantities, and relationships between variables (Hegarty et al., 1995). This is also the role for schematic diagrams, assuming that they are appropriately integrated with text and they are taught to learners. As mentioned earlier in this chapter, studies where diagrams are merely embedded with word problems without added instruction (Booth & Koedinger, 2012; Pantziara et al., 2009; Reuter et al., 2015) may have resulted in what Sweller (1994) calls the "split-attention effect." That is, even greater demands are made upon working memory simply because subjects constantly shift their attention between the word problem and the diagram.

Teaching students how to translate textual material into a well-structured, schematic diagram not only builds appropriate mental models or schema over time, but it reduces cognitive load. Multimedia (e.g., Mayer, 2009) frameworks and learning theory (Paivio, 1986) support this integration of text and visual imagery because of its positive effects on understanding and retention. The challenge for teachers and curriculum designers is to construct an appropriate fit between a range of word problems (e.g., routine, inconsistent, nonroutine, multistep) and a diagram or discrete set of diagrams that enable students to map text onto diagrams in an integrated manner.

SCHEMA-BASED INSTRUCTION (SBI): INTEGRATED INSTRUCTION IN WORD PROBLEMS AND VISUAL REPRESENTATIONS

In the final portion of this chapter, we will offer an extended example of schema-based instruction (SBI), a research-based approach that integrates schematic diagrams with word problems for ratios, proportions, and percent. Several randomized controlled studies have provided empirical evidence of the effectiveness of SBI (e.g., Jitendra et al., 2015; Jitendra, Harwell, Karl, et al., 2017; Jitendra, Star, Dupuis, & Rodriguez, 2013; Jitendra, Star, Rodriguez, Lindell, & Someki, 2011). Many of the critical features of the SBI curriculum are highlighted in recent syntheses of high-quality research on students in the intermediate and middle grades (see Gersten et al., 2009; Woodward et al., 2012).

The SBI approach incorporates a number of pedagogical and curricular features that attend to concerns described in the previous section. Perhaps the most distinguishing feature of their materials is their attention to extraneous factors that affect cognitive load and schema construction. Each schematic diagram

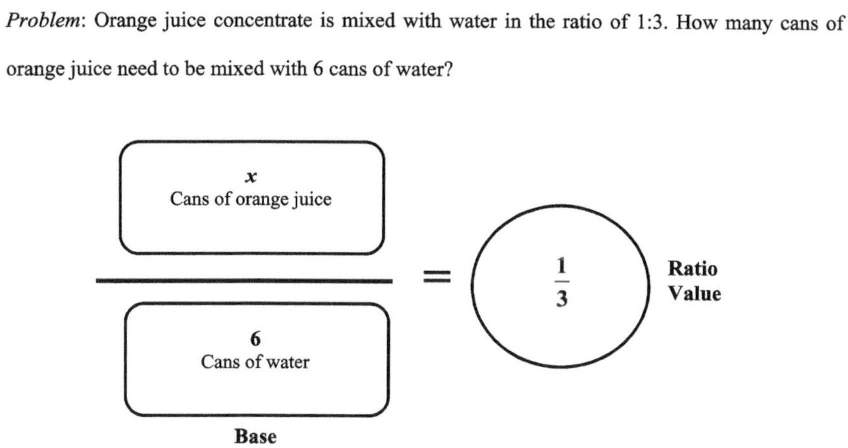

FIG. 5 SBI ratio diagram.

in their program has been designed to accentuate the underlying relationships found in the type of word problem being taught. Fig. 5 shows an example of ratio word problem and the associated diagram in the SBI curriculum.

This kind of word problem follows a *restate* schema where a relation exists between two things (e.g., twice as much as, nth times) and there is a, "restatement of this relationship using values different from those involved in the initial relation statement" (Kalyuga, 2006, p. 6). Moreover, as a ratio it involves a known relation between two things (1:6 or 6 times as much), and the same relation applies to two different values describing the same things: unknown amount of orange juice and 6 cans of water, which should be 6 times more than the amount of orange juice concentrate expressed in cans. Thus $1{:}3 = x{:}\ 6$, or 2 cans of orange juice are needed to mix 6 cans of water.

The schematic diagram in Fig. 5 is constructed in a way to draw explicit attention to the relationship between variables. Teacher instruction adds valuable meaning to the labels in each diagram. The labels *compare* and *base* labels reinforce the multiplicative relationship between orange juice concentrate and the water. Teacher also presents how the *ratio value* label highlights the fact that the quantities being compared are of the *same type*. Comparisons of quantities that are the same type are what distinguish ratios from proportions.

Fig. 6 presents a second type of diagram used in the SBI curriculum. Word problems used for this diagram comply with underlying relationships that follow a *vary* schema, which represents a situation in which, "a systematic relationship exists between the amounts of two different things. If the amount of one thing varies (decreases or increases), the amount of the second thing changes in a fixed way. These situations could be redescribed as involving

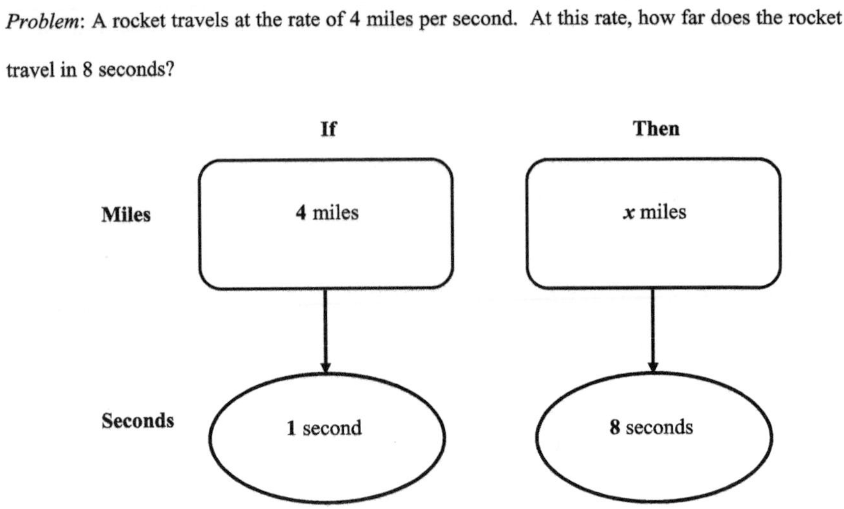

FIG. 6 SBI proportion diagram.

'IF … THEN' relationship" (Kalyuga, 2006, p. 6). The world problem in Fig. 6 describes a relation between miles traveled and the time taken. If the amount of time increases (8 times, in this instance), then the number of miles traveled will also increase by 8 times. The structure of the diagram along with teacher guidance helps students unpack the intensive quantities found in these kinds of rate relationships. It is well documented that intensive quantities are particularly difficult for students (Howe, Nunes, & Bryant, 2010), perhaps due in part to the kind of subtle semantic structure that Boonen et al. (2016) describes in their research.

The overall organization and flow of the SBI curriculum also attends to extraneous factors that both integrate the diagram with the word problem type and control for excessive cognitive load. The different types of ratios, proportions, and percent problems are introduced gradually over time with sufficient opportunities for practice and mastery. This structure requires students to contrast different diagrams and their associated word problems in a cumulative manner, thus necessitating a student's active attention to the quantitative relationships that underlie a given problem type.

A further curriculum design feature of the SBI curriculum is the way the solution to a problem is segmented into four distinct steps. Students begin by reading and rereading the problem to look for relationships that signify the problem type. The proportion word problem in Fig. 6 contains an "if … then" structure as well as the fact that different kinds of quantities are involved (i.e., miles and minutes). After identifying these features, the student is in a position to complete the second step, which involves translating the key numbers and variables into the schematic diagram. This step is critical because it separates the identification of quantitative relationships and a complete representational model of the problem from the next step, which involves estimating as well as computing the answer.

SBI curriculum designers intentionally created problems that afforded opportunities for estimation and number sense development. The rocket example in Fig. 6 can be written symbolically as $\frac{4}{1}=\frac{x}{8}$. Working vertically to solve this problem (i.e., 1 times 4 = 4, so 8 times 4 = 32) follows a unit rate strategy. Students can also solve this problem working horizontally (i.e., 1 times 8 = 8, so $4 \times 8 = 32$). This is an instance of an equivalent fraction strategy. The curriculum developers used simple numbers in their problems such as these to reinforce the multiplicative relationships found in ratios, proportions, and percent problems. The final step in the process is to check the calculations to see if the estimate generated for the answer in Step 3 is close to the exact answer. All of these steps have been codified in an acronym *DISC* (Discover, Identify, Solve, and Check) that students learn to apply as they solve proportion problems in word problem contexts. Additionally, SBI emphasizes metacognition skills and students are guided to "think about what they are doing and why they are doing it, evaluate the steps they are taking to solve the problem, and connect new concepts to what they already know" (Woodward et al., 2012, p. 17).

The readymade diagrams used in the SBI curriculum differ significantly from tape diagrams, tables, and double number lines recommended in CCSS-M (National Governors Association Center for Best Practices, Council of Chief State School Officers, 2010). Presumably, tape diagrams, double number lines, and/or tables are to be generated by students as they read and solve ratios and proportions word problems.

SBI curriculum developers created their own readymade diagrams for two reasons. First, it obviated the considerable amount of time it would take to teach and solidify each of the kinds of diagrams encouraged in the CCSS-M. This, in turn, helped control for extraneous cognitive load and the split attention effect. Second, the SBI diagrams were specifically designed for their mathematical correctness. As previously mentioned, the structure of the ratio diagram in Fig. 5 affords teachers the opportunity to draw attention to the fact that ratios involve the same kinds of quantities. The diagram for percent change problems shown in Fig. 7—the most complex of the SBI diagrams—assists teachers in reminding students that percent (captured by the circle and the label *ratio value*) is always a part-to-whole relationship. The triangular symbol for change is a direct mapping of the delta symbol used in mathematics to signify changeable quantities. It is unlikely that these kinds of mathematical dimensions would be captured in student-generated tape diagrams, tables, or double number lines or that they would arise in subsequent classroom discussions of these kinds of diagrams.

Problem: The movie theater is offering a 10% discount off of their regular $14 tickets to the first 100 people in line to see the new *Amazing Spiderman* film. How much will you pay if you are one of the first people in line?

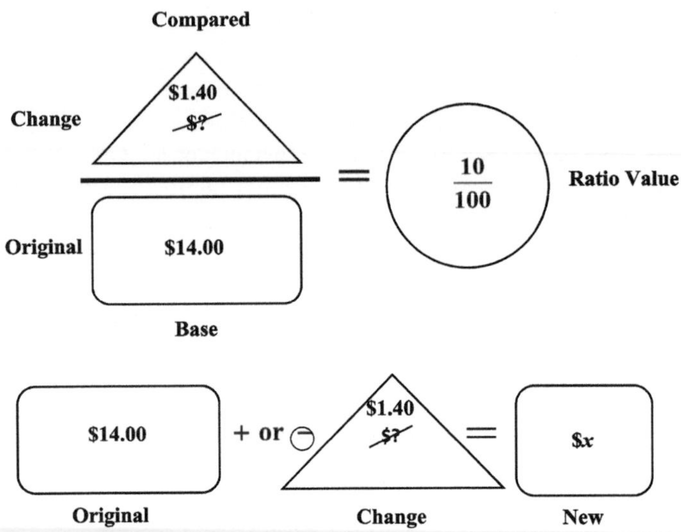

FIG. 7 Percent change problem.

Finally, the pedagogical component of the SBI curriculum is designed to help modulate cognitive load. Teachers spend the majority of a class lesson analyzing only two or three problems by working students through each step of the DISC strategy. The first step of DISC involves the teacher and the class reading and rereading each problem for relevant and irrelevant information, important numbers, variable names, and relational terms. This practice allows teachers to meet the needs of a range of student abilities and in doing so, mitigates cognitive load issues by moving the reading portion of the problem solving from the individual working alone to the classroom as a whole. To be sure, this is unlike the more common experimental practices in previously discussed research where subjects are required to read and solve word problems on their own (e.g., Boonen et al., 2013; Booth & Koedinger, 2012; Hegarty et al., 1995; Hegarty & Kozhevnikov, 1999; Reuter et al., 2015).

Teachers are also able to assist students in learning how to use each of the four schematic diagrams during the remaining three steps in DISC. Classroom video analyses conducted as part of the SBI studies indicated that teachers and students often participated in lively and ad hoc classroom discussions as they translated the word problem into the diagram, created estimations, calculated the solution, and checked to determine if the estimate aligned with the answer. These interactions are also informal assessment opportunities for teachers to probe areas where students are weak or exhibited misconceptions. These practices are consistent with Klein et al. (1998), who describe the importance of classroom discussions as students learn how to use a novel representation such as the empty number line and develop flexible number sense strategies at the same time.

Further research questions remain that might help clarify the long-term effects of SBI visual representations on word problem solving. Taken together, the studies conducted by Jitendra and her research team have met the high-quality design standards for educational research, which encourages randomized controlled trials. Now that the SBI curricular materials and instructional practices have been refined and validated through these methods, it would be important to return to clinical interviews where a range of individual students are assessed for the methods that they use to solve routine as well as multistep ratio and proportion problems in the maintenance phase of a study.

Of particular interest would be how well students can retrieve and use the schematic diagrams under conditions where these representations are not present in readymade form as part of the clinical interview. Given the unique design of SBI diagrams, one wonders how (or if) they are retained in long-term memory. Perhaps the primary purpose of these diagrams is simply to help students construct robust mental models of word problems as described in early word problem research (e.g., Hegarty et al., 1995; Kintsch & Greeno, 1985) and that the diagrams function as a scaffold for achieving this kind of cognitive framework. Contrasts with simpler diagrams such as double number lines that use instructional design and pedagogy similar to the SBI curriculum may help determine which kinds of diagrams are retained in long-term memory. This might further isolate the extent to which diagrams, when retrieved and used under circumstances where they are not in a ready-made form for the learner, impact problem-solving competence.

CONCLUSIONS AND FUTURE DIRECTIONS

Well-designed visual representations hold the promise of assisting students in constructing the kind of mental models needed to solve routine as well as multistep word problems. One cannot make too strong of an inference from many of the education research studies summarized in this chapter because they tend to be multicomponent in nature. The SBI curriculum described earlier, for example, included diagrams, metacognitive strategies, and extensive teacher-student dialogs. Nonetheless, carefully orchestrated instruction—teaching that helps students align a diagram with an associated word problem—appears to be essential for both the construction of mental models and the modulation of cognitive load.

At the same time, successful applications of schematic diagrams to complex multimedia environments or lengthy real world problems remain an area for further research. Returning to issues Aldous (2007) raised in the beginning of this chapter, a central goal of visual representations is to *internalize* them so that they become another tool for thinking about mathematical problems. Integration of text and imagery has long been proffered as benefiting understanding and retention (Mayer, 2009; Paivio, 1986; Reed, 2006; Schnotz & Bannert, 2003). The extent to which the kinds of schematic diagrams discussed in this chapter (e.g., line or network diagrams, tape diagrams, SBI diagrams) ultimately reside in long-term memory as tools to be used internally or as a strategy applied on pencil and paper in everyday contexts merits further consideration. Siegler et al. (2011) have noted that specific types of diagrams, particularly the number line, appear necessary for constructing the kind of mental images fundamental to long-term success in mathematics.

REFERENCES

Ainsworth, S. (2006). DeFT: a conceptual framework for considering learning with multiple representations. *Learning and Instruction*, *16*, 183–198.

Aldous, C. (2007). Creativity, problem solving and innovative science: insights from history, cognitive psychology and neuroscience. *International Education Journal*, *8*(2), 176–186.

Baddeley, A. D., & Hitch, G. (1974). Working memory. *Psychology of Learning and Motivation*, *8*, 47–89. https://doi.org/10.1016/S0079-7421(08)60452-1.

Baxter, J., Woodward, J., & Olson, D. (2001). Effects of reform-based mathematics instruction in five third grade classrooms. *The Elementary School Journal*, *101*(5), 529–548.

Beitzel, B., & Staley, R. (2015). The efficacy of using diagrams when solving probability word problems in college. *Journal of Experimental Education*, *83*(1), 130–145.

Boonen, A., de Koning, B., Jolles, J., & van der Schoot, M. (2016). Word problem solving in contemporary math education: a plea for reading comprehension skills training. *Frontiers in Psychology*, *7*, 1–10.

Boonen, A., van der Schoot, M., van Wesel, F., De Vries, M., & Jolles, J. (2013). What underlies successful word problem solving? A path analysis in sixth grade students. *Contemporary Educational Psychology*, *38*, 271–279.

Boonen, A., van Wesel, F., Jolles, J., & van der Schoot, M. (2014). The role of visual representation type, spatial ability, and reading comprehension in word problem solving: an item-level analysis in elementary school children. *American Journal of Educational Research*, *68*, 15–26.

Booth, J., & Koedinger, K. (2012). Are diagrams always helpful tools? Developmental and individual differences in the effect of presentation format on students' solutions of algebra problems. *British Journal of Educational Psychology*, *82*, 492–511.

Campbell, K., Collis, K., & Watson, J. (1995). Visual processing during mathematical problem solving. *Educational Studies in Mathematics*, *28*(2), 177–194.

Dark, V., & Benbow, C. (1990). Enhanced problem translation and short-term memory: components of mathematical talent. *Journal of Educational Psychology*, *82*, 420–429.

Dehaene, S., Spelke, E., Pinel, P., Stanescu, R., & Tsivkin, S. (1999). Sources of mathematical thinking: behavioral and brain-imaging evidence. *Science*, *284*(5416), 970–974.

De Jong, T. (2010). Cognitive load theory, educational research, and instructional design: some food for thought. *Instructional Science*, *38*, 105–134.

Dewolf, T., Van Dooren, W., Cimen, E., & Verschaffel, L. (2014). The impact of illustrations and warnings on solving mathematical word problems realistically. *Journal of Experimental Education*, *82*(1), 103–120.

Diezmann, C., & English, L. (2001). Promoting the use of diagrams as tools for thinking. In A. Cuoco & F. Curcio (Eds.), *The roles of representation in school mathematics* (pp. 77–89). Reston, VA: National Council of Teachers of Mathematics.

Elia, I., & Philippou, G. (2004). In M. J. Hoines & A. B. Fuglestad (Eds.), *The functions of pictures in problem solving. Proceedings of the 28th conference of the International Group for the Psychology of Mathematics Education* (Vol. 2, pp. 327–334). Bergen, Norway: University College.

Fagnant, A., & Vlassis, J. (2013). Schematic representations in arithmetical problem solving: analysis of their impact on grade 4 students. *Educational Studies in Mathematics*, *84*(1), 149–168.

Fang, H., Herron, S., Zhou, Q., Hartsell, T., & Mohn, R. (2015). The effects of simplified schema-based instruction on elementary students' mathematical word problem solving performance. *Journal of Mathematics Education*, *8*(1), 37–55.

Garcia-Retamero, R., & Hoffrage, U. (2013). Visual representation of statistical information improves diagnostic inferences in doctors and their patients. *Social Science & Medicine*, *83*, 27–33.

Geary, D., Hoard, M., Byrd-Craven, J., & DeSoto, M. (2004). Strategy choices in simple and complex addition: contributions of working memory and counting knowledge for children with mathematical disability. *Journal of Experimental Child Psychology*, *88*, 121–151.

Gersten, R., Beckmann, S., Clarke, B., Foegen, A., Marsh, L., Star, J. R., et al. (2009). *Assisting students struggling with mathematics: Response to Intervention (RtI) for elementary and middle schools (NCEE 2009-4060)*. Washington, DC: National Center for Education Evaluation and Regional Assistance, Institute of Education Sciences, U.S. Department of Education. Retrieved from http://ies.ed.gov/ncee/wwc/publications/practiceguides/.

Hegarty, M., & Kozhevnikov, M. (1999). Types of visual-spatial representations and mathematical problem solving. *Journal of Educational Psychology*, *91*(4), 684–689.

Hegarty, M., Mayer, R., & Monk, C. (1995). Comprehension of arithmetic word problems: a comparison of successful and unsuccessful problem solvers. *Journal of Educational Psychology*, *87*, 18–32.

Howe, C., Nunes, T., & Bryant, P. (2010). Intensive quantities: why they matter to developmental research. *British Journal of Developmental Psychology*, *28*, 307–329.

Izard, V., Sann, C., Spelke, E., & Streri, A. (2009). Newborn infants perceive abstract numbers. *Proceedings of the National Academy of Sciences of the United States of America*, *106*, 10382–10385.

Jitendra, A., Star, J., Dupuis, D., & Rodriguez, M. (2013). Effectiveness of schema-based instruction for improving seventh-grade students' proportional reasoning: a randomized experiment. *Journal of Research on Educational Effectiveness*, *6*(2), 114–136.

Jitendra, A. K., Dupuis, D. N., Rodriguez, M. C., Zaslofsky, A. F., Slater, S., Cozine-Corroy, K., et al. (2013). A randomized controlled trial of the impact of schema-based instruction on mathematical outcomes for third grade students with mathematics difficulties. *The Elementary School Journal*, *114*, 252–276.

Jitendra, A. K., Dupuis, D. N., Star, J. R., & Rogriguez, M. C. (2016). The effects of schema-based instruction on the proportional thinking of students with mathematics difficulties with and without reading difficulties. *Journal of Learning Disabilities*, *49*, 354–367.

Jitendra, A. K., Harwell, M. R., Dupuis, D. N., & Karl, S. R. (2017). A randomized trial of the effects of schema-based instruction on proportional problem solving for students with mathematics problem-solving difficulties. *Journal of Learning Disabilities*, *50*, 322–336.

Jitendra, A. K., Harwell, M. R., Dupuis, D. N., Karl, S. R., Lein, A. E., Simonson, G., et al. (2015). Effects of a research based mathematics intervention to improve seventh-grade students' proportional problem solving: a cluster randomized trial. *Journal of Educational Psychology*, *107*, 1019–1034.

Jitendra, A. K., Harwell, M. R., Karl, S. R., Simonson, G. R., & Slater, S. C. (2017). Investigating a Tier 1 intervention focused on proportional reasoning: a follow-up study. *Exceptional Children, 83*(4), 340–358.

Jitendra, A. K., Star, J. R., Rodriguez, M., Lindell, M., & Someki, F. (2011). Improving students' proportional thinking using schema-based instruction. *Learning and Instruction, 21*, 731–745.

Johnson, M. (1987). *The body in the mind: The bodily basis of meaning, imagination and reason.* Chicago: University of Chicago Press.

Kalyuga, S. (2006). Rapid cognitive assessment of learners' knowledge structures. *Learning & Instruction, 16*(1), 1–11.

Kintsch, W., & Greeno, J. G. (1985). Understanding and solving word arithmetic problems. *Psychological Review, 92*, 109–129.

Klein, T., Beishuizen, M., & Treffers, A. (1998). The empty number line in Dutch second grades: 'realistic' versus 'gradual' program design. *Journal for Research in Mathematics Education, 29*, 443–464.

Kribbs, E., & Rogowsky, B. (2016). A review of the effects of visual-spatial representations and heuristics on word problem solving in middle school mathematics. *International Journal of Research in Education and Science, 2*(1), 65–74.

Lefevre, J., Fast, L., Skwarchuk, S., Smith-Chant, B., Bisanz, J., Kamawar, D., et al. (2010). Pathways to mathematics: longitudinal predictors of performance. *Child Development, 81*(6), 1753–1767.

Li, Y., & Geary, D. C. (2017). Children's visuospatial memory predicts mathematics achievement through early adolescence. *PLoS One, 12*(2), e0172046.

Mayer, R. E. (2009). *Multimedia learning* (2nd ed.). New York: Cambridge University Press.

McCrink, K., & Wynn, K. (2004). Large-number addition and subtraction in infants. *Psychological Science, 15*, 776–781.

Meyer, M. L., Salimpoor, V. N., Wu, S. S., Geary, D. C., & Menon, V. (2010). Differential contribution of specific working memory components to mathematics achievement in 2nd and 3rd graders. *Learning and Individual Differences, 20*(2), 101–109.

Murata, A. (2008). Mathematics teaching and learning as a mediating process: the case of tape diagrams. *Mathematical Thinking and Learning, 10*(4), 374–406.

Nathan, M., Kintsch, W., & Young, E. (1992). A theory of algebra-word-problem comprehension and its implications for the design of learning environments. *Cognition and Instruction, 9*, 329–389.

National Council of Teachers of Mathematics (NCTM). (1989). *Curriculum and evaluation standards for school mathematics.* Reston, VA: Author.

National Council of Teachers of Mathematics (NCTM). (2000). *Principles and standards for school mathematics.* Reston, VA: Author.

National Governors Association Center for Best Practices, Council of Chief State School Officers. (2010). *Common core state standards in mathematics.* Washington, DC: Author.

Paivio, A. (1986). *Mental representations: A dual coding approach.* Oxford: Oxford University Press.

Pantziara, M., Gagatsis, A., & Elia, I. (2009). Using diagrams as tools for the solution of non-routine mathematical problems. *Educational Studies in Mathematics, 72*(1), 39–60.

Quilici, J., & Mayer, R. (1996). Role of examples in how students learn to categorize statistics word problems. *Journal of Educational Psychology, 88*, 144–161.

Raghubar, K., Barnes, M., & Hecht, S. (2010). Working memory and mathematics: a review of developmental, individual difference, and cognitive approaches. *Learning and Individual Differences, 20*(2), 110–122.

Reed, S. (1999). *Word problems: Research and curriculum reform*. Mahwah, NJ: Lawrence Erlbaum Associates.

Reed, S. (2006). Cognitive architectures for multimedia learning. *Educational Psychologist, 41*(2), 87–98.

Reuter, T., Schnotz, W., & Rasch, R. (2015). Drawings and tables as cognitive tools for solving non-routine word problems in primary school. *American Journal of Educational Research, 3*, 1187–1197.

Schnotz, W., & Bannert, M. (2003). Construction and interference in learning from multiple representations. *Learning and Instruction, 13*, 141–156.

Schuler, A., Scheiter, K., Rummer, R., & Gerjets, P. (2012). Explaining the modality effect in multimedia learning: is it due to a lack of temporal contiguity with written text and pictures? *Learning and Instruction, 22*, 92–102.

Schwonke, R., Berthold, K., & Renkl, A. (2009). How multiple external representations are used and how they can be made more useful. *Applied Cognitive Psychology, 23*(9), 1227–1243.

Sfard, A. (1994). Reification as the birth of metaphor. *For the Learning of Mathematics, 14*, 44–55.

Siegler, R., Fazio, L., Bailey, D., & Zhou, X. (2013). Fractions: the new frontier for theories of numerical development. *Trends in Cognitive Sciences, 17*, 13–19.

Siegler, R., Thompson, C., & Schneider, M. (2011). An integrated theory of whole number and fractions development. *Cognitive Development, 62*, 273–296.

Stylianou, D. (2002). On the interaction of visualization and analysis: the negotiation of a visual representation in expert problem solving. *The Journal of Mathematical Behavior, 21*, 303–317.

Swanson, H., & Beebe-Frankenberger, M. (2004). The relationship between working memory and mathematical problem solving in children at risk and not at risk for serious math difficulties. *Journal of Educational Psychology, 96*, 471–491.

Sweller, J. (1994). Cognitive load theory, learning difficulty, and instructional design. *Learning and Instruction, 4*, 295–312.

Terwel, J., van Oers, B., van Dijk, I., & van den Eeden, P. (2009). Are representations to be provided or generated in primary mathematics education? Effects on transfer. *Educational Research and Evaluation, 15*(1), 25–44.

Tweney, R. D. (2009). Mathematical representations in science: a cognitive-historical case history. *Topics in Cognitive Science, 1*, 758–776.

van den Heuvel-Panhuizen, M. (2008). Learning from "didactikids": an impetus for revisiting the empty number line. *Mathematics Education Research Journal, 20*(3), 6–31.

van Garderen, D. (2007). Teaching students with LD to use diagrams to solve mathematical word problems. *Journal of Learning Disabilities, 40*, 540–553.

van Garderen, D., & Montague, M. (2003). Visual-spatial representation, mathematical problem solving, and students of varying abilities. *Learning Disabilities Research and Practice, 18*(4), 246–254.

Verschaffel, L., Greer, B., & De Corte, E. (2000). *Making sense of word problems*. Lisse [Netherlands]: Swets & Zeitlinger Publishers.

Weaver, C., & Kintsch, W. (1992). Enhancing students' comprehension of the conceptual structure of algebra word problems. *Journal of Educational Psychology, 84*, 419–428.

Woodward, J. (2001). In *Mathematical problem solving through performance assessments. Paper presented at the annual Pacific Coast research conference, San Diego, CA*. February.

Woodward, J., & Baxter, J. (1997). The effects of an innovative approach to mathematics on academically low achieving students in mainstreamed settings. *Exceptional Children, 63*(3), 373–388.

Woodward, J., Beckmann, S., Driscoll, M., Franke, M., Herzig, P., Jitendra, A., et al. (2012). *Improving mathematical problem solving in grades 4 through 8: A practice guide (NCEE 2012-4055).* Washington, DC: National Center for Education Evaluation and Regional Assistance, Institute of Education Sciences, U.S. Department of Education. Retrieved from http://ies.ed.gov/ncee/wwc/publications_reviews.aspx#pubsearch/.

Woodward, J., Monroe, K., & Baxter, J. (2001). Enhancing student achievement on performance assessments in mathematics. *Learning Disabilities Quarterly*, *24*(1), 33–46.

Wynn, K. (1992). Addition and subtraction by human infants. *Nature*, *358*, 749–750.

Chapter 12

The Role of Cognitive Processes in Treating Mathematics Learning Difficulties☆

Lynn S. Fuchs, Douglas Fuchs, Amelia S. Malone, Pamela M. Seethaler and Caitlin Craddock
Department of Special Education, Vanderbilt University, Nashville, TN, United States

Despite important advances in intervention for students with learning difficulties, the dominant approach to intervention—explicit skills instruction—fails to meet the needs of a substantial proportion of these students. For example, Fuchs, Geary, et al. (2013) investigated the effects of a first-grade number knowledge intervention, a form of explicit skills instruction focused on the foundational number knowledge necessary for early mathematics development. The intervention occurred three times per week for 16 weeks. In each 30-minute session, the major emphasis (25 minutes) was number knowledge; the other 5 minutes provided one of two forms of practice on solving simple number sentences: nonspeeded practice, which reinforced the relations and principles addressed in the number knowledge portion of the intervention, or speeded strategic practice, which was designed to integrate number knowledge with efficient counting procedures.

Children who participated in this study were identified as at risk for mathematics learning disabilities based on delayed mathematics performance at the start of first grade. They were randomly assigned to remain in their school

☆This research was supported by 2 R01 HD053714 from the Eunice Kennedy Shriver National Institute of Child Health & Human Development to Vanderbilt University and by Grant R324C100004 from the Institute of Education Sciences in the U.S. Department of Education to the University of Delaware with a subcontract to Vanderbilt University. The content is solely the responsibility of the authors and does not necessarily represent the official views of the Eunice Kennedy Shriver National Institute of Child Health & Human Development, the National Institutes of Health, the National Center on Special Education Research, or the Institute of Educational Sciences.

Mathematical Cognition and Learning, Vol. 5. https://doi.org/10.1016/B978-0-12-815952-1.00012-8

program without the research-based intervention ($n = 206$) or participate in the research-based number knowledge intervention (190 in nonspeeded practice, 195 in speeded practice). Results indicated that the number knowledge intervention with either form of practice produced stronger learning than occurred for children who continued in their school program without research-based intervention. This was the case on all four mathematics outcomes: arithmetic, 2-digit calculations, number knowledge, and word problems. Also, speeded practice produced stronger learning than nonspeeded practice on arithmetic and 2-digit calculations. The effects favoring both intervention conditions over the typical school program were large, suggesting strong practical value for improving the course of mathematics learning for these children.

Nevertheless, as with all interventions that include research-based explicit skills interventions and are implemented as intended, not all children responded. Consider arithmetic, one of the study's major outcomes, to which we applied the following definition of responsiveness: improvement over the 16 weeks of intervention commensurate with the average increase of not-at-risk classmates' improvement (i.e., at least 1 standard error of measurement above the not-at-risk classmates' mean). Using this definition, 94% of at-risk students were classified as adequately responsive to intervention, the other 6% classified as unresponsive. This rate of unresponsiveness (i.e., 6% of at-risk students whose mathematics performance at the start of first grade fell below the 40th percentile) represents 2.4% of the general population or approximately 40% of students with mathematics learning disabilities.

This rate of unresponsiveness, combined with the heterogeneity of learning disabilities (Fletcher, Lyon, Fuchs, & Barnes, in press), suggests the need to identify innovative approaches that target the specific needs of subgroups of students with learning disabilities. One strategy for identifying subgroups involves focusing on the cognitive, linguistic, or socioemotional processes associated with learning. (In this chapter, we use the term *cognitive processes* broadly, to denote cognitive and linguistic processes.) Yet, as shown in the working memory training literature, although cognitive processing may be strengthened, transfer to academic skill is limited in at-risk students (Kearns & Fuchs, 2013; Melby-Lervag & Hulme, 2013). This may be due to the difficulty students with or at risk for disabilities have with transfer (Haskell, 2001; National Research Council, 2000) or their weak foundational academic skill.

The question addressed in this chapter is whether we can combine what is known about explicit skills intervention with what is known about the role of cognitive processes in mathematics development to identify methods for expanding the efficacy of mathematics intervention. We address two possible approaches. The first involves embedding cognitive training within direct skills intervention. Here, students' cognitive resources are strengthened while intervention builds a stronger foundation of mathematics skill and facilitates transfer from the embedded cognitive training to mathematics performance. In the first section of this chapter, we illustrate this approach by embedding language comprehension instruction within first-grade word-problem intervention and demonstrate how such embedded instruction can offer added value over word-problem intervention without embedded instruction in strengthening word-problem outcomes.

The second approach involves allocating alternative forms of explicit skills intervention to subgroups of learners with different cognitive profiles. Here, the goal is to compensate within direct skills intervention for students' cognitive limitations. In the second section of this chapter, we describe how responsiveness to two variants of a fourth-grade fractions intervention depends on different preexisting cognitive processes. We explain how such research findings can be used to allocate alternative forms of direct skills interventions to students with different cognitive profiles and to strengthen the design of interventions to address a broader range of students.

These two approaches form the organizational structure of this chapter. Within each of these two major sections, we first present information about the context for the study: the theoretical framework for the subdomain of mathematics performance addressed and for the instructional approaches on which we relied. Then we outline the study methods, followed by an overview of findings. We conclude the chapter with a summary of key points.

EMBEDDING COGNITIVE TRAINING WITHIN EXPLICIT SKILLS INTERVENTION TO FACILITATE TRANSFER TO ACADEMIC PERFORMANCE

Embedding cognitive training within explicit skills intervention is an approach for combining what is known about direct, explicit skills intervention with what is known about the role of cognitive processes in mathematics development. The goal is to expand the range of students whose needs are addressed effectively in mathematics intervention. With embedded instruction, the goal is to strengthen cognitive processing while building a foundation of mathematics skill and facilitating transfer from the embedded cognitive training to mathematics performance.

Study Context

Why Embed Cognitive Process Training within Direct Skills Intervention?

As already noted, the literature suggests promise for some forms of cognitive process training, such as working memory. For example, in a meta-analysis conducted by Melby-Lervag and Hulme (2013), the mean effect size on verbal working memory tasks was 0.55. This indicates that students' working memory improves an average of half a standard deviation, typically operationalized as complex span tasks. Yet, as this meta-analysis also revealed, the effect size for transfer to calculation skill among at-risk learners was 0.25, lower than the effect of many explicit skills interventions. Few studies have addressed transfer to other math outcomes.

Research on language therapy indicates a similar pattern of effects for at-risk learners (Catts & Kamhi, 2017): Oral language comprehension improves, but transfer to academic tasks that rely on oral language comprehension, such as word-problem solving or reading comprehension, is disappointing. This lack of transfer from working memory training or oral language therapy to academic performance is not surprising because at-risk learners experience substantial challenges with transfer (Haskell, 2001; National Research Council, 2000). Moreover, expecting cognitive process training to boost academic performance in this population seems unrealistic because at-risk learners, by definition, have an inadequate foundation of academic skill (Geary, Nicholas, Li, & Sun, 2017). This argues for conducting cognitive process training in the context of academic intervention (to facilitate transfer), while building a stronger foundation of academic skill via that explicit skills intervention.

Why Focus on Word Problems?

In this section of the chapter, we describe preliminary findings from an ongoing study in which oral language instruction is embedded within word-problem intervention at first grade (Fuchs, Fuchs, Craddock, Seethaler, & Geary, 2017). We focus on word problems at first grade for the following reason. Although individual differences in word-problem performance increase at a rapid pace during first grade (Fuchs, Geary, et al., 2013), early mathematics research has focused predominantly on number knowledge and arithmetic, with only incidental attention to word problems to support understanding of number and operations.

This was the case in the Fuchs, Geary, et al. (2013) study, described at the start of this chapter. Results showed that although children who received number knowledge intervention experienced superior arithmetic and word-problem outcomes than did at-risk control group students, effect sizes were substantially smaller on word problems than on arithmetic. Moreover, while the arithmetic achievement gap with respect to not-at-risk classmates narrowed substantially by the end of intervention, the word-problem achievement gap increased substantially: Year-end word-problem performance of students who received the number knowledge intervention was 0.69–0.73 standard deviations below their not-at-risk classmates.

This widening word-problem achievement gap in the face of a narrowing arithmetic achievement gap for at-risk learners is alarming for three reasons. First, word problems are the best school-age predictor of employment and wages in adulthood (Bynner, 1997; Every Child a Chance Trust, 2009; Parsons & Bynner, 1997). Second, word problems can be a persistent deficit even when arithmetic skill is adequate (Swanson, Jerman, & Zheng, 2008), and the cognitive processes involved in word problems differ from and are more numerous than those underlying arithmetic skill (e.g., Fuchs et al., 2006, 2008, 2010a, 2010b; Geary, Hoard, Nugent, & Bailey, 2012). So word-problem and arithmetic skill may represent distinct components of mathematics competence, and word-problem difficulty may be more difficult to prevent. Third, word problems represent a major emphasis in almost every strand of the mathematics curriculum from kindergarten through high school. So word-problem difficulty limits school as well as occupational success.

Why Embed Instruction on Language Comprehension?

With this context in mind, we designed the Fuchs et al. (2017) ongoing study not only to directly target explicit word-problem intervention at first grade but also to incorporate a strong focus on language comprehension. We embedded language comprehension instruction in the word-problem materials used for explicit word-problem intervention. The goal is to facilitate transfer from the language instruction to word-problem skill, while simultaneously building children's word-problem skill.

We target language comprehension for embedded cognitive training, because oral language comprehension is a specific predictor of word-problem skill. Language comprehension predicts word problems but not calculations, at least beyond the earliest stages of calculation learning (Fuchs et al., 2006, 2010a, 2010b), and the indirect effects of language comprehension on prealgebraic knowledge occur entirely through word problems, not calculations (Fuchs et al., 2012). Also, the specific language comprehension demands within word problems differ from everyday language comprehension demands (Schleppegrell, 2007), such that language therapy outside the context of word problems is unlikely to transfer to word problems: Language therapy is broad, whereas word problems involve focused academic language knowledge ("WP language"), and the linguistic redundancy and nonlinguistic support for language comprehension in typical conversations is not available in word-problem language.

Conceptual Model for Linking Language Comprehension and Our Approach to Explicit Skills Word-Problem Intervention

Our conceptual model for linking language comprehension with explicit skills word-problem intervention is based on the seminal work of Kintsch and colleagues (e.g., Cummins, Kintsch, Reusser, & Weimer, 1988; Kintsch & Greeno, 1985; Nathan, Kintsch, & Young, 1992). In this model, word-problem solving involves an interaction between language comprehension processes and problem-solving strategies that rely on working memory and reasoning. Based on theories of text comprehension and discourse processing (e.g., Perfetti, Landi, & Oakhill, 2005), the model assumes that general features of the text comprehension process apply across stories, essays, and word-problem statements, but that the comprehension strategies, the nature of required knowledge structures, and the form of resulting macrostructures, situation models, and problem models differ by task.

The conceptual model is visualized in Fig. 1 and assumes that memory representations of word problems have three components. The first involves constructing a coherent microstructure and deriving a hierarchical macrostructure to capture the text's essential ideas. The second component, the situation model, requires supplementing the text with inferences based on the child's world knowledge, including informal knowledge about conceptual relations among quantities. The problem solver coordinates this information with the third component—problem models or schema—to formalize the conceptual relations among quantities. The schema guides application of solution strategies. At first grade, whole-number combine, compare, and change problem types are the major schema.

It follows from the model that this process of building the propositional text structure, inferencing, identifying schema, and applying solution strategies makes strong demands on three cognitive resources. In terms of working memory and reasoning ability, consider a combine problem (Part 1 plus Part 2 equals Total or $P1+P2=T$): *Joe has 3 marbles. Tom has 5 marbles. Tom also has 2 balls. How many marbles do the boys have in all?* A competent problem solver processes sentence 1's propositional text base to identify object = marbles; quantity = 3; actor = Joe; but Joe's role = to be determined. This is represented in short-term memory. In sentence 2, propositions are similarly coded and held in memory. In sentence 3, *balls* fails to match the object code in sentences 1 and 2, signaling 2 as perhaps irrelevant; this is added to memory. In sentence 4, the question, the quantitative proposition *how many marbles* and the phrase *in all* cues the child to identify the combine schema; assign the role of superset (Total) to the question; assign subset roles (Parts 1 and 2) to the TBDs in memory; and reject 2 balls. Filling in these slots of the schema triggers a set of problem-solving strategies.

With typical school instruction, children gradually construct the combine schema on their own, just as they devise their own strategies for handling

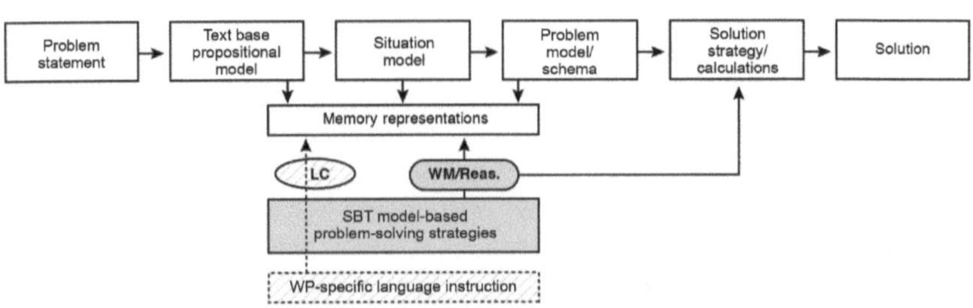

FIG. 1 Model linking intervention with cognitive processes and improved word-problem skill (LC is language comprehension; WM is working memory; Reas. is reasoning; SBT is schema-based intervention; WP is word-problem).

the demands on working memory and reasoning involved in this problem-solving sequence. Errors are viewed as failures to produce the intended mental representations (see the three components in the preceding paragraph) or to manage demands on working memory and reasoning. Our word-problem intervention, referred to as schema-based intervention, explicitly teaches the at-risk child step-by-step strategies that begin with identifying word-problem statements as combine, compare, or change schema. Then they are taught how to build the propositional text structure. Schema-based intervention facilitates connections among the situation model, schema, and productive solution strategies by making these connections explicit, while reducing demands on working memory and reasoning.

With schema-based intervention, the child RUNs through the problem: Reads it, Underlines the question in which the object code (marbles) is revealed, and Names the explicitly taught combine schema. This prompts the child to write the combine schema's number sentence, P1 + P2 = T (Part 1 plus Part 2 equals Total). The child then rereads the problem statement and while rereading, writes replacements for P1 and P2 (quantities for each relevant "part") and crosses out irrelevant objects and numbers. This is designed to reduce the burden on working memory and reasoning and provide the equation for problem solving, which prompts the counting-up adding strategy ("5, 6, 7, 8," for this problem) to find the solution (also taught in schema-based intervention).

As Kintsch and colleagues discussed and as Cummins et al. (1988) demonstrated, word-problem solving also relies heavily on multiple processes that contribute to language comprehension. As per Kintsch and Greeno (1985), children "understand important vocabulary and language constructions prior to school entry" (p. 111) and "through instruction in arithmetic and word problems, learn to treat these words in a special, task-specific way, including extensions to ordinary usage for terms (e.g., *all or more*) to more complicated constructions involving sets (*in all* and *more than*)" (p. 111). The assumption is that "students have the necessary language comprehension to understand problem statements and … form an appropriate problem model" (Nathan et al., 1992, p. 330).

But for at-risk children in a typical school program, this assumption is tenuous. Cummins et al. (1988) computationally modeled incorrect problem solving with two types of errors: incorrect math problem-solving processing versus incorrect language comprehension processing. Correct problem representation depended more on language comprehension, and changing wording in minor ways dramatically affected accuracy. As Nathan et al. (1992) concluded, "The study prescribes that instruction focus on language comprehension processes as well as the mathematical aspects of word-problem solving" (p. 332).

To illustrate how word-problem solving taxes language comprehension, consider this combine problem: *Joe has 3 cats. Tom has 5 dogs. Tom has 2 plants. How many pets do the boys have in all?* Objects in this text increase demands on language comprehension for assigning roles for the propositional text structure (despite similar demand for inducing the schema), due to more sophisticated representations of vocabulary involving taxonomic relations at superordinate levels and subtle distinctions among categories (dogs+cats= pets; plants are not pets). Embedding language comprehension instruction within schema-based intervention explicitly strengthens word-problem language for building propositional, situational, and schema representations.

In the ongoing study (Fuchs et al., 2017), we therefore hypothesize that embedding language comprehension will compensate for limitations in at-risk children's language comprehension (the study's compensatory moderation hypothesis) and improve word-problem solving not only directly but also indirectly by strengthening word-problem specific language (the study's mediation hypothesis). When we present preliminary results later in this chapter, we describe main effect outcomes among study conditions as well as mediation effects. Sample size still needs to accumulate to test the moderation hypothesis. (In the second approach described later in this chapter, we explain how compensatory moderation effects operate.)

Study Overview

In this ongoing study (Fuchs et al., 2017), we define risk at the start of first grade as low arithmetic skill and low performance on mathematics concepts and applications. At-risk students are randomly assigned, stratifying by classroom, to four conditions: schema-based intervention, schema-based intervention with embedded language comprehension instruction, number knowledge intervention, and control (the schools' business-as-usual program, which includes classroom instruction and, for most participants, intervention). We include the number knowledge intervention condition to assess whether transfer from arithmetic to word problems is sufficient to support word-problem development. Each active intervention condition lasts 15 weeks, with three sessions per week and 30 min per session. All three active intervention conditions include 5 minutes of speeded, strategic arithmetic practice. A total of 400 students will enter the study, 20% each year. In this chapter, we report findings on 159 students.

As already explained, with schema-based instruction, the tutor explicitly teaches step-by-step strategies to reduce demands on reasoning and working memory. This includes strategies for understanding word problems as belonging to word-problem types (i.e., schemas) and for building the word-problem model. As the child processes and solves a word problem, he/she uses these strategies to name the problem type, to represent that problem model with a number sentence, to enter relevant quantities from the word-problem statement into the problem model while crossing out "extra" (irrelevant) numbers, and to solve for the missing quantity. Fig. 2 shows a screenshot of a first-grader's work completing a combine problem (we teach the children to use the word *total* (T) instead of *combine*).

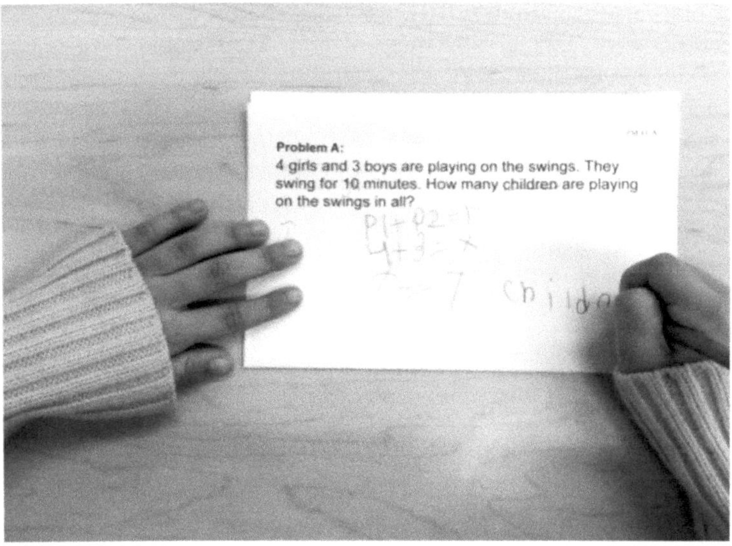

FIG. 2 Screenshot of first-grader applying explicit word-problem instructional strategies to a combine (Total = T) problem.

The embedded language comprehension component addresses word-problem language relevant to the first-grade word-problem schemas. Some are specific to problem types. For example, in combine problems, joining words (e.g., *altogether, in all*) and superordinate categories (e.g., *animals = dogs + cats*) are critical. In compare problems, compare words (e.g., *more, fewer, than*, *-er* words) and adjective *-er* versus verb *-er* words (e.g., *bigger* vs. *teacher*) are essential. In change problems, cause-effect conjunctions (e.g., *then, because, so*), implicit quantity change verbs (e.g., *cost, ate, found*), and time passage phrases (e.g., *3 hours later, the next day*) are important. Others apply across problem types and include, for example, confusing cross-problem constructions (e.g., *more than* vs. *then … more*) and "tricky" labels (e.g., questions with superordinate category words, without a label, noun that's the wrong label [as in money questions]).

With number knowledge intervention, explicit instruction is provided on number knowledge, relations, and principles. This includes a strong focus on number families and decomposition of number sets (7 = 6 + 1, 5 + 2, 4 + 3, etc.) and relies on the number line, manipulatives, and games to reinforce concepts.

Preliminary Results

We report on results to date for three outcomes: arithmetic, word-problem solving, and understanding of word-problem language. With respect to arithmetic, the three active conditions performed comparably at posttest, with each group outperforming the control group. The effect size for number knowledge intervention versus control was 0.81; the effect size for word-problem intervention versus control, 1.27; and the effect size for word-problem with embedded language comprehension, 1.68. These are all large effect sizes, indicating strong positive effects.

On the word-problem solving outcomes, the pattern of effects differs. The effect between the two word-problem conditions, with and without embedded language comprehension, is not significantly different, but is nearly so ($P = .06$). If the pattern of effects holds throughout the study cohorts, then an advantage for the embedded language comprehension component will be significant. Importantly, children in both word-problem conditions perform significantly stronger than number knowledge intervention children and stronger than control children, who perform comparably to each other. The effect size for word-problem intervention with embedded language comprehension instruction versus control is 1.16; for word-problem intervention versus control, 0.79; and for number knowledge intervention versus control 0.10. The effect size between the two word-problem conditions is 0.37.

So, both word-problem intervention conditions are producing significantly better word-problem outcomes than number knowledge intervention and control, but differences between the two word-problem conditions are not significant. To date, number knowledge intervention appears comparable to control on word-problem solving, despite that number knowledge intervention is producing significantly better arithmetic performance (effect size = 0.81). This raises questions about whether transfer from arithmetic to word problems is sufficient to support word-problem development among at-risk children. It also suggests added value for explicit word-problem intervention over number knowledge intervention on first-grade word-problem outcomes.

In terms of understanding of word-problem language, children in the word-problem intervention with embedded language comprehension instruction condition completed intervention with superior understanding compared to children in each of the other three conditions. The effect size contrasting number knowledge intervention with control was 0.01; the effect size contrasting word-problem intervention with control was 0.08. By contrast, the effect size contrasting word-problem intervention with embedded language comprehension with control was 0.50. The effect size comparing the two word-problem conditions was 0.42.

We also asked whether end-of-intervention understanding of word-problem language mediates condition effects on the word-problem solving outcome. In Fig. 3, *condition effect* refers to each active condition versus control (i.e., three models were run, one for each of the three active conditions versus control). Each model controls for pretest word-problem and arithmetic skill. Each model includes posttest word-problem language understanding (the potential mediator) and the condition effect: number knowledge intervention versus control in the upper left panel; word-problem intervention versus control, in the upper right panel; and word-problem intervention with embedded language comprehension instruction versus control in the lower panel.

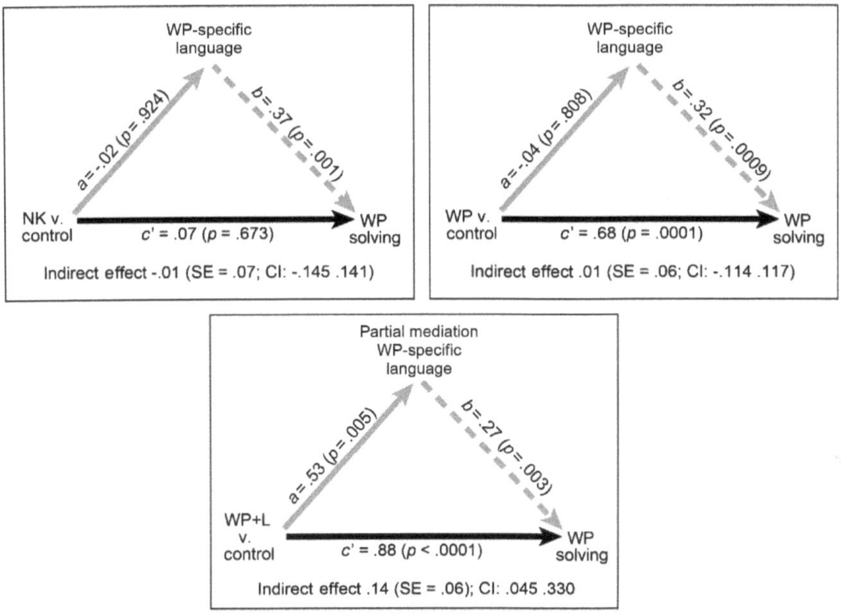

FIG. 3 Mediation effects for word-problem specific language understanding.

In all three models, the effect of word-problem language on the word-problem outcome (the *b* path shown on the descending side of each triangle) was significant. This indicates that stronger word-problem language is related to stronger word-problem solving. The *a* path (the ascending side of each triangle) shows whether the intervention had a positive effect on word-problem language. The mediation effect is the product of the *a* path and the *b* path. Given the similarly strong *b* paths for all three models, the strength of the mediation effect depends on the strength of *a* path. The *a* path was not significant for number knowledge intervention versus control: The effect of number knowledge intervention on word-problem language was weak, with an *a* path coefficient of .02 ($P = .924$). The *a* path was also not significant for word-problem intervention versus control: The effect of word-problem intervention on word-problem language is weak, with an *a* path coefficient of .04 ($P = .808$). By contrast, the *a* path was significant for word-problem intervention with embedded language instruction versus control: The effect of word-problem intervention with embedded language instruction on word-problem language is strong, with an *a* path coefficient of .53 ($P = .005$). Therefore in the word-problem intervention condition with embedded language comprehension instruction, additional power accrues on the word-problem outcome via the language component's unique effect on word-problem language, which in turn supports word-problem solving.

Some Tentative Conclusions

Based on results to date, we offer the following conclusions, while emphasizing the tentative nature of these findings pending the completion of the randomized control trial. First, word-problem intervention with embedded language comprehension instruction provides added value over word-problem intervention without language comprehension on word-problem outcomes. Moreover, only in the embedded language condition does word-problem language exert an indirect effect on the word-problem outcomes. This strengthens evidence for the role of language comprehension in word problems, within an experimental paradigm. It also suggests promise for an expanded intervention framework embedding cognitive training in direct skills intervention.

Second, on arithmetic outcomes, the added value for both word-problem conditions over number knowledge intervention suggests the importance of contextualizing number knowledge instruction with word problems to promote arithmetic learning. It also suggests the efficiency of word-problem intervention for promoting both word problems and arithmetic outcomes in the same amount of intervention time. Third, the added value of both word-problem conditions over number knowledge intervention on word-problem learning indicates the need for an explicit focus on word-problem instruction in first-grade intervention. If results hold across the study cohorts, then findings should have implications for first-grade mathematics intervention and would suggest potentially productive avenues for research in which the effects of embedded cognitive process training are tested for improving other mathematics outcomes.

One example is to address the link between working memory and fraction magnitude understanding by providing working memory span practice on fraction comparison tasks (Fuchs, Schumacher, et al., 2013; Fuchs, Schumacher, Malone, & Fuchs, 2014; Fuchs, Schumacher, Sterba, et al., 2014; Hansen et al., 2015; Jordan et al., 2013; Seethaler, Fuchs, Star, & Bryant, 2011). Strengthening working memory in the context of explicit instruction on fraction magnitudes may have similarly synergistic effects as appear with the embedding language instruction within word problems. We next turn our attention to fraction magnitude understanding. However, instead of illustrating embedded cognitive training, we focus on the second approach by which cognitive processes may provide direction for improving intervention efficacy for at-risk learners.

ALLOCATING VARYING FORMS OF EXPLICIT SKILLS INTERVENTION TO SUBGROUPS OF LEARNERS WITH DIFFERENT COGNITIVE PROFILES

The second approach for relying on what we know about cognitive processes to expand intervention efficacy for at-risk learners involves allocating different forms of explicit skills intervention to students with different cognitive profiles to improve overall intervention efficacy. We illustrate this approach by describing a study conducted by Fuchs, Malone, et al. (2016).

Study Context

Why Focus on Fractions at Fourth Grade?

The focus of the Fuchs, Malone, et al. (2016) study was improving fourth-grade students' understanding of fraction magnitudes. Competence with fractions is an important foundational mathematics domain for success with advanced mathematics learning and in the American workforce (Geary et al., 2012; National Mathematics Advisory Panel [NMAP], 2008; Siegler et al., 2012). Yet, many students experience difficulty in understanding about fractions and operating with fractions (NMAP). The NMAP therefore assigned high priority to improving fraction learning.

Why Focus Specifically on Fraction Magnitude Comparisons and Word Problems?

The multicomponent intervention we developed focuses mainly on fraction magnitude understanding. Such understanding can be linked to children's experiences with measuring, but depends largely on formal instruction that explicates the conventions of symbolic fraction notation (e.g., what the 3 and 4 mean in ¾; the inversion property of fractions in which fractions with the same numerator become smaller as denominators increase), and the infinite density of fractions on any segment of the number line. Comparing fraction magnitudes, ordering fraction magnitudes, and placing fractions on number lines are central tasks representing fraction magnitude understanding.

The other form of understanding fractions relevant at fourth grade is part-whole. This involves understanding a fraction as one or more equal parts of a single object (e.g., two of eight equal parts of a cake) or a subset of a group of objects (e.g., two of eight cakes). Such understanding is typically represented using an area model, in which a region of a shape or a subset of objects is shaded. It is intuitive, based on children's experiences with sharing, and is apparent in young children (Mix, Levine, & Huttenlocher, 1999).

The NMAP (2008) hypothesized that fraction magnitude understanding, which is often contextualized in terms of comparing, ordering, and placing fractions on number lines, is more critical than the part-whole understanding to help students learn about fractions. Support for this idea comes from Fuchs, Schumacher, et al. (2013), who showed that improvement in fraction magnitude but not improvement in part-whole understanding mediated the effects of a fraction intervention. Support also comes from Siegler and Ramani's (2009) demonstration that a linear game board mapped more effectively than a circular game board to the linear mental representation of numerical magnitude. Part-whole understanding encourages separate counting of numerator and denominator segments, increasing children's tendency to conceptualize a fraction as two independent whole numbers. By contrast, fraction magnitude understanding encourages relational thinking about numerators and denominators as determinants of a single number. The Fuchs, Malone, et al. (2016) study was therefore conducted in the context of a multicomponent fraction intervention focused predominantly on fraction magnitude understanding.

Why Incorporate a Self-Explaining or Word-Problem Instructional Component Into the Multicomponent Fractions Intervention?

A major purpose of the Fuchs, Malone, et al. (2016) study was to isolate the effects of an intervention component teaching children to construct sound explanations about why fraction magnitudes differ. This was motivated in part because having students provide explanations is a broadly recommended instructional strategy and represents a strong focus in the most recent wave of mathematics standards reform, such as Common Core State Standards (National Governors Association Center for Best Practices, 2013).

Three types of self-explaining are described in the literature. *Spontaneous self-explaining* occurs when learners generate explanations for the to-be-learned material without being prompted to do so. Studies demonstrate that individuals who spontaneously engage in self-explaining experience superior learning (e.g., Chi & Van Lehn, 1991; Siegler, 2002). However, not all learners spontaneously engage in explanations.

With *elicited self-explaining*, learners are prompted to invent explanations. Enhanced learning with elicited self-explaining can occur for a broad range of ages and domains, but results are mixed. Rittle-Johnson (2006) provided insight for the inconsistency in findings. In this study of typically achieving third through fifth graders, elicited self-explaining promoted stronger procedural accuracy on whole-number mathematical equivalency problems than a no-explanation condition. However, elicited self-explaining did not produce more sophisticated procedures or understanding, perhaps because children's self-explanations rarely included a conceptual focus. Instead, the children tended to state whether an answer was correct or describe procedures to obtain answers. Their self-explanations also sometimes led to incorrect procedures. So inventing sound explanations appears challenging, and its success may depend on cognitive processes associated with strong learning, such as reasoning, working memory, and language comprehension.

These findings also suggest that the key ingredient in self-explaining is processing and expressing high-quality explanations, not inventing explanations. This is in line with Crowley and Siegler (1999) and Rittle-Johnson, Fyfe, Loehr, and Miller (2015). Thus with the third form of self-explaining, *supported self-explaining*, learners operate on high-quality explanations already created for them. Because the focus in Fuchs, Malone, et al. (2016) was students with a history of poor mathematics learning, who may be vulnerable to inventing subpar explanations, we adopted supported explaining. With this approach, tutors model high-quality explanations, and children practice analyzing and applying those explanations, as they elaborate on and discuss important features of those explanations.

Fuchs, Malone, et al. (2016) hypothesized more accurate fraction magnitude comparisons for the explaining condition, compared to a contrast condition that received the same multicomponent fraction program without the explaining component. This hypothesis was based on the generative model of learning (Wittrock, 1990), in which learning requires active processing and elaboration of the to-be-learned information while meaningfully connecting new information to existing knowledge.

To control for intervention time, the contrast condition received a previously validated intervention component focused on fraction word problems that require multiplicative reasoning. We chose word problems as the focus of the contrast intervention for three reasons. First, as noted previously, accuracy in solving word problems strongly predicts employment and wages in adulthood (e.g., Parsons & Bynner, 1997). Second, multiplicative thinking, a major stumbling block for many students, is central to fraction knowledge (e.g., finding equivalent fractions requires multiplying or dividing the numerator and denominator by the same quantity). Third, magnitude comparisons and word problems are both conceptually challenging, and the cognitive processes involved in fraction magnitude comparisons appear similar to those that support word problems.

The word-problem intervention was based on schema theory, as described earlier in this chapter: Children are taught to conceptualize word problems as problem types (schemas); once they have identified the word-problem type, they apply a taught solution strategy for that problem type. Based on Fuchs, Schumacher, et al. (2016), who showed positive effects for this fractions word-problem intervention, we expected stronger word-problem solutions for the word-problem condition over the explaining condition. Contrasting the explaining condition to a condition that received the same multicomponent fraction program without the explaining component ensured that the contrast condition had high-quality relevant intervention on the same content.

Does Supported Self-Explaining Compensate for Limitations in Cognitive Processes?

In this chapter, our interest in Fuchs, Malone, et al. (2016) is the moderation effects, which provide the basis for allocating different intervention components to children with differing cognitive strengths or weaknesses. We focused on three cognitive processes: working memory, reasoning, and language comprehension.

Both fraction magnitude comparisons and word problems transparently demand reasoning ability. They also tax working memory in that (a) fraction magnitude comparisons require children to store and access information across a series of steps that include finding equivalent fractions that need to be compared to benchmark fractions, while (b) word problems require children to process text describing a series of quantities, which must be sequentially evaluated and iteratively considered to create a coherent problem-solving model. Also, both tasks involve language comprehension because teachers use language to convey new ideas and procedures, because students use language when self-explaining, and because word problems are presented linguistically. In longitudinal studies aimed at predicting children's development of fraction and word-problem knowledge, roles have been demonstrated for reasoning (Fuchs, Geary, et al., 2013; Fuchs, Schumacher, et al., 2013; Seethaler et al., 2011), working memory (Fuchs, Schumacher, et al., 2013; Fuchs, Schumacher, Malone, et al., 2014; Fuchs, Schumacher, Sterba, et al., 2014; Hansen et al., 2015; Jordan et al., 2013; Seethaler et al., 2011), and language comprehension (Fuchs, Geary, et al., 2013; Fuchs, Schumacher, et al., 2013; Jordan et al., 2013).

Fuchs, Malone, et al. (2016) extended this descriptive prediction literature by asking whether the effects of supported self-explaining occur, at least in part, by compensating for limitations in cognitive processes. To address this question, we assessed whether individual differences in reasoning, working memory, or language comprehension moderate the effect of the explaining versus word-problem condition on fraction magnitude comparisons (when all other variables in the model, such as performance on other cognitive processes and pretest performance, are held constant).

We hypothesized a *compensatory moderator effect*, which occurs in the following way. On the fraction magnitude comparison outcome, we expected outcomes for students in the word-problem condition to correlate with cognitive processing scores, such that students with severe cognitive deficits experience poorer outcomes than those with more adequate cognitive processing. By contrast, on this outcome, we expected the supported self-explaining intervention to compensate for cognitive limitations, such that students achieve similarly well regardless of where they fall along the cognitive processing distribution. Such a compensatory moderation effect is visualized in Fig. 4. Finding a compensatory moderator effect would provide further evidence for the role of these cognitive processes in self-explaining and would indicate that supported self-explaining increases learning by decreasing cognitive load.

We hypothesized a parallel moderation for the word-problem outcomes, in which the correlation between cognitive capacity and word-problem outcomes is less strong for students who receive the word-problem intervention component than for students who do not receive that component (who instead receive the explaining component). Moderator analysis can specify where along the distribution of cognitive process scores the effect of supported self-explaining transitions from significant to nonsignificant. Moderator analysis may therefore help schools identify which learners are more and less likely to benefit from which intervention component.

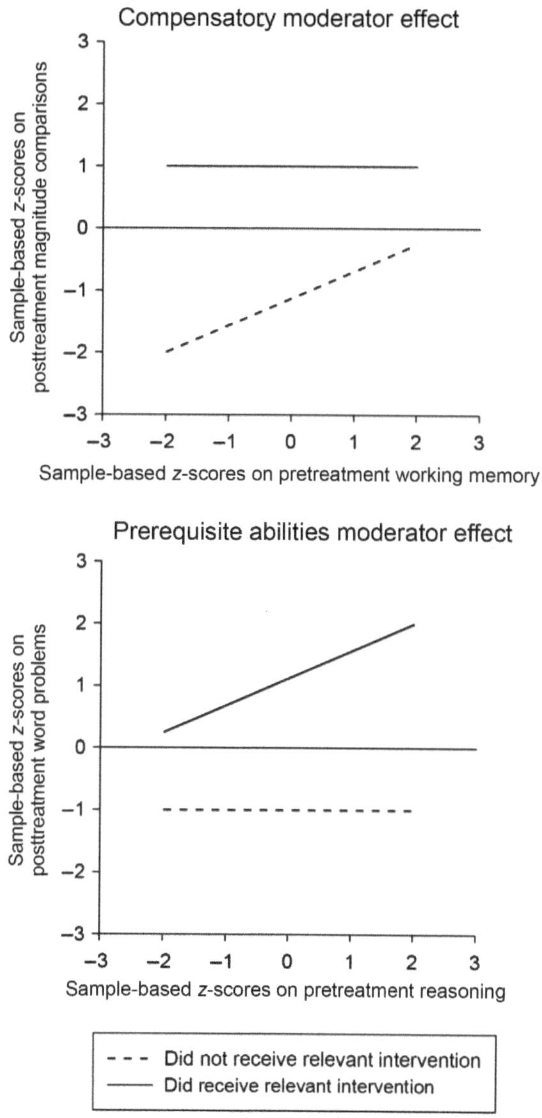

FIG. 4 Visualization of two types of moderator effects. The *top panel* illustrates a *compensatory moderator effect*, in which students who receive relevant explicit intervention perform similarly well on the mathematics outcome regardless of their pretest cognitive performance (see *flat solid line*). By contrast, for students who did not receive relevant explicit intervention, posttest performance is correlated with their pretest cognitive performance (see *angled broken line*). The *bottom* panel illustrates a *prerequisite ability moderator effect*, in which the correlation between pretest cognitive performance and the mathematics outcome is stronger for students who did receive relevant explicit intervention (see *angled solid line*) compared to the correlation for students who did not receive relevant explicit intervention (see *flat broken line*).

Study Overview

In Fuchs, Malone, et al. (2016), risk was defined as low mathematics calculation skill at the start of fourth grade: below the 34th percentile, with half the sample scoring below the 15th percentile and half drawn from between the 15th and 34th percentile. Students were pretested on working memory, reasoning, language comprehension (the three potential moderators), and fractions performance. Stratifying by classroom and the bands of risk severity, 236 at-risk students from 52 classrooms in 14 schools were randomly assigned to three conditions: two variants of the multicomponent fractions intervention and a business-as-usual control group. Eighteen students, representing the three conditions comparably, moved out of the study's reach during the academic year, with 218 students completing the study. Intervention was delivered three times per week for 12 weeks, with 35 minutes per session. We then posttested children's fraction performance. Students in the three conditions were comparable on pretest whole-number and fractions performance, pretest moderator variables (cognitive processes), and demographics.

The multicomponent intervention, known as *Fractions Faceoff!* (Fuchs, Schumacher, Malone, et al., 2014) includes 36 sessions and is delivered in small groups (groups of two in Fuchs, Malone, et al., 2016). Each session is divided into five segments: "Warm-Up" (beginning on Lesson 7), which is where the supported self-explaining or the word-problem variant occurs (depending on randomly assigned condition); "Training," where new concepts and strategies are introduced; "The Relay," where group work on that day's concepts and strategies are practiced in cooperative fashion with students assisting each other as needed; "The Sprint" (beginning in Lesson 4), which is where speeded practice occurs cooperatively to build fluency on foundational fraction tasks; and "The Individual Contest," where individual practice on that day's topics and cumulative review occurs, with individual responsibility for the quality and accuracy of work. The main focus of segments two through five is why and how fractional values differ.

As reflected in segments two through five, in the two intervention variants, fractions intervention was identical for 28 minutes of each of the 35-minute sessions, focused largely on understanding about fraction magnitudes (e.g., comparing, ordering, and placing fractions on number lines). The other (Warm-Up) segment, which lasted 7 minutes each session, differed. In one variant ($n=79$ completed), students were taught strategies for producing written explanations for why fractional values differ; in the other ($n=79$ completed), students were taught strategies for solving word problems involving multiplication and division fraction concepts. As already noted, our main focus in this chapter is the contrast between these two variants of the multicomponent fractions intervention and the associated outcomes: fraction magnitude comparisons and word-problem solving.

In the explaining variant, students were taught and explicitly practiced strategies for providing quality explanations about why a fraction is greater than another (e.g., Which is greater: 5/8 or 1/2?). They were taught to draw a representation of each fraction, to label representations in ways that explicate thinking about the fractions' relative value, to write their answer (which is greater), and to explain why answer makes sense. One explaining student's posttest response on an item from the fraction magnitudes comparison condition is shown in Fig. 5. As shown, he correctly entered the greater than sign between 3/4 and 3/6; he drew tenable representations of each fraction (each with approximately the same area, each divided into the right number of approximately same-sized parts); labeled each representation correctly; and explained, "Same number of parts but fourths are bigger."

The multiplicative fraction word-problem intervention variant component focused on "splitting" and "grouping" word-problem types. An example of the splitting problem type is, Matthew has 2 watermelons. He cuts each watermelon into fifths, How many pieces of watermelon does Matthew have? An example of the grouping problem type is, Keisha wants to make 8 necklaces for her friends. For each necklace, she needs ½ of a yard of string. How many yards of string does Keisha need? The instructional approach was schema based (e.g., Fuchs, Zumeta, et al., 2010; Jitendra & Star, 2012). As with the first-grade study described previously, students learn to identify word problems as belonging to word-problem types that share structural features (splitting and grouping word-problem types). Students are also taught to represent the underlying structure of the word-problem type with a number sentence (e.g., Fuchs et al., 2003, 2009; Fuchs, Zumeta, et al., 2010) or visual display (e.g., Jitendra et al., 2009; Jitendra, Star, Rodriguez, Lindell, & Someki, 2011) and to use a structured step-by-step strategy to organize their solution strategy.

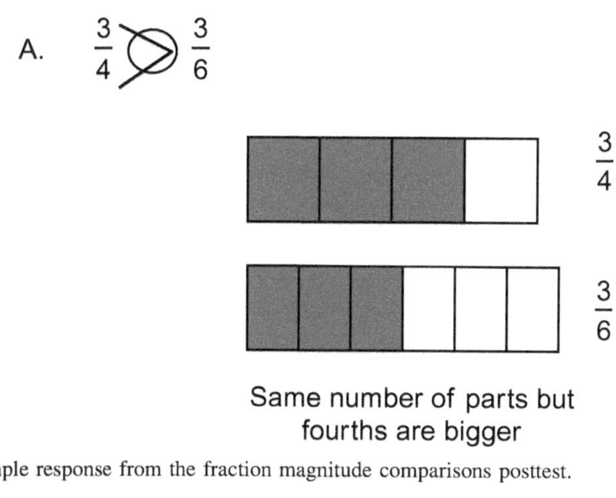

FIG. 5 Sample response from the fraction magnitude comparisons posttest.

Results

Main effect results were as expected. On understanding of fraction magnitude, which we assessed via the accuracy with which students identify larger and smaller fractions, students in the explaining condition outperformed those in the word-problem condition with a moderate effect size of 0.43. On a measure of the quality of explanations about why fraction magnitudes differ (scored for the explanations' conceptual content), effects more dramatically favored the explaining over the word-problem condition, with an effect size of 0.93. These outcomes for the explaining condition over word-problem condition are noteworthy because the multicomponent fraction intervention provided children in *both* conditions the same instruction on the essential ideas and strategies for comparing fraction magnitudes. What distinguished the explaining condition from the word-problem condition was supported self-explaining, which involved explicit instruction.

On the word-problem outcome, students in the word-problem variant significantly and substantially outperformed the explaining students, with an effect size of 1.43. As with the first-grade study, this demonstrates the importance of addressing word problems explicitly within intervention, rather than assuming the effects of conceptual instruction about number (or calculations) will transfer to word-problem skill. It also speaks to the power of schema-based instruction.

Of greater interest for the present chapter, however, is that individual differences in the cognitive processes associated with fractions development helped determine individual student responsiveness to the two intervention variants. In this study, as explained previously, we targeted working memory, reasoning ability, and language comprehension.

On the fraction magnitudes outcome, we found an interaction between students' working memory capacity and the effect between the two intervention variants, while controlling for student's reasoning ability and oral language comprehension. This interaction indicated that the explanation variant compensated for limited working memory capacity. That is, in the multicomponent fraction intervention with*out* the explaining variant, students' working memory performance, measured at the start of the study, correlated with their magnitude comparisons outcome scores ($r = .32$). This is visualized in the top panel of Fig. 4 with the angled broken line (with negative value sample-based z-scores for this condition). By contrast, in the multicomponent fraction intervention *with* the explaining variant, the correlation between students' working memory performance and their magnitude comparisons outcome scores approximated zero ($r = -.04$). This compensatory moderation effects is visualized in Fig. 4's flat solid line (with positive value sample-based z-scores for this condition).

Thus students in the explaining variant did comparably well in learning about fraction magnitudes regardless of their working memory capacity;

by contrast, the fraction magnitudes outcomes of students in the contrasting (word-problem) variant reflected their working memory capacity. The effect favoring the explaining variant over the word-problem variant was robust across the sample's distribution of working memory scores *except* for students whose working memory scores exceeded the sample's 86th percentile. At this relatively intact level of working memory capacity (approximating average working memory performance in an unselected sample), students in the word-problem variant did relatively well with*out* the explaining variant of the fractions intervention. This compensatory moderator effect indicates that explicit instruction on generating strong explanations about why fractional values differ decreases cognitive load. More practically, by revealing where along the distribution of cognitive process scores the effect of the explaining variant transitions from significant to nonsignificant, results may help schools identify which learners are more and less likely to benefit from this intervention variant.

On the word-problem outcome, however, we found a different kind of moderator effect. We already noted that the main effect favoring the word-problem variant over the explaining variant on the word-problem outcome was strong (as was the advantage for the explaining variant over the word-problem variant on the fraction magnitudes outcome). Our word-problem instructional approach, which is rooted in schema theory, is designed to compensate for limitations in students' cognitive processes by teaching them to identify word problems as belonging to categories that share structural features and to represent the underlying structure of the word-problem category with a number sentence or visual display.

Because schema-based instruction is designed to compensate for limitations in students' cognitive resources, we had expected a compensatory moderator effect involving the word-problem variant. Although we found a significant moderator effect (while controlling for the effects of pretest working memory, language comprehension, and word-problem performance), it did not indicate compensation. Instead, it suggests that responsiveness to the word-problem variant grew stronger as reasoning ability *increased.* That is, word-problem outcomes correlated with reasoning ability more strongly for students in the word-problem variant ($r = .36$, see angled solid line in the bottom panel of Fig. 4, with positive value sample-based z-scores) than for students in the non-word-problem (explaining) variant ($r = .05$, see flat broken line in Fig. 4, with negative value sample-based z-score). The effect favoring word-problem over the explaining variant transitioned from significant to nonsignificant at 1.81 *SD*s (4th percentile) below the sample's mean on the reasoning measure. This suggests that the word-problem intervention's instructional methods demanded stronger reasoning than did the explanation variant on the word-problem outcome.

This makes sense for the following reason. The explaining variant provided no word-problem instruction. Therefore explaining variant students called upon strategies taught in their classrooms/school interventions to solve word problems. These strategies required superficial analysis (such as key words) rather than deep reasoning. By contrast, in relying on the schema-based strategies, students in the word-problem variant exercised more analytical strategies. As such, students with more severe reasoning deficits benefited less from word-problem variant than students with more adequate reasoning. This prerequisite ability moderator effect therefore reveals the need to develop the intervention further to address the full range of learners' reasoning ability.

Implications for Practice

Both types of moderator effects qualify the main effects between the two variants, to identify cognitive processes that make students more or less responsive to the intervention. Results suggest that students with adequate working memory do not require the explaining variant of the larger multicomponent intervention but instead benefit more from the word-problem variant. By contrast, students with very low reasoning ability require additional support to benefit from the word-problem component. The hope is that identifying such moderator effects may eventually help schools deploy interventions in ways that improve effectiveness and efficiency. The Fuchs, Malone, et al. (2016) and Fuchs, Schumacher, et al. (2016) study's moderator effects illustrate how this might occur.

SUMMARY AND FUTURE DIRECTIONS

The question addressed in this chapter was whether we can combine what is known about explicit skills intervention with what is known about the role of cognitive processes in mathematics development to identify methods for expanding the range of students whose needs are effectively addressed by intervention. We addressed two possible approaches. The first involves embedding cognitive training within direct skills intervention to strengthen cognitive resources while building a foundation of mathematics skill and while facilitating transfer from the embedded cognitive training to mathematics performance.

We illustrated this approach with an ongoing study in which language comprehension instruction was embedded within first-grade word-problem intervention. Results showed that embedded language comprehension instruction offers added value over word-problem intervention without that embedded language comprehension for improving first graders' word-problem solving. Results also showed that language comprehension mediates that effect because the embedded instruction improves language comprehension, which in turn contributes to word-problem solving. This encourages the idea that other forms of cognitive process training, when embedded within explicit skills intervention, may prove effective. More specifically, results to date suggest promise of embedding language comprehension instruction within word-problem intervention.

The second approach involves identifying which cognitive processes moderate (or help determine students' responsiveness to) the effects of a specific intervention. We presented a study that illustrates such moderator effects, in which students with relatively intact working memory capacity can forego the supported self-explaining component of a larger fractions intervention. At the same time, students with extremely weak reasoning ability will need special attention to profit from a schema-based word-problem component of the same larger fractions intervention.

Information about moderator effects can eventually help school personnel personalize intervention, by identifying an individual child's profile and cognitive strengths and weaknesses and making research-informed decisions about how to individualize an intervention package to his/her needs. This should make the intervention process more effective and efficient. At the present time, research on moderator effects is evolving. Making progress toward the goal of using research to personalize mathematics intervention will require an extended program of research that identifies which cognitive processes moderate interventions in which subdomains of mathematics performance.

Before closing, we offer one additional observation, drawn from findings across the two studies presented in this chapter. In both studies, conducted at different grade levels (grade 1 vs. grade 4) and on different subdomains of mathematics (whole numbers vs. fractions), results argue for the importance of addressing word problems explicitly within intervention, rather than assuming that the effects of instruction about number or calculations will transfer to word-problem skill. This represents a finding that schools can enact immediately.

REFERENCES

Bynner, J. (1997). Basic skills in adolescents' occupational preparation. *Career Development Quarterly*, *45*, 305–321.

Catts, H. W., & Kamhi, A. G. (2017). Prologue: reading is not a single ability. *Language, Speech, and Hearing Services in Schools*, *48*, 73–76.

Chi, M. T. H., & Van Lehn, K. A. (1991). The content of physics self-explanations. *The Journal of the Learning Sciences*, *1*, 69–105. https://doi.org/10.1207/s15327809jls0101_4.

Crowley, K., & Siegler, R. S. (1999). Explanation and generalization in young children's strategy learning. *Child Development*, *70*, 304–316. https://doi.org/10.1111/1467-8624.00023.

Cummins, D. D., Kintsch, W., Reusser, K., & Weimer, R. (1988). The role of understanding in solving word problems. *Cognitive Psychology*, *20*, 405–438.

Every Child a Chance Trust. (2009). *The long-term costs of numeracy difficulties*. Retrieved 14 August 2009. http://www.everychildachancetrust.org/counts/index.cfm.

Fletcher, J.M., Lyon, G.R., Fuchs, L.S., & Barnes, M.A. (in press). Learning disabilities: From identification to intervention (2nd ed.). New York: Guilford.

Fuchs, L. S., Compton, D. L., Fuchs, D., Powell, S. R., Schumacher, R. F., Hamlett, C. L., et al. (2012). Contributions of domain-general cognitive resources and different forms of arithmetic development to pre-algebraic knowledge. *Developmental Psychology*, *48*, 1315–1326. https://doi.org/10.1037/a0027475.

Fuchs, L. S., Fuchs, D., Compton, D. L., Powell, S. R., Seethaler, P. M., Capizzi, A. M., et al. (2006). The cognitive correlates of third-grade skill in arithmetic, algorithmic computation, and arithmetic word problems. *Journal of Educational Psychology*, *98*, 29–43.

Fuchs, L. S., Fuchs, D., Craddock, C., Seethaler, P. M., & Geary, D. C. (2017). In *Embedding word-problem specific language comprehension within word-problem intervention. Preliminary results presented at the 2017 math cognition conference.* Available from L.S. Fuchs (lynn.fuchs@vanderbilt.edu).

Fuchs, L. S., Fuchs, D., Prentice, K., Burch, M., Hamlett, C. L., Owen, R., et al. (2003). Explicitly teaching for transfer: effects on third-grade students' mathematical problem solving. *Journal of Educational Psychology*, *95*, 293–305. https://doi.org/10.1037/0022-0663.95.2.293.

Fuchs, L. S., Fuchs, D., Stuebing, K., Fletcher, J. M., Hamlett, C. L., & Lambert, W. E. (2008). Problem-solving and computation skills: are they shared or distinct aspects of mathematical cognition? *Journal of Educational Psychology*, *100*, 30–47.

Fuchs, L. S., Geary, D. C., Compton, D. L., Fuchs, D., Hamlett, C. L., & Bryant, J. V. (2010b). The contributions of numerosity and domain-general abilities to school readiness. *Child Development*, *81*, 1520–1533.

Fuchs, L. S., Geary, D. C., Compton, D. L., Fuchs, D., Hamlett, C. L., Seethaler, P. M., et al. (2010a). Do different types of school mathematics development depend on different constellations of numerical and general cognitive abilities? *Developmental Psychology*, *46*, 1731–1746.

Fuchs, L. S., Geary, D. C., Compton, D. L., Fuchs, D., Schatschneider, C., Hamlett, C. L., et al. (2013). Effects of first-grade number knowledge tutoring with contrasting forms of practice. *Journal of Educational Psychology*, *105*, 58–77.

Fuchs, L. S., Malone, A., Schumacher, R. F., Namkung, J. M., Hamlett, C. L., Jordan, N. C., et al. (2016). Supported self-explaining during fraction intervention. *Journal of Educational Psychology*, *108*, 493–508.

Fuchs, L. S., Powell, S. R., Seethaler, P. M., Cirino, P. T., Fletcher, J. M., Fuchs, D., et al. (2009). Remediating number combination and word problem deficits among students with mathematics difficulties: a randomized control trial. *Journal of Educational Psychology*, *101*, 561–576. https://doi.org/10.1037/a0014701.

Fuchs, L. S., Schumacher, R. F., Long, J., Namkung, J., Hamlett, C. L., Cirino, P. T., et al. (2013). Improving at-risk learners' understanding of fractions. *Journal of Educational Psychology*, *105*, 683–700. https://doi.org/10.1037/a0032446.

Fuchs, L. S., Schumacher, R. F., Long, J., Namkung, J., Malone, A., Wang, A., et al. (2016). Effects of intervention to improve at-risk fourth graders' understanding, calculations, and word problems with fractions. *The Elementary School Journal*, *116*, 625–651.

Fuchs, L. S., Schumacher, R. F., Malone, A., & Fuchs, D. (2014). *Fractions face-off!* Available at www.peerassistedlearningstrategies.net.

Fuchs, L. S., Schumacher, R. F., Sterba, S. K., Long, J., Namkung, J., Malone, A., et al. (2014). Does working memory moderate the effects of fraction intervention? An aptitude-treatment interaction. *Journal of Educational Psychology*, *106*, 499–514.

Fuchs, L. S., Zumeta, R. O., Schumacher, R. F., Powell, S. R., Seethaler, P. M., Hamlett, C. L., et al. (2010). The effects of schema-broadening instruction on second graders' word-problem performance and their ability to represent word problems with algebraic equations: a randomized control study. *The Elementary School Journal*, *110*, 446–463. https://doi.org/10.1086/651191.

Geary, D. C., Hoard, M. K., Nugent, L., & Bailey, D. H. (2012). Mathematical cognition deficits in children with learning disabilities and persistent low achievement: a five year prospective study. *Journal of Educational Psychology*, *104*, 206–223.

Geary, D. C., Nicholas, A., Li, Y., & Sun, J. (2017). Developmental change in the influence of domain-general abilities and domain-specific knowledge on mathematics achievement: an eight-year longitudinal study. *Journal of Educational Psychology*, *109*, 680–693. https://doi.org/10.1037/edu0000159.

Hansen, N., Jordan, N. C., Fernandez, E., Siegler, R. S., Fuchs, L. S., Gersten, R., et al. (2015). General and math-specific predictors of sixth-graders' knowledge of fractions. *Cognitive Development*, *35*, 34–49. https://doi.org/10.1016/j.cogdev.2015.02.001.

Haskell, R. E. (2001). *Transfer of learning: Cognition, instruction, and reasoning*. Cambridge, MA: Academic Press.

Jitendra, A. K., & Star, J. A. (2012). An exploratory study contrasting high- and low-achieving students' percent word problem solving. *Learning and Individual Differences*, *22*, 151–158. https://doi.org/10.1016/j.lindif.2011.11.003.

Jitendra, A. K., Star, J. A., Rodriguez, M., Lindell, M., & Someki, F. (2011). Improving students' proportional thinking using schema-based instruction. *Learning and Instruction*, *21*, 731–745. https://doi.org/10.1016/j.learninstruc.2011.04.002.

Jitendra, A. K., Star, J. A., Starosta, K., Leh, J. M., Sood, S., Caskie, G., et al. (2009). Improving seventh-grade students' learning of ratio and proportion: the role of schema-based instruction. *Contemporary Educational Psychology*, *34*, 250–264. https://doi.org/10.1016/j.cedpsych.2009.06.001.

Jordan, N. C., Hansen, N., Fuchs, L. S., Siegler, R. S., Gersten, R., & Micklos, D. (2013). Developmental predictors of fraction concepts and procedures. *Journal of Experimental Child Psychology*, *116*, 45–58. https://doi.org/10.1016/j.jecp.2013.02.001.

Kearns, D. M., & Fuchs, D. (2013). Does cognitive-focused instruction improve the academic performance on low-achieving children? *Exceptional Children*, *79*, 263–290.

Kintsch, W., & Greeno, J. G. (1985). Understanding and solving word arithmetic problems. *Psychological Review*, *92*, 109–129.

Melby-Lervag, M., & Hulme, C. (2013). Is working memory training effective? A meta-analytic review. *Developmental Psychology*, *49*, 270–291.

Mix, K. S., Levine, S. C., & Huttenlocher, J. (1999). Early fraction calculation ability. *Developmental Psychology*, *35*, 164–174. https://doi.org/10.1037/0012-1649.35.1.164.

Nathan, M. J., Kintsch, W., & Young, E. (1992). A theory of algebra-word-problem comprehension and its implications for the design of learning environments. *Cognition and Instruction*, *9*, 329–389.

National Governors Association Center for Best Practices, Council of Chief State School Officers. (2013). *Common core state standards (mathematics standards)*. Washington, DC: National Governors Association Center for Best Practices. http://www.corestandards.org/.

National Mathematics Advisory Panel. (2008). *Foundations for success: Final report of the national mathematics advisory panel*. Washington, DC: U.S. Department of Education.

National Research Council. (2000). *How people learn: Brain, mind, experience, and school*. Washington, DC: The National Academies Press.

Parsons, S., & Bynner, J. (1997). Numeracy and employment. *Education and Training*, *39*, 43–51.

Perfetti, C. A., Landi, N., & Oakhill, J. (2005). The acquisition of reading comprehension skill. In M. J. Snowling & C. Hulme (Eds.), *The science of reading: A handbook* (pp. 227–247). Oxford: Blackwell.

Rittle-Johnson, B. (2006). Promoting transfer: effects of self-explanation and direct instruction. *Child Development*, *77*, 1–15. https://doi.org/10.1111/j.1467-8624.2006.00852.x.

Rittle-Johnson, B., Fyfe, E. R., Loehr, A. M., & Miller, M. R. (2015). Beyond numeracy in preschool: adding patterns to the equation. *Early Childhood Research Quarterly*, *31*, 101–112. https://doi.org/10.1016/j.ecresq.2015.01.005.

Schleppegrell, M. J. (2007). The linguistic challenges of mathematics teaching and learning: a research review. *Reading & Writing Quarterly: Overcoming Learning Difficulties*, *23*, 139–159.

Seethaler, P. M., Fuchs, L. S., Star, J. R., & Bryant, J. (2011). The cognitive predictors of computational skill with whole versus rational numbers: an exploratory study. *Learning and Individual Differences*, *21*, 536–542. https://doi.org/10.1016/j.lindif.2011.05.002.

Siegler, R. S. (2002). Microgenetic studies of self-explanation. In N. Garnott & J. Parziale (Eds.), *Microdevelopment: A process-oriented perspective for studying development and learning* (pp. 31–58). Cambridge: Cambridge University Press.

Siegler, R. S., Duncan, G. J., Davis-Kean, P. E., Duckworth, K., Claessens, A., Engel, M., et al. (2012). Early predictors of high school mathematics achievement. *Psychological Science*, *23*, 691–697. https://doi.org/10.1177/0956797612440101.

Siegler, R. S., & Ramani, G. B. (2009). Playing linear number board games—but not circular ones—improves low-income preschoolers' numerical understanding. *Journal of Educational Psychology*, *101*, 545–560. https://doi.org/10.1037/a0014239.

Swanson, H. L., Jerman, O., & Zheng, X. (2008). Growth in working memory and mathematical problem solving in children at risk and not at risk for serious math difficulties. *Journal of Educational Psychology*, *100*, 343–379.

Wittrock, M. C. (1990). Generative processes of comprehension. *Educational Psychologist*, *24*, 345–376. https://doi.org/10.1207/s15326985ep2404_2.

Chapter 13

Explanations and Implications of Diminishing Intervention Impacts Across Time

Drew Bailey
School of Education, University of California, Irvine, Irvine, CA, United States

INTRODUCTION

Math cognition researchers sometimes orient their readers to the big-picture importance of their research by citing evidence of robust longitudinal associations between children's early mathematical skills and their math achievement many years later. Some also cite evidence from experimental evaluations of early childhood education programs such as Perry Preschool program that produced impressive impacts on achievement longer after the program ended and sometimes on educational attainment and earnings in adulthood as well. However, these patterns of longer run impacts paint a much more optimistic picture of the long-term academic impacts of early academic interventions than is typically found in the all-too-rare cases when children are randomly assigned to receive an effective early math intervention and followed for years following the end of treatment: quickly diminishing, and often null, impacts a year or more following the end of treatment.

The storied Perry program as well as the longer running Abecedarian program both produced lasting effects on achievement and on adult life outcomes, such as educational attainment and lack of incarceration (Campbell et al., 2014; Schweinhart et al., 2005). However, these programs differ in important ways from early math interventions. Neither targeted math skills, and both were intensive, with Perry lasting one or two years and Abecedarian running for five years. Both may have produced long-term effects through a variety of pathways (e.g., literacy, social-emotional development, parent–child interactions; for review, see Elango, García, Heckman, & Hojman, 2015) that are not primary targets of most early math interventions.

Mathematical Cognition and Learning, Vol. 5. https://doi.org/10.1016/B978-0-12-815952-1.00013-X

In a meta-analysis of early childhood educational intervention programs, Li et al. (2017) found that typical impacts on academic outcomes all but disappeared in the 3 years following the conclusion of a successful intervention. The reasons for these diminishing impacts are not well understood and are difficult (but possible, I will argue) to reconcile with views of mathematical development that assume basic skills are crucial for learning much more advanced skills.

PATTERNS OF EFFECTS ACROSS TIME AND THEORIES OF CHILDREN'S MATHEMATICAL DEVELOPMENT

In this chapter, the term "fadeout" is used to refer to a pattern of diminishing effects following the end of an effective math intervention. Importantly, effects ("impacts") of programs are always defined in comparison with some other group, usually a randomly assigned control group. For this reason, the extent of fadeout always depends on what happens in the control group (which, given successful random assignment, shows what *would* have happened to children who received an effective intervention, had they not received that intervention). Because control-group children in studies of early math interventions are always receiving *some* sort of educational experience—and often a fair amount of math instruction—this review is about the effects of adding more, or better, or different math instruction on children's math achievement rather than about the effects of learning vs. not learning any math at all. Therefore it would not be appropriate to extend the arguments in this chapter to a discussion of the possibility of *eliminating* formal mathematics instruction for children, a policy I and most others in the field would certainly not recommend. The term fadeout is an empirical one not inherently tied to any kind of value judgment. The fact that an intervention's effects fade out does not necessarily eliminate the utility of the intervention; utility depends on values, not just facts. Further, it is possible that we will discover effective ways to extend short-term effects into the long term by using complementary interventions (an approach I will discuss later). However, given the long-term goals interventionists and policymakers have for the children affected by early math interventions, I argue that fadeout is an important regularity worthy of more theoretical and empirical attention.

The purpose of this chapter is to introduce readers to explanations for fadeout, the implications of these explanations for research on children's mathematical cognition and development, and their implications for developing more effective intervention strategies. Many of these implications are speculative, because fadeout has not been a major topic of study in math cognition research. However, my hope is that the testing of these ideas will lead to a fuller understanding of children's mathematical development and better approaches for teaching mathematics.

Possible Explanations of Fadeout

A common theme I have noticed in discussing fadeout with educational researchers is the wide range of ideas they propose to explain and/or prevent it, the intuitive appeal of many of these ideas, and the incommensurability of many of them. For example, after giving the talk that served as the basis for this chapter at the Math Cognition Conference, I received two consecutive comments from individuals in the audience. The first questioned whether the faster growth for control-group children responsible for fadeout would generalize to current conditions in the United States, in which kindergarteners are exposed to an increasingly complex set of materials (Bassok, Latham, & Rorem, 2016). Perhaps, the commenter proposed, children who enter school behind their peers—the kind of children who comprise control groups in existing experiments—will never catch up. In this scenario, early interventions that bring these children up to a level similar to their peers would be more likely to produce lasting effects now than in past decades.

The second commenter noted that as early childhood math instruction becomes more advanced, children in the control group are now receiving a higher level of instruction than was the case in previous decades. This faster growth should lead control-group children to catch up even more quickly than before, and thus for fadeout to happen more quickly than in past decades.

Both comments were thoughtful and intuitively appealing and deserve careful consideration when contemplating the practical implications of the research I will discuss in this chapter. But there is a problem: these hypotheses cannot simultaneously be true. Increases in the level of early math instruction cannot cause early math intervention to have longer *and* shorter lasting effects than it used to.

In any case, the point of this example is not to criticize these comments. Indeed, speculating about the generalizability of findings under a changing set of educational conditions is a productive way to generate plausible theories of fadeout and persistence. However, the incompatibility of these intuitive ideas underscores the need for a deeper scientific understanding of the processes underlying fadeout and persistence. In the following section, I will describe some of the explanations for fadeout of the effects of early childhood math interventions and evaluate them in light of existing evidence. My evaluations should be taken as tentative: while some explanations have more support than others, none can be fully ruled out as a contributor to fadeout on the basis of the existing evidence.

Measurement-Based Explanations

Some explanations for fadeout focus on the ways that we measure it. Perhaps fadeout reflects problems in the ways that researchers measure learning across development, and not a true decrease in the impact of early interventions across time.

Instrumentation: On this view, fadeout reflects the use of different instruments or tests at different points in time:

> *Math achievement is not a unidimensional construct. Effective math interventions teach material that is developmentally appropriate for children recruited for the study; it is unfair to evaluate their longer-run impacts with assessments designed to measure aspects of children's math achievement taught in subsequent years.*

This explanation is certainly true to some extent. Despite high correlations between knowledge across some mathematical domains (Purpura & Lonigan, 2013; Schenke, Rutherford, Lam, & Bailey, 2016), mastery of one domain does not logically imply mastery of another. Indeed, the existence of fadeout proves this: children who outperform their peers on short-run measures of math achievement must outperform their peers less on measures of math achievement administered at a later point. For this instrumentation argument to account for fadeout, though, impacts on measures of knowledge taught in the intervention should persist, while impacts on other measures should be persistently smaller or null.

A schematic depiction of the instrumentation explanation, along with a structural equation model (SEM) representation, appears in Fig. 1. A persistent impact occurs on a hypothetical test (Panel A, SEM), while no impact is found

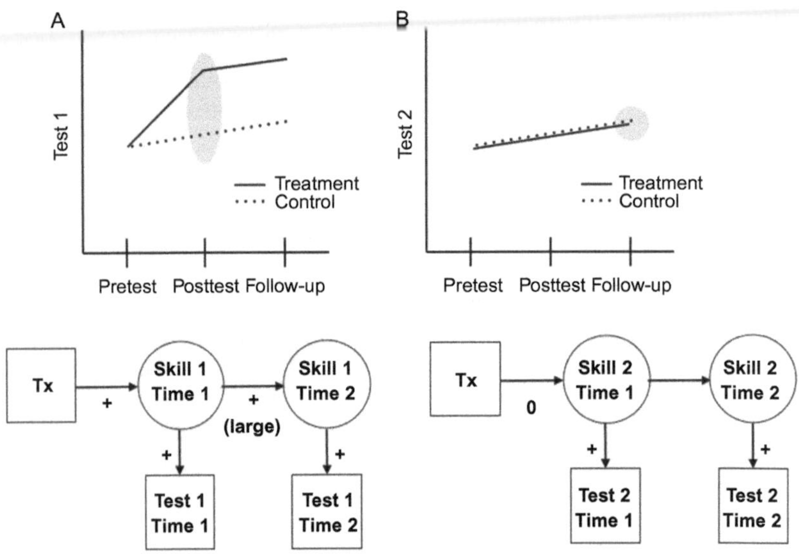

FIG. 1 Possible explanations for fadeout: instrumentation. Note: For instrumentation to fully account for fadeout, there must be a persistent effect on Test 1 (Panel A) but no effect on Test 2 (Panel B). In the underlying SEMs, Tx denotes treatment (1 = treatment, 0 = control). There is no fadeout on Skill 1 or Skill 2, but fadeout may be falsely inferred if Skill 1 is assessed at posttest (Panel A) and Skill 2 is assessed at follow-up (Panel B). This model implies a large effect of Skill 1 at Time 1 on Skill 1 at Time 2 (SEM, bottom left). Because there is no effect of the treatment on Skill 2, the effect of Skill 2 at Time 1 on Skill 2 at Time 2 is unknown (SEM, bottom right).

on another hypothetical test (Panel B, SEM). Using one test at the posttest and another at a later follow-up assessment would create the illusion of fadeout (highlighted test-wave combinations in Panels A and B).

An example in which the instrumentation hypothesis cannot be ruled out comes from a recent randomized controlled trial of a math intervention, which found substantial impacts on a variety of math measures in the spring of kindergarten, but no effects on a standardized math achievement test in the winter of first grade (Clarke et al., 2016). Because the same achievement test was not administered at the posttest in kindergarten, we cannot know whether this null effect is attributable to the later time at which the test was administered or the different content on the test. However, evaluations of math interventions that have used the same vertically scaled math achievement test (i.e., a test designed to be administered to children across ages, for which differences are intended to be comparable across the score distribution) have also yielded diminishing impacts across time (Clements, Sarama, Wolfe, & Spitler, 2013; Hofer, Lipsey, Dong, & Farran, 2013; Smith, Cobb, Farran, Cordray, & Munter, 2013). Children may indeed maintain an advantage years later on content emphasized during an effective math intervention (see Dillon, Kannan, Dean, Spelke, & Duflo, 2017, for a possible example), but there is very little evidence assessing this idea.

Scaling: Another possibility is that fadeout effects are due to the way in which achievement tests are developed. The associated explanation is that fadeout is a scaling artifact:

> *Fadeout is a statistical artifact of the way we scale achievement tests. Children who receive an effective early math intervention may learn the same amount or more than children in the control group in subsequent years, but the same amount of learning is worth fewer points on achievement tests relative to learning in earlier years, creating the illusion of fadeout.*

This hypothesis is a more specific version of the instrumentation hypothesis and is worthy of serious consideration. Two empirical regularities support it. First, variance on achievement tests increases with age, so a standard deviation of knowledge is worth more points for older children than for younger children. Cascio and Staiger (2012) noted that this could make impacts appear to fadeout across time when expressed in standard deviation units. However, given realistic estimates of changes in impacts and test variance across time, they found that the increase in test variance was not large enough to account for more than 10% of the fadeout effect in the year following an intervention nor more than 20% of the fadeout effect many years after the intervention.

The second empirical regularity that makes scaling a possible explanation for fadeout is that children's learning, as expressed in standard deviation units on vertically scaled tests, decelerates from year to year. Children improve by over a standard deviation worth of achievement from kindergarten to first grade, but by under one-third of a standard deviation per year in the middle school years (Hill, Bloom, Black, & Lipsey, 2008). Perhaps scales place too much weight on items learned in the early school years. In less technical terms, a standard deviation of math knowledge may consist of many more facts in fifth grade than in kindergarten. If so, interpreting changes in impacts across years is difficult: Trends in gaps on achievement tests between groups with different average scores can be quite sensitive to how the tests are scaled (Bond & Lang, 2013).

A schematic depiction of the scaling explanation, along with a SEM representation, appears in Fig. 2. A persistent impact occurs on children's latent underlying skill (Panel B, SEM), but test score impacts fade (Panel A). However, the fadeout interpretation is misleading: Because changes in true skill have a weaker influence on test scores on the follow-up assessment than on the posttest assessment (SEM), persistent skill effects create the illusion of fadeout.

To address the problem of comparing changes in groups with different levels, one might compare children who received an effective math intervention to children with the same level of *posttest* math achievement, but who did

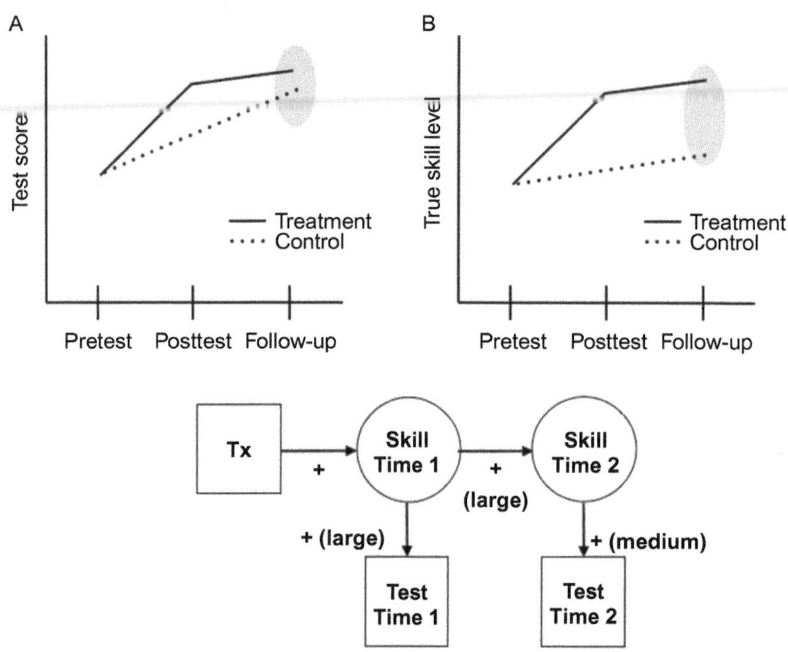

FIG. 2 Possible explanations for fadeout: scaling. Note: For scaling to fully account for fadeout, there must be a persistent effect on some skill (Panel B, SEM), but partial or full fadeout is observed because the impact of higher skill on test score is lower at Time 2 than at Time 1 (Panel A, SEM). In the underlying SEM, Tx denotes treatment (1 = treatment, 0 = control). Test scores when scaled normally will misleadingly show fadeout and smaller effects of prior skill on later skill than truly exist.

not receive the intervention. If the groups show the same level of subsequent learning, with unmatched lower-achieving control children catching up to both treatment children and higher-achieving matched control children, this would be consistent with the scaling hypothesis. In contrast, if children who received the effective mathematics intervention are subsequently outperformed by posttest-matched controls, this would suggest a theoretically meaningful difference in postintervention learning, favoring the control group. My colleagues and I made this comparison, matching control children to children who had received an effective pre-k intervention on their posttest achievement scores and comparing their growth in the following year. Children in the control group learned significantly more in the following year, and the difference was close to the size of the fadeout effect (Bailey et al., 2016). While this comparison implies that scaling is not *sufficient* to explain the fadeout effect, it was hardly precise enough to rule out any influence of scaling on fadeout, and more research into this possibility is clearly warranted.

Cognitive Processing Explanations

In addition to methodological explanations of fadeout, there are several plausible cognitive explanations.

Forgetting: Fadeout may be caused by children in the treatment group forgetting information they learned from the treatment:

> *After the completion of an effective early mathematics intervention, children stop accessing the concepts and procedures they learned, which leads them to forget this content.*

Looking at patterns of the decay of treatment effects of early math interventions (Bailey, Duncan, Watts, Clements, & Sarama, 2018), a cognitive psychologist cannot help but be reminded of forgetting curves. An exponential-like pattern of decay—that is, rapid forgetting soon after the end of the intervention and then a more modest decline in retention thereafter—is a strong empirical regularity in research on forgetting (Cepeda, Vul, Rohrer, Wixted, & Pashler, 2008). Additionally, children at risk for persistently low mathematics achievement (those who are often targeted by early math interventions) may show faster decay of problem-relevant information (Geary, 1993) or hypersensitivity to interference in retrieving arithmetic facts (De Visscher & Noël, 2014) than typically achieving children, perhaps predisposing them to forget information learned during brief mathematics interventions. Finally, forgetting is an intuitively sensible explanation for fadeout: Once the math intervention is removed, children stop rehearsing the math they were learning, and they begin to lose the ability to recall it.

A schematic depiction of the forgetting explanation, along with a SEM representation, appears in Fig. 3. Whether children in the treatment group experience a net skill loss in the posttreatment period (Panel A) or a net skill

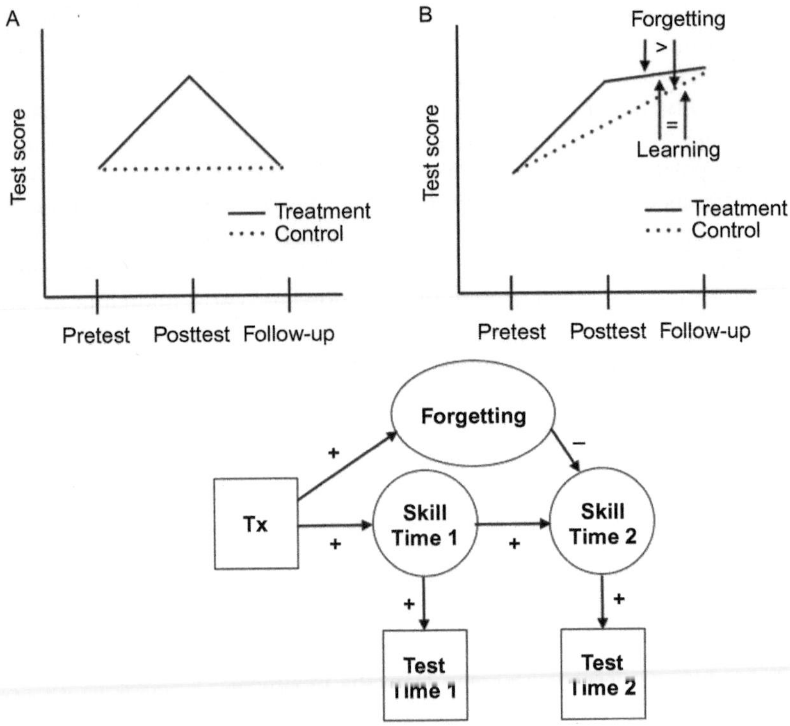

FIG. 3 Possible explanations for fadeout: forgetting. Note: For forgetting to fully account for fadeout, an indirect effect of the treatment on skill at time 2 must be offset by an indirect effect of the treatment on forgetting (SEM). Panel A shows an example in which the treatment group undergoes a net skill loss between the posttest and follow-up. This is less likely in early childhood than the case illustrated in Panel B, in which both groups experience net skill gain, but the treatment group experiences more forgetting than the control group. In the underlying SEM, Tx denotes treatment (1 = treatment, 0 = control).

gain (Panel B), the control group experiences a higher net skill gain in the posttreatment period. For forgetting to fully account for fadeout, an indirect effect of the treatment on skill at the follow-up assessment must be offset by an indirect effect of the treatment on forgetting (SEM).

However, this explanation overlooks important theoretical and empirical considerations. Children probably do not stop practicing the math they learn from an effective early math intervention, because in the early years, math is hierarchically organized: For example, children practice addition as they learn multiplication (Lemaire & Siegler, 1995). Further, children learn a full standard deviation worth of material from kindergarten to grade 1, but math intervention impacts on vertically scaled achievement tests rarely approach that magnitude. Thus children are likely to be learning content from early math interventions that they will be rehearsing regularly in the following year. In long-term postintervention assessments, children who received a successful intervention do not perform worse at long-term follow-up assessments than they did on the posttest in the absolute sense; they merely learn less during the time since the end of the intervention than children in the control group (Clements et al., 2013; Elango et al., 2015). In other words, results resemble Fig. 3, Panel B much more than Panel A. Thus fadeout is often reasonably reframed in the context of early interventions as "catchup."[1]

But while forgetting is unlikely to be the primary cognitive process underlying fadeout in early math interventions, it may still play a role: Children who receive an effective math intervention could, in principle, experience the same amount of learning as the control group following the end of treatment with more forgetting. My colleagues and I are beginning to look into this possibility and hope to report estimates of the role forgetting plays in fadeout in the near future.

Teaching to the Test: If interventions target test content in ways that do not also build fundamental supporting skills, gains may be short-lived:

> *Fadeout is a predictable consequence of teaching to the test. Children's learning from early math interventions tends to be shallow and superficial, and thus unlikely to persist or transfer to more advanced skills.*

It is not difficult to understand why teaching to the test implies fadeout. If children gain only the most superficial understanding of the content to which they are exposed, it is unlikely that this understanding will transfer to more advanced knowledge in the future. Indeed, the "hollowness" of test score gains has been invoked for decades as an explanation for fadeout in experiments that attempt to increase children's general cognitive ability (Jensen, 1998; but see Protzko, 2016).

A schematic depiction of the teaching to the test explanation, along with a SEM representation, appears in Fig. 4. Simply, there is an effect on test scores (Panel A, SEM), but this effect does reflect true changes in the underlying latent skill (Panel B, SEM). Test-specific effects, in this example, do not transfer to later test scores (SEM).

Is teaching to the test a fair characterization of existing early mathematics interventions? The question deserves a full treatment, including a systematic review of observer reports of early math interventions and of psychometric analyses of the domains in which children learn most when receiving such interventions compared to business-as-usual control conditions. However, based on my unsystematic evaluation of the evidence, including discussions with early math interventionists, reading reports, and reanalyzing data, I am skeptical that this hypothesis accounts for much of the fadeout we observe.

Developers of these interventions are often leading researchers in math cognition, development, and education, and they monitor the fidelity with

1. All catchup will result in fadeout, but fadeout is theoretically possible without catchup, in cases in which the treatment group experiences a net skill loss and the control group exhibits no improvement.

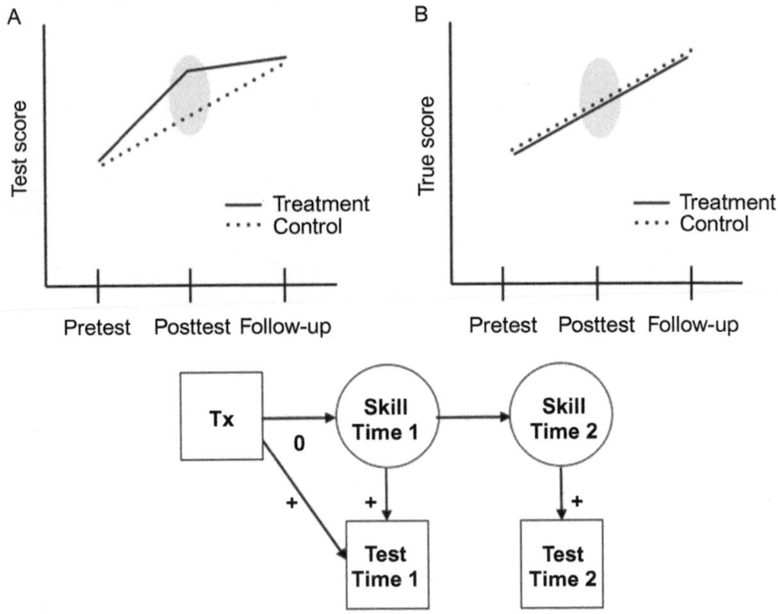

FIG. 4 Possible explanations for fadeout: teaching to the test. Note: For teaching to the test to fully account for fadeout, there must be an impact on test-specific variance at the posttest (Panel A, SEM), but no impact of the treatment on children's true score on some skill (Panel B, SEM). Because test-specific variance in this example does not transfer across testing occasions, there is no effect of the treatment on test score at the follow-up (SEM). In the underlying SEM, Tx denotes treatment (1 = treatment, 0 = control).

which their interventions are implemented. Their descriptions of their math interventions often highlight the extent to which they target children's conceptual understanding of number (e.g., Clarke et al., 2016), and their interventions show impacts on a diverse set of tests at the end of treatment (Clarke et al., 2016; Smith et al., 2013). Furthermore, interventions often occur in the same or similar contexts as those in which children receive their typical math instruction and extend the dosage of instruction by a substantial amount.

All of these suggest to me that early math interventions are providing children with knowledge and understanding no more shallow and superficial than that which they would develop under business-as-usual conditions. However, to reiterate, this is not a systematic review of the extent to which early math interventions provide children with deeper or shallower understanding than they would otherwise receive, and doing so would make a useful contribution to the literature.

Constraining Content: This hypothesis posits that children are not exposed to more advanced content than they learned during an effective intervention, leading to fadeout:

After the conclusion of an effective early math intervention, students return to an instructional environment where teachers tend to target lower-achieving students and instruction does not build on the knowledge they received in the intervention, which leads to stagnation in the treatment group and catchup by the control group.

This explanation for fadeout is intuitively appealing, perhaps because it implies a simple solution: teach more difficult material (or, when it can be done effectively, differentiate instruction). Consistent with this explanation are (1) observations that kindergarten teachers report spending a substantial proportion of math instruction time on content that most of their students already know (Engel, Claessens, & Finch, 2013), and (2) the ubiquitous pattern of "catchup" described previously, where fadeout occurs as the treatment group's posttreatment trajectory slows while the control group's does not. Further, it seems highly likely under some conditions. For example, in an extreme case in which children would never encounter any content more advanced than the content to which they had been exposed following an effective early math intervention, these children would only make progress by spontaneously generating it.

However, as noted previously, children learn very quickly in the early school years (when learning is expressed in standard deviation units on a vertically scaled achievement test); they learn more in the year following the intervention than the posttest impact of the most effective early childhood interventions of which I am aware. Thus the extreme scenario in which children in the treatment group are not exposed to new content in the year following an effective early intervention is not realistic.

Still, it remains possible that the content to which children are exposed following an effective early intervention contributes to fadeout. My colleagues and I used the posttest matching approach described previously to test this possibility (Bailey et al., 2016). If the constraining content hypothesis fully explained fadeout, children in the treatment group and posttest-matched higher-achieving children in the control group should have similar trajectories in the following year (Fig. 5, Panel B and SEM), while the unmatched lower-achieving control children who were not matched to children in the treatment group catch up to both groups. We found that, consistent with Fig. 6, Panel B and contrary to predictions of the constraining content hypothesis, children in the treatment group fell behind their posttest-matched peers from the control group in the following year at approximately the same rate that the control group caught up to the treatment group. This pattern held for higher- and lower-achieving children in the treatment group, consistent with the possibility that preexisting differences between children in the treatment group and higher-achieving posttest-matched controls continued to favor the latter group following the conclusion of the intervention.

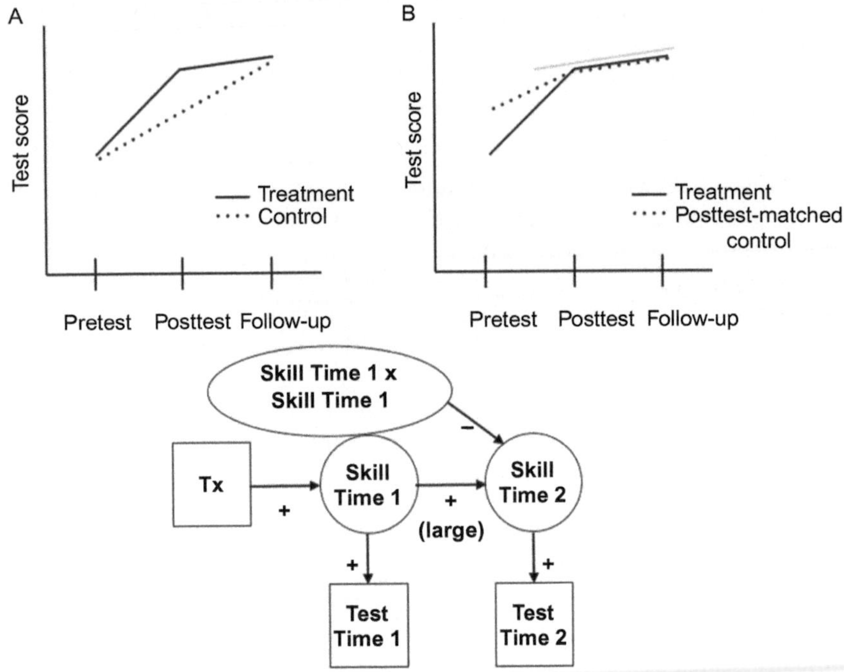

FIG. 5 Possible explanations for fadeout: constraining content. Note: Panel A shows the standard catchup pattern accompanying fadeout of early math intervention effects. For constraining content to fully account for fadeout, children in the treatment group and similarly high achieving children in the control group must be equally constrained by the low level of content to which they are exposed in the period following the end of the intervention (denoted by the gray line in Panel B). The effect of skill at time 1 on skill at time 2 is thus constrained at the higher end of skill at time 1 for both groups, indicated by the negative effect of skill at time 1^2 on skill at time 2 (SEM). In the underlying SEM, Tx denotes treatment (1=treatment, 0=control). *Figure adapted from Bailey et al. (2016).*

Modest Transfer: The Modest Transfer explanation posits that causal effects of earlier skills on later learning are real positive smaller than many assume, so that changes to early skills lead to diminishing effects on more advanced skills:

> *Early math skill boosts are insufficient to generate long-term effects on children's much later math achievement. Catchup occurs when transfer from basic skills to more sophisticated skills is modest and development under counterfactual conditions is rapid.*

The explanation for fadeout I find most consistent with the current literature, which has survived several tests since my colleagues and I first described it (Bailey, Watts, Littlefield, & Geary, 2014), is that although children's early math skills are *necessary* for their later math learning, they are not *sufficient*

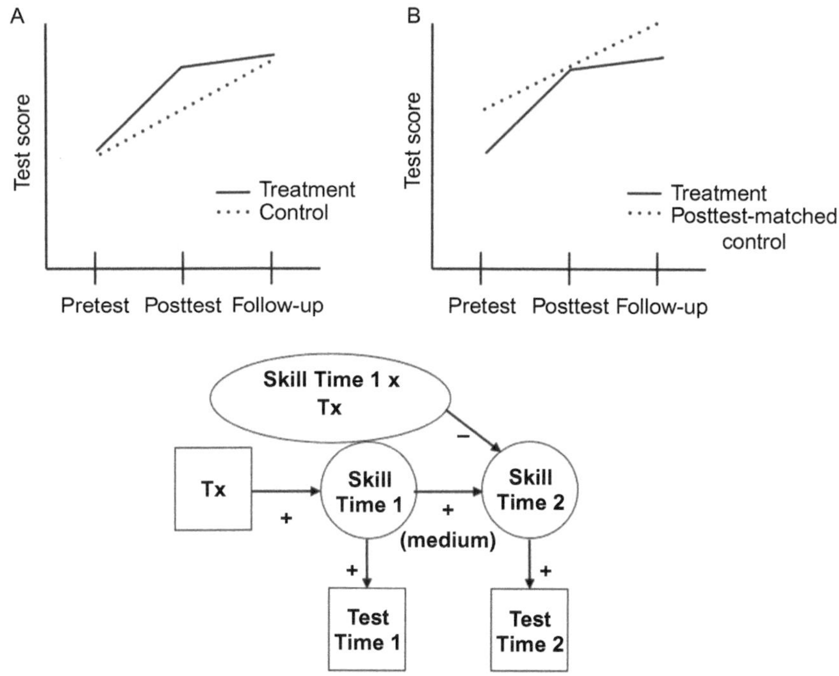

FIG. 6 Possible explanations for fadeout: modest transfer. Note: Panel A shows the standard catchup pattern accompanying fadeout of early math intervention effects. If modest transfer accounts for fadeout, children in the treatment group and similarly high achieving children in the control group will diverge following the end of treatment (Panel B) at the same rate as the treatment effect in Panel A decays. In this case, the predicted effect of skill at time 1 on skill at time 2 is smaller in the treatment group than in the control group, indicated by the higher slope in the control group in the posttreatment interval in the graphs and the negative treatment by skill interaction in the SEM. In the underlying SEM, Tx denotes treatment (1 = treatment, 0 = control). *Figure adapted from Bailey et al. (2016).*

to produce substantial long-term effects on children's math achievement. Before outlining evidence for this hypothesis, I note why I provide it. Several math cognition researchers have expressed to me that the minimal transfer idea is overly pessimistic and that advancing it will do harm, either to the field or to children who would otherwise benefit from early math intervention. My hope is that these ideas will benefit the field and children in one of two ways. First, the strongest way to falsify this hypothesis would be to design an early math intervention that shows a large initial impact with an impact of about the same magnitude several years later. I do not think or hope that this chapter will convince all readers of this hypothesis, and I think it would be a mistake to conclude that designing better early math interventions is no longer a good use of anyone's time. Second, if the hypothesis continues to survive attempts at falsification, I hope it will help motivate math cognition researchers to pursue research on complementary approaches to improving children's academic outcomes (see Fuchs et al., this volume), which I will focus on later in this chapter.

The idea that transfer of learning would be modest in math development is counterintuitive. As described previously under *Teaching to the Test*, basic skills are frequently reemployed in service of learning and performance of more advanced skills. Consistent with this idea, children's math skills around school entry are robust statistical predictors of their much later mathematics achievement, statistically controlling for a wide range of cognitive and contextual covariates (e.g., Aunola, Leskinen, Lerkkanen, & Nurmi, 2004; Duncan et al., 2007; Geary, Hoard, Nugent, & Bailey, 2013; Jordan, Kaplan, Ramineni, & Locuniak, 2009; Siegler et al., 2012; Watts, Duncan, Siegler, & Davis-Kean, 2014).

How can one reconcile these findings with evidence of fadeout from experimental studies? One possibility is that the set of statistical controls employed in these studies is insufficient to capture all of the factors influencing children's math learning throughout their development. If these omitted variables are positively related to both early and later math achievement, a failure to control for them would lead us to overestimate the causal effect of changes in children's early math achievement on their much later math achievement. Consistent with this hypothesis, we found in two longitudinal datasets that factors influencing children's math learning throughout development accounted for more of the long-term stability in children's math achievement scores during schooling, and that commonly used control variables failed to account for between 1/3 and 1/2 of the variance in these factors (Bailey et al., 2014).

To rule out the hypothesis that preexisting math knowledge per se constituted much of the residual variation in this stable unmeasured factor, we tested whether an effective early math intervention acted on children's math achievement via (1) a factor with similar influences on children's math achievement throughout development, (2) an autoregressive process, where children's math achievement at time t affects their math achievement at time $t+1$, which in turn affects their math achievement at time $t+2$, or some combination of these pathways. Only hypothesis 2 was supported (Watts et al., 2017).

Accounting for the unmeasured persistent variation in children's academic achievement, in these two studies, we estimate an effect of children's prior achievement of approximately .35 *SD* following a standard deviation boost in children's math achievement in the previous year, and .12 (=.35*.35) *SD* following a standard deviation boost in children's math achievement two years ago. In other words, a 1 SD gain in math achievement is 1st grade is predicted to result in a .35 SD gain in 2nd grade, but only a .12 SD gain in third grade. This pattern is close to what has been reported in experimental studies that boosted children's math achievement near school entry and followed children for at least a year subsequent to the end of treatment

(Bailey et al., 2018). It is important to note that the highly significant .35 *SD* effect size is fully consistent with the hypothesis that important math transfer take place from one year to the next, but because longer-run transfer estimates are a *product* of their .35 *SD* annual building blocks, they rapidly diminish in size. Transfer matters, but math achievement in a given year is the product of many other influences as well.

Because these models reconcile robust relations between early and later math achievement with observed patterns of fadeout, it is not clear that explanations specific to early math interventions (relative to other factors leading to specific changes in children's math achievement) are necessary. To the extent that this pattern of diminishing impacts reflects overlap between the items assessed in consecutive years, transfer of learning contributes even less to persistence. Still, these findings do not rule out the possibility of the other explanations proposed previously. Indeed, some of the ideas discussed previously are likely complementary explanations for modest transfer (e.g., the heterogeneity of math achievement across development implies that early gains will not necessarily lead to later gains; forgetting and constraining content could impede transfer).

So far, this explanation is more technical than conceptual. How might one make sense of the possibility that boosts in children's early math skills would not lead to lasting advantages relative to children's business-as-usual educational experiences, despite the strong evidence for the importance of early math skills for later math learning and performance? Two important factors require consideration: (1) Although early skills are necessary for more advanced math learning and performance, they are not sufficient. For example, addition fact fluency is not sufficient for learning multiplication. To know how to multiply, one must at least have some familiarity with the multiplication symbol and a procedure for multiplication. (2) As noted previously, in the early school years, children in the control group who receive business-as-usual math instruction progress quickly (this point applies to pre-k and the early school years and may vary across grades, a possibility I will discuss later).

Taken together, these factors limit the potential for large, long-lasting transfer effects. Hypothetically, imagine there are three sequential math skills, S_1, S_2, and S_3, and that learning each skill is a necessary but not sufficient condition for learning the subsequent skill. Receiving effective instruction on a skill increases the probability of learning the next skill before a peer with the same socioeconomic status, reading achievement, IQ, and subsequent schooling by one half. However, importantly, in the time it takes the child to learn the more advanced skill, peers also master the previous skill. In this case, a child who receives an effective intervention, which changes her from not knowing S_1 to knowing S_1, will see her probability of learning S_2 before an otherwise equal peer increase from .5 (an equal chance of either student learning S_2 first) to .5+(50% * .5), or .75. If she learns S_2 before her peer, she has a probability of .75 of learning S_3 before her peer. Therefore the probability of learning S_3 before her peer is the probability of learning S_2 first and S_3 first ($.75^2$), plus the probability of learning S_2 second but still learning S_3 first ($.25^2$), or .625. As the number of skills in the sequence increases, the initial boost in the probability that the student treated on S_1 will outperform her matched peer will decay by one half, approaching an asymptote of .5 (equally likely to learn; see Fig. 7).

Obviously, this is an oversimplified scenario. For example, there is not a single route to learning a specific skill, and it is possible that more basic skills continue to influence more advanced skill learning after intermediate skills are learned. However, because of the rapid learning rates for control-group children discussed previously, the assumption that catchup occurs on the more basic skill during the period in which transfer occurs may not differ substantially from what happens. Consistent with this simplified model, approximately exponential decay of treatment effects (rapid decline after the end of treatment followed by more modest declines thereafter) has been observed in math and other early academic interventions (Bailey et al., 2018; Li et al., 2017).

Some plausible objections to the ubiquity of the Modest Transfer explanation remain. Perhaps skills that have been (or can be) targeted by early math interventions do not develop quickly under counterfactual conditions and

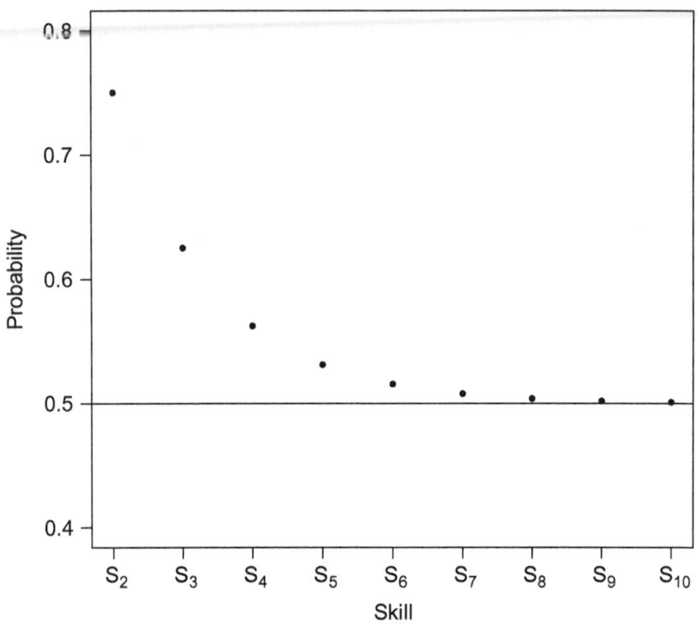

FIG. 7 Hypothetical pattern of skill learning impacts. Note: This pattern assumes a successful intervention on an individual's skill S_1, that skill learning is influenced only by the prior skill, and that the probability of learning a skill before a matched peer, conditional on learning the prior skill, is 75%.

might affect children's math learning throughout development. Further, the hypothesis that treatment effects regularly approach an asymptote of exactly zero, rather than some small but positive value that the average early intervention study is not powered to detect, deserves further empirical attention. Some meta-analytic estimates of long-term impacts of early childhood educational interventions (which include more than just math instruction, but often show a similar pattern of declining impacts after the end of treatment) approach an asymptote of approximately .05 *SD* (Li et al., 2017); to what extent this small but positive estimate reflects cognitive skill building effects, long-term effects of environmental enrichment via other processes, publication bias in reported outcomes, or sampling error, is not clear. However, while an effect of .05 may strike the reader as quite small, it is important to bear in mind that the average initial effect in these studies was just over .20 *SD*; to the extent that the long-term asymptote scales with the size of the initial effect (i.e., if 1/4 of the initial effect is regularly maintained years later), substantial long-term effects may be possible. The relative influences of mediating and moderating cognitive and educational processes across children's long-term mathematical development are not well understood and deserve additional empirical attention.

IMPLICATIONS FOR THE STUDY OF CHILDREN'S MATHEMATICAL DEVELOPMENT

Understanding fadeout has important implications for the study of children's mathematical development. I have reviewed these implications elsewhere (Bailey, 2018), but summarize the main points here.

First, the disappointing but important regularity of fadeout conflicts with common understandings of mathematical development. Specifically, as reviewed previously, analyses of nonexperimental longitudinal datasets have yielded estimates of the effects of early math achievement on much later math achievement that are higher than most corresponding estimates from experimental studies (Bailey et al., 2018). Fortunately, relevant high quality experimental and quasi-experimental studies exist. Resolving the discrepancy between experimental and nonexperimental findings in research on children's mathematical development in a robust and replicable way is a worthy endeavor for some math cognition and educational researchers (especially those doing nonexperimental research) to pursue. In short, theories about the long-term effects of math knowledge on other math knowledge must, at the very least, be consistent with regularities in the existing literature.

For this to happen, researchers conducting nonexperimental work on cognitive development need to provide clear and thoughtful theoretical interpretations of their results. For example, consider the empirical regularity that children's earlier math achievement is a robust statistical predictor of their much later math achievement, controlling for a variety of child- and family-level covariates. If we estimate a partial correlation of, say, .4 between children's grade 1 and grade 5 math achievement, what is the proper interpretation of this estimate? Is it the causal effect of a 1 *SD* boost in children's grade 1 math achievement on their grade 5 math achievement which, given the nonexperimental source of variation, may still suffer from lingering bias? Or is it only useful for prediction, perhaps indicating that achievement scores are a useful way to identify first graders who are at risk for low grade 5 math achievement, because of many causal factors in addition to their low grade 1 math achievement (e.g., working memory constraints)? The parameter underlying the .4 estimate is rarely discussed in method and results sections. On one hand, it is admirable that researchers do not wish to overstep their data. On the other hand, the scientific and educational value of this work is constrained by the extent to which it can inform theories that will make specific predictions about the effects of changes to children's grade 1 math achievement in the real world (Borsboom, 2013). Making clear the assumptions of the model and the interpretation of the parameter of interest exemplify what Meehl (1990) described as "precision in the derivation chain."

This is not to argue that researchers ought to naïvely proclaim their results to be unbiased causal estimates. On the contrary, the goal is to use prior knowledge yielded from causally informative designs along with a series of robustness checks and falsification tests based on this knowledge to gauge, perhaps imprecisely, the likely magnitude and direction of lingering biases. The approach, known as "coherent pattern matching," is detailed in Shadish, Cook, and Campbell (2002). For example, if our models yield similar estimates of effects of early math achievement on reading achievement, or on math achievement a year later and math achievement 5 years later, these might be signs of model misspecification. Some of the models I described in the section on the Modest Transfer explanation appear to recover the long-run observed effects of experimentally induced changes in early math achievement on children's math achievement several years later (Bailey et al., 2018), but this should be seen as only one of possibly many models that are consistent with the fadeout data. Alternative specifications, replications, extensions, and robustness checks of these methods in the context of children's mathematical development would make useful contributions to the literature. Being able to predict the long-term effects of changes in one math skill on others reasonably accurately, and to understand how these vary over time (Geary, Nicholas, Li, & Sun, 2017), would be even more useful.

Finally, more practically, math cognition theory and real-world practice will benefit from larger field experiments that target specific skills and include long-term follow-up assessments after the intervention. My discussion of plausible explanations of fadeout was largely informed by a handful of RCTs of math interventions in pre-k or the early school years that assessed children at least a year after the end of treatment. This work, along with field studies in developing countries (Barner et al., 2016; Dillon et al., 2017), provides a rich combination of reasonably large sample size and well-measured constructs, along with a source of exogenous variation not present in nonexperimental longitudinal studies. Generating models that fit well in nonexperimental longitudinal datasets, replicate across datasets, and reliably reproduce the patterns of treatment effects observed in experimental datasets would allow for a better integration of cognitive developmental theory, longitudinal data analysis, and educational practice.

How can Researchers of Mathematical Cognition Help Produce Long-Lasting Effects?

In reanalyses of experimental and nonexperimental longitudinal datasets, and in the design of early math interventions, it is worth considering whether some types of target skills produce more persistent effects than other skills. My colleagues and I have hypothesized that these skills share at least three important characteristics (Bailey, Duncan, Odgers, & Yu, 2017): First, and most obvious, is that the skill must be *malleable* by the early intervention of interest. Long-term impacts in the absence of any short-term impacts (sometimes called "sleeper effects") are intuitively unlikely.

Second, the skill must be *fundamental* for academic success. Ideally, the skill will be something that is used in the learning or performance of more advanced math. Third, skills most likely to produce persistent effects are those that do not develop quickly under counterfactual conditions. This is especially important for interventions targeting early achievement skills because, as discussed earlier, young children learn malleable and fundamental skills quickly—even in the absence of some prior intervention.

Unfortunately, even assuming its accuracy, this framework also does not clearly identify the early skills meeting all three conditions (which Bailey et al. call "trifecta" skills) that might be targeted via early intervention. Schools ought to be promoting the learning of all fundamental and malleable skills, but doing so makes these skills develop more rapidly in the absence of intervention. Further, skills that do not develop readily under counterfactual conditions but may be fundamental for academic success (e.g., solving systems of equations) may not be readily malleable (i.e., teachable) for young children. Following, I highlight some possible cases in which all three criteria might be met. I caution that this is speculative, and again, hope that in attempting to falsify these ideas, the field will learn some practically useful principles.

Targeting At-Risk Children

Knowledge of the characteristics of children at risk for persistently low mathematics achievement may allow us to target children who under business-as-usual conditions would take several years to learn as much as they would learn from a successful early math intervention (e.g., Berch & Mazzocco, 2007; National Mathematics Advisory Panel, 2008). This has been an appealing approach for interventionists, who often disproportionately target at-risk children. The long-term effects of interventions aimed at at-risk children may vary substantially depending on the targeted skill and age of intervention. I propose that a useful metric worth reporting in math intervention research is the ratio of the treatment effect on the targeted skill to growth in the control group during the same year. Based on the framework described previously, interventions with ratios substantially higher than 1 would be predicted to produce the longest lasting effects.

Targeting Advanced Skills in Older Children

Another way to approach this would be to target older children. As noted previously, children progress far more slowly in middle and high school than in elementary school (Hill et al., 2008). This is not purely a psychometric artifact. For example, numerous children and adults struggle to understand fractions many years after they are introduced during schooling (NMAP, 2008; Schneider & Siegler, 2010). Children's knowledge of fractions is fundamental to their ability to learn more advanced math, and it can be changed in children at risk for persistently low math achievement (Fuchs et al., 2013). I am not aware of a study that has examined the effects of an effective fraction intervention many years later, but based on the framework I outlined previously, this seems like a promising direction.

Math cognition randomized controlled trials with multiyear follow-up intervals for older children are not common. However, two relevant and important examples come from large causally informative regression discontinuity designs, in which children who scored below a set threshold on a math achievement placement test were required to take two periods of math instead of one.

In Chicago, children who performed below a set threshold on a math test were required to take a second period of algebra in ninth grade. Children who scored just below this threshold showed persistent benefits several years later relative to children who scored just above the threshold (Cortes, Goodman, & Nomi, 2015). Most notably, the effect of being double dosed on high school graduation was 12 percentage points (the graduation rate at the cutoff was 58%). This is certainly an important study for helping to understand the factors that might improve at-risk children's educational outcomes. However, the pattern of impacts of the program implies that math cognition may not have been the primary active ingredient. In grade 10, children took a standardized math test, and the effect of being double dosed was a nonsignificant .09 *SD*. In grade 11, there was a statistically significant impact on students' ACT Math scores (.18 *SD*), but oddly, it was smaller than the impact on their ACT Verbal scores (.27 *SD*). Elsewhere, we hypothesized that this effect might tell us more about the important effects of receiving credit for a high school algebra class on students' persistence in high school (Bailey et al., 2017); being double dosed increased children's chances of receiving a C or better in algebra by 12 percentage points. The extent to which an even more effective algebra intervention would lead to persistent effects on children's real world outcomes, and whether this knowledge influences children's later outcomes directly, rather than simply through course completion, is not clear, but are important relevant questions for math interventionists.

Taylor (2014) evaluated a similar program in Miami that assigned sixth graders who performed below a set threshold on a math test to take a second period of math in sixth grade. The effect on grade six math achievement was .16–.18 *SD*, which was reduced to approximately one-third of this size by the end of grade eight. Impacts in high school were null. Perhaps the difference in the persistence of these two interventions was attributable to the systemic advantage afforded to the Chicago children who would have failed high school algebra if not for the intervention. In contrast, perhaps the link between middle school math performance and educational attainment is less mechanistic. The possible impacts of interventions on "staying on track" in educational settings warrant consideration by interventionists and by developmentalists interpreting the long-term effects of interventions (Bailey et al., 2017).

The end-of-treatment effects in both of these studies were much smaller than some obtained by early math interventions, offsetting any potential benefit of targeting children with slower learning under counterfactual conditions. The promise of the approach of targeting older children depends on the inevitability of this pattern. Given the strong focus of math cognition research on younger children (for a review of the current state of the field, see Alcock et al., 2016), perhaps a somewhat increased focus on older children's learning and cognition would be sensible.

Complementary Follow-Through Interventions

Several interventionists have told me that they do not intend their early math interventions to be inoculations against later low achievement, and that children must receive better instruction following the conclusion of effective early interventions for effects to be sustained. Based on the patterns of effects observed following the end of effective early math interventions, this seems like a reasonable stance.

Before considering what an effective complementary follow-through intervention might look like, it is important to consider what it means for an intervention to be complementary. In economics, complementarity refers to the condition under which the causal effect of one variable (early math intervention, in this case) is larger for people who also receive some other treatment (subsequent schooling environment, in this case). To establish that a second intervention has sustained the effect of the first intervention, one would ideally rerandomize children from both the treatment and control groups from an RCT of an early math intervention into a treatment or control group for a second intervention. A positive interaction between receiving the first intervention and the second intervention (i.e., whereby the effect of the earlier intervention is more persistent for children who received the second intervention than for children who did not) would be evidence of complementarity, which would mitigate fadeout. I have not seen this design used in published RCTs of early math programs, but identifying complementary interventions would be a very useful step toward raising the long-term achievement of children at risk for persistently low math achievement with targeted intervention.

Proof of concept for the idea of complementary interventions comes from recent work that estimates the effect of attending Head Start, the effect of attending a school receiving increased funding, and the interaction between these variables (Johnson & Jackson, 2017). Findings suggest positive effects of both on adult income, but larger effects of Head Start attendance for children who subsequently attended schools that received more funding. This work is clever and causally informative, but our ability to draw conclusions about complementary math cognition interventions is limited, because the independent variables of interest (spending) are so distally related to math instruction. Importantly, complementarity is a hypothesis, not a law. The opposite pattern of substitutability is also possible. For example, a study of Danish children found positive effects on educational attainment for children who received pre-k and positive effects of receiving a nurse home visiting program in infancy, but the effect of pre-k disappeared almost completely for the children who had received the nurse home visiting program, meaning that the programs were nearly perfect substitutes for each other (Rossin-Slater & Wüst, 2016).

What would an effective complementary follow-through intervention look like? As reviewed in the summary of the constraining content hypothesis previously, I am skeptical as to whether merely teaching more advanced content would disproportionately benefit children who received an effective early intervention (except in the unrealistic case that the early intervention was so effective that there was almost no overlap in achievement between the groups by the end of treatment). Jenkins et al. (2018) found that the long-term effects of receiving an effective pre-k math intervention did not interact with classroom quality in kindergarten and first grade. Perhaps complementary intervention is most likely when early and later interventions are more directly aligned in pedagogy, terminology, instructional materials, and teacher knowledge. Of course, this is also speculative.

The Possibility of Different Effects of Improving Early Math Intervention at Scale?

Finally, while the constraining content hypothesis has limited support under existing conditions, it is likely that at some level of scale and effective implementation, effective early intervention could push up the level of later instruction for children across the elementary school years. For example, changes in county level spending for early childhood education programs in North Carolina across years were associated with changes in children's math and reading scores at age 11 (Dodge, Bai, Ladd, & Muschkin, 2017). Importantly, effects were positive even for children who were not eligible for participation in these programs, consistent with the possibility of "spillover" effects on nonparticipants.

Perhaps implementing a universal program for young children that improved their school-entry math skills would allow instruction in later grades to change. Of course, to the extent this would become economically and practically feasible to implement, math cognition researchers would likely favor such a program. In the meantime, perhaps the impacts of one-year math interventions assigned at the school or district level on children's achievement many years later would be a theoretically and practically useful set of interventions to study.

CONCLUSIONS AND FUTURE DIRECTIONS

Explanations for fadeout, their implications for research practices, and their implications for intervention are conceptually intertwined. However, while interventionists are concerned with designing treatments that will produce persistent effects in the field, lab research on mathematical cognition rarely addresses the problem of persistence. This is a shame, because math cognition researchers possess important knowledge about learning, transfer, and memory, along with (although not usually expressed in these terms) the malleability, fundamentality, and development under counterfactual conditions of aspects of children's math knowledge. All of these ideas feature prominently into explanations for fadeout and implications for practice. The purpose of this chapter is to attempt to introduce math cognition researchers to the regularity of fadeout in field studies that follow children for a long time following the conclusion of an effective early math intervention and to get them thinking about reasons why, what this means for their research, and whether they might be able to improve math intervention by combining this kind of information with their knowledge of developmental changes in mathematical cognition and learning.

In the short-term, thoughtful theoretical interpretation of nonexperimental research findings, comparison of these findings to experimental findings when possible will help us make more accurate predictions about the long-term effects of changes to early math skills on important outcomes. Combining these theories with tests involving variation in targeted skills, child baseline achievement and age, and complementary interventions, would further develop these theories and yield practically important information as well.

ACKNOWLEDGMENTS

I am grateful to the Eunice Kennedy Shriver National Institute of Child Health & Human Development of the National Institutes of Health under award number P01-HD065704. I also thank Dan Berch, Greg Duncan, Dave Geary, and Kathy Mann Koepke for helpful comments on a previous draft.

REFERENCES

Alcock, L., Ansari, D., Batchelor, S., Bisson, M. J., De Smedt, B., Gilmore, C., et al. (2016). Challenges in mathematical cognition: a collaboratively-derived research agenda. *Journal of Numerical Cognition*, *2*(1), 20–41. https://doi.org/10.5964/jnc.v2i1.10.

Aunola, K., Leskinen, E., Lerkkanen, M.-L., & Nurmi, J.-E. (2004). Developmental dynamics of math performance from pre-school to Grade 2. *Journal of Educational Psychology*, *96*, 699–713.

Bailey, D. H. (2018). Correlational data analysis in cognitive development: the primacy of risky tests. In P. Lemaire (Ed.), *Cognitive development from a strategy perspective: A Festschrift for Robert Siegler* (pp. 194–206). New York, NY: Routledge.

Bailey, D. H., Duncan, G., Odgers, C., & Yu, W. (2017). Persistence and fadeout in the impacts of child and adolescent interventions. *Journal of Research on Educational Effectiveness*, *10*, 7–39.

Bailey, D. H., Duncan, G. J., Watts, T., Clements, D., & Sarama, J. (2018). Risky business: correlation and causation in longitudinal studies of skill development. *American Psychologist*, *73*(1), 81–94.

Bailey, D. H., Nguyen, T., Jenkins, J. M., Domina, T., Clements, D. H., & Sarama, J. S. (2016). Fadeout in an early mathematics intervention: constraining content or preexisting differences? *Developmental Psychology*, *52*, 1457–1469.

Bailey, D. H., Watts, T. W., Littlefield, A. K., & Geary, D. C. (2014). State and trait effects on individual differences in children's mathematical development. *Psychological Science*, *25*, 2017–2026.

Barner, D., Alvarez, G., Sullivan, J., Brooks, N., Srinivasan, M., & Frank, M. C. (2016). Learning mathematics in a visuospatial format: a randomized, controlled trial of mental abacus instruction. *Child Development*, *87*(4), 1146–1158.

Bassok, D., Latham, S., & Rorem, A. (2016). Is kindergarten the new first grade? *AERA Open*, *2*(1). https://doi.org/10.1177/2332858415616358.

Berch, D. B., & Mazzocco, M. M. M. (Eds.), (2007). *Why is math so hard for some children? The nature and origins of mathematical learning difficulties and disabilities*. Baltimore, MD: Brookes.

Bond, T. N., & Lang, K. (2013). The evolution of the Black-White test score gap in Grades K-3: the fragility of results. *The Review of Economics and Statistics*, *95*(5), 1468–1479.

Borsboom, D. (2013). *Theoretical amnesia*. Open Science Collaboration Blog [November 20].

Campbell, F., Conti, G., Heckman, J. J., Moon, S. H., Pinto, R., Pungello, E., & Pan, Y. (2014). Early childhood investments substantially boost adult health. *Science*, *343*, 1478–1485. https://doi.org/10.1126/science.1248429.

Cascio, E. U., & Staiger, D. O. (2012). *Knowledge, tests, and fadeout in educational interventions* (No. w18038). Cambridge, MA: National Bureau of Economic Research.

Cepeda, N. J., Vul, E., Rohrer, D., Wixted, J. T., & Pashler, H. (2008). Spacing effects in learning: a temporal ridgeline of optimal retention. *Psychological Science*, *19*(11), 1095–1102.

Clarke, B., Doabler, C., Smolkowski, K., Kurtz Nelson, E., Fien, H., Baker, S. K., et al. (2016). Testing the immediate and long-term efficacy of a Tier 2 kindergarten mathematics intervention. *Journal of Research on Educational Effectiveness*, *9*(4), 607–634.

Clements, D. H., Sarama, J., Wolfe, C. B., & Spitler, M. E. (2013). Longitudinal evaluation of a scale-up model for teaching mathematics with trajectories and technologies: persistence of effects in the third year. *American Educational Research Journal*, *50*(4), 812–850.

Cortes, K. E., Goodman, J. S., & Nomi, T. (2015). Intensive math instruction and educational attainment long-run impacts of double-dose algebra. *Journal of Human Resources*, *50*(1), 108–158.

De Visscher, A., & Noël, M.-P. (2014). Arithmetic facts storage deficit: the hypersensitivity-to-interference in memory hypothesis. *Developmental Science*, *17*, 434–442.

Dillon, M. R., Kannan, H., Dean, J. T., Spelke, E. S., & Duflo, E. (2017). Cognitive science in the field: a preschool intervention durably enhances intuitive but not formal mathematics. *Science*, *357*(6346), 47–55.

Dodge, K. A., Bai, Y., Ladd, H. F., & Muschkin, C. G. (2017). Impact of North Carolina's early childhood programs and policies on educational outcomes in elementary school. *Child Development*, *88*(3), 996–1014.

Duncan, G. J., Dowsett, C. J., Claessens, A., Magnuson, K., Huston, A. C., Klebanov, P., et al. (2007). *Developmental Psychology*, *43*(6), 1428–1446. School readiness and later achievement. https://doi.org/10.1037/0012-1649.43.6.1428.supp.

Elango, S., García, J. L., Heckman, J. J., & Hojman, A. (2015). *Early childhood education* (No. w21766). Cambridge, MA: National Bureau of Economic Research.

Engel, M., Claessens, A., & Finch, M. A. (2013). Teaching students what they already know? The (mis) alignment between mathematics instructional content and student knowledge in kindergarten. *Educational Evaluation and Policy Analysis*, *35*, 157–178.

Fuchs, L. S., Schumacher, R. F., Long, J., Namkung, J., Hamlett, C. L., Cirino, P. T., et al. (2013). Improving at-risk learners' understanding of fractions. *Journal of Educational Psychology*, *105*(3), 683–700. https://doi.org/10.1037/a0032446.

Geary, D. C. (1993). Mathematical disabilities: cognitive, neuropsychological, and genetic components. *Psychological Bulletin*, *114*, 345–362.

Geary, D. C., Hoard, M. K., Nugent, L., & Bailey, D. H. (2013). Adolescents' functional numeracy is predicted by their school entry number system knowledge. *PLoS One*, *8*(1), e54651.

Geary, D. C., Nicholas, A., Li, Y., & Sun, J. (2017). Developmental change in the influence of domain-general abilities and domain-specific knowledge on mathematics achievement: an eight-year longitudinal study. *Journal of Educational Psychology*, *109*, 680–693.

Hill, C. J., Bloom, H. S., Black, A. R., & Lipsey, M. W. (2008). Empirical benchmarks for interpreting effect sizes in research. *Child Development Perspectives*, *2*(3), 172–177.

Hofer, K. G., Lipsey, M. W., Dong, N., & Farran, D. C. (2013). *Results of the early math project—scale-up cross-site results (working paper)*. Nashville, TN: Vanderbilt University, Peabody Research Institute.

Jenkins, J. M., Watts, T. W., Magnuson, K., Gershoff, E. T., Clements, D. H., Sarama, J., & Duncan, G. J (2018). Do high-quality kindergarten and first-grade classrooms mitigate preschool fadeout? *Journal of Research on Educational Effectiveness*, *11*, 339–374.

Jensen, A. R. (1998). *The g factor: The science of mental ability*. Westport, CT: Praeger.

Johnson, R. C., & Jackson, C. K. (2017). *Reducing inequality through dynamic complementarity: Evidence from head start and public school spending* [No. w23489]. National Bureau of Economic Research.

Jordan, N. C., Kaplan, D., Ramineni, C., & Locuniak, M. N. (2009). Early math matters: kindergarten number competence and later mathematics outcomes. *Developmental Psychology*, *45*(3), 850–867. https://doi.org/10.1037/a0014939.

Lemaire, P., & Siegler, R. S. (1995). Four aspects of strategic change: contributions to children's learning of multiplication. *Journal of Experimental Psychology: General*, *124*, 83–97.

Li, W., Leak, J., Duncan, G. J., Magnuson, K., Schindler, H., & Yoshikawa, H. (2017). In *Is timing everything? How early childhood education program impacts vary by starting age, program duration and time since the end of the program. Working paper, national forum on early childhood policy and programs, meta-analytic database project. Center on the Developing Child, Harvard University.*

Meehl, P. E. (1990). Why summaries of research on psychological theories are often uninterpretable. *Psychological Reports*, *66*, 195–244.

National Mathematics Advisory Panel. (2008). *Foundations for success: The final report of the National Mathematics Advisory Panel*. Washington, DC: U.S. Department of Education.

Protzko, J. (2016). Does the raising IQ-raising g distinction explain the fadeout effect? *Intelligence*, *56*, 65–71.

Purpura, D. J., & Lonigan, C. J. (2013). Informal numeracy skills: the structure and relations among numbering, relations, and arithmetic operations in preschool. *American Educational Research Journal*, *50*(1), 178–209.

Rossin-Slater, M., & Wüst, M. (2016). *What is the added value of preschool? Long-term impacts and interactions with a health intervention* [No. w22700]. National Bureau of Economic Research.

Schenke, K., Rutherford, T., Lam, A. C., & Bailey, D. H. (2016). Construct confounding among predictors of mathematics achievement. *AERA Open*, *2*(2). https://doi.org/10.1177/2332858416648930.

Schneider, M., & Siegler, R. S. (2010). Representations of the magnitudes of fractions. *Journal of Experimental Psychology. Human Perception and Performance*, *36*(5), 1227.

Schweinhart, L. J., Montie, J., Xiang, Z., Barnett, W. S., Belfield, C. R., & Nores, M. (2005). *Lifetime effects: The high/scope perry preschool study through age 40 (monographs of the high/scope educational research foundation, 14)*. Ypsilanti, MI: HighScope Press.

Shadish, W. R., Cook, T. D., & Campbell, D. T. (2002). *Experimental and quasi-experimental designs for generalized causal inference*. Boston, MA: Houghton Mifflin.

Siegler, R. S., Duncan, G. J., Davis-Kean, P. E., Duckworth, K., Claessens, A., Engel, M., et al. (2012). Early predictors of high school mathematics achievement. *Psychological Science*, *23*, 691–697.

Smith, T. M., Cobb, P., Farran, D. C., Cordray, D. S., & Munter, C. (2013). Evaluating math recovery: assessing the causal impact of a diagnostic tutoring program on student achievement. *American Educational Research Journal*, *50*(2), 397–428.

Taylor, E. (2014). Spending more of the school day in math class: evidence from a regression discontinuity in middle school. *Journal of Public Economics*, *117*, 162–181.

Watts, T. W., Clements, D. H., Sarama, J., Wolfe, C. B., Spitler, M. E., & Bailey, D. H. (2017). Does early mathematics intervention change the processes underlying children's learning? *Journal of Research on Educational Effectiveness*, *10*, 96–115.

Watts, T. W., Duncan, G. J., Siegler, R. S., & Davis-Kean, P. E. (2014). What's past is prologue: relations between early mathematics knowledge and high school achievement. *Educational Researcher*, *43*(7), 352–360. https://doi.org/10.3102/0013189X14553660.

Index

Note: Page numbers followed by "*f*" indicate figures, and "*t*" indicate tables.

D

Z

FSC
www.fsc.org
MIX
Papier | Fördert
gute Waldnutzung
FSC® C083411

Kolibri 360 GmbH
Ferdinand-Jühlke-Straße 7
99095 Erfurt, Deutschland
produktsicherheit@kolibri360.de